An Industrial IoT Approach for Pharmaceutical Industry Growth

An Industrial IoT Approach for Pharmaceutical Industry Growth

Volume 2

Edited by

Valentina Emilia Balas
"Aurel Vlaicu" University of Arad, Arad, Romania

Vijender Kumar Solanki
CMR Institute of Technology (Autonomous), Hyderabad, TS, India

Raghvendra Kumar
GIET University, Gunupur, OR, India

ELSEVIER

ACADEMIC PRESS
An imprint of Elsevier

Academic Press is an imprint of Elsevier
125 London Wall, London EC2Y 5AS, United Kingdom
525 B Street, Suite 1650, San Diego, CA 92101, United States
50 Hampshire Street, 5th Floor, Cambridge, MA 02139, United States
The Boulevard, Langford Lane, Kidlington, Oxford OX5 1GB, United Kingdom

British Library Cataloguing-in-Publication Data
A catalogue record for this book is available from the British Library

Library of Congress Cataloging-in-Publication Data
A catalog record for this book is available from the Library of Congress

ISBN: 978-0-12-821326-1

For Information on all Academic Press publications
visit our website at https://www.elsevier.com/books-and-journals

Publisher: Andre Gerhard Wolff
Acquisitions Editor: Erin Hill-Parks
Editorial Project Manager: Pat Gonzalez
Production Project Manager: Punithavathy Govindaradjane
Cover Designer: Mark Rogers

Typeset by MPS Limited, Chennai, India

Contents

List of contributors

T. Lucia Agnesbeena Department of Information Technology, St. Josephs College, Tiruchirappalli, India

Rashmi Agrawal Manav Rachna Institute of Research and Studies, Faridabad, India

Balamurugan Balusamy School of Computing Science and Engineering, Galgotias University, Greater Noida, India

Saakshi Bhargava Department of Physical Sciences and Engineering, Banasthali Vidyapith, Tonk, India

Devender Bhushan Manav Rachna Institute of Research and Studies, Faridabad, India

Pushpa Choudhary Department of Information Technology, G. L. Bajaj Institute of Technology and Management, Greater Noida, India

K. Eswari School of Pharmaceutical Sciences, Vels Institute of Science, Technology and Advanced Studies (VISTAS), Chennai, India

Neda Firoz Departments of Computational Biology and Bioinformatics, Sam Higginbottom University of Agriculture, Technology and Sciences, Allahabad, India

Prachi Gupta Department of Information Technology, Netaji Subhas University of Technology (formerly known as Netaji Subhas Institute of Technology), New Delhi, India

Sandeep Singh Jagdev Ellen Technology (P) Ltd., Jalandhar, Punjab, India,

Anjali Jain Department of Instrumentation and Control, Netaji Subhas University of Technology (formerly known as Netaji Subhas Institute of Technology), New Delhi, India

S. Janarthanan School of Computing Science and Engineering, Galgotias University, Greater Noida, India

Long Cu Kim School of Information Communication Technology, Hanoi University of Science and Technology, Hanoi, Vietnam

Chandrakanta Mahanty Department of CSE, GIET University, Gunupur, India

Brojo Kishore Mishra Department of CSE, GIET University, Gunupur, India

P.K. Pattnaik School of Computer Engineering, KIIT Deemed to be University, Bhubaneswar, India

T. Poongodi School of Computing Science and Engineering, Galgotias University, Greater Noida, India

Priety Department of Information Technology, Netaji Subhas University of Technology (formerly known as Netaji Subhas Institute of Technology), New Delhi, India

C. Priya Department of Information Technology, Vels Institute of Science, Technology and Advanced Studies (VISTAS), Chennai, India

Vikram Puri Duy Tan University, Da Nang, Vietnam

Jay Sarraf School of Computer Engineering, KIIT Deemed to be University, Bhubaneswar, India

Deepak Kumar Sharma Department of Information Technology, Netaji Subhas University of Technology (formerly known as Netaji Subhas Institute of Technology), New Delhi, India

Arun Kumar Singh Department of Information Technology, G. L. Bajaj Institute of Technology and Management, Greater Noida, India

K.K. Singh Department of ECE, KIET Group of Institutions, Ghaziabad, India

Kartik Singhal Department of Manufacturing Process and Automation Engineering, Netaji Subhas University of Technology (formerly known as Netaji Subhas Institute of Technology), New Delhi, India

Ashish Tripathi Department of Information Technology, G. L. Bajaj Institute of Technology and Management, Greater Noida, India

Vaibhaw School of Computer Engineering, KIIT Deemed to be University, Bhubaneswar, India

Chung Van Le Duy Tan University, Da Nang, Vietnam

Prem Chand Vashist Department of Information Technology, G. L. Bajaj Institute of Technology and Management, Greater Noida, India

A. Vidhyalakshmi Department of Computer Science, Vels Institute of Science, Technology and Advanced Studies (VISTAS), Chennai, India

About the editors

Valentina Emilia Balas, PhD, is currently Full Professor in the Department of Automatics and Applied Software at the Faculty of Engineering, "Aurel Vlaicu" University of Arad, Romania. She holds a PhD in Applied Electronics and Telecommunications from Polytechnic University of Timisoara, Romania. She is the author of more than 270 research papers in refereed journals and international conferences. Her research interests are in intelligent systems, fuzzy control, soft computing, smart sensors, information fusion, and modeling and simulation. She is the Editor-in-Chief to *International Journal of Advanced Intelligence Paradigms* (IJAIP) and *International Journal of Computational Systems Engineering* (IJCSysE), a member of the editorial board of several national and international journals, and is an evaluator expert for national and international projects. She served as General Chair of the International Workshop Soft Computing and Applications in seven editions 2005–16 held in Romania and Hungary. She has participated in many international conferences as Organizer, Session Chair, and member in International Program Committees. She is currently working in a national project with EU funding support: BioCell-NanoART = Novel Bio-inspired Cellular Nano-Architectures—For Digital Integrated Circuits, 2M Euro from National Authority for Scientific Research and Innovation. She is a member of EUSFLAT, ACM, and a Senior Member of IEEE, member in TC—Fuzzy Systems (IEEE CIS), member in TC—Emergent Technologies (IEEE CIS), and member in TC—Soft Computing (IEEE SMCS). She was Vice-President (Awards) of the IFSA International Fuzzy Systems Association Council (2013–15) and is a Joint Secretary of the Governing Council of the Forum for Interdisciplinary Mathematics (FIM), a multidisciplinary academic body based in India.

Vijender Kumar Solanki, Ph.D. is an Associate Professor in Computer Science & Engineering, CMR Institute of Technology (Autonomous), Hyderabad, TS, India. He has more than 10 years of academic experience in network security, IoT, Big Data, Smart City, and IT. Prior to his current role, he was associated with Apeejay Institute of Technology, Greater Noida, UP, KSRCE (Autonomous) Institution, Tamilnadu, India, and Institute of Technology & Science, Ghaziabad, UP, India. He is a member of ACM, COPE, and IEEE.

He has attended an orientation program at UGC-Academic Staff College, University of Kerala, Thiruvananthapuram, Kerala, and a refresher course at the Indian Institute of Information Technology, Allahabad, UP, India.

He has authored or co-authored more than 50 research articles that have been published in various journals, books, and conference proceedings. He has edited or co-edited 14 books and Conference Proceedings in the area of soft computing.

He received a Ph.D. in Computer Science and Engineering from Anna University, Chennai, India, in 2017, an ME and MCA from Maharishi Dayanand University, Rohtak, Haryana, India, in 2007 and 2004, respectively, and a bachelor's degree in Science from JLN Government College, Faridabad Haryana, India, in 2001.

He is the Book Series Editor of *Internet of Everything (IoE): Security and Privacy Paradigm*, CRC Press, Taylor & Francis Group, USA; *Artificial Intelligence (AI): Elementary to Advanced Practices Series*, CRC Press, Taylor & Francis Group, USA; *IT, Management & Operations Research Practices*, CRC Press, Taylor & Francis Group, USA; *Bio-Medical Engineering: Techniques and Applications*, Apple Academic Press, USA; and *Computational Intelligence and Management Science Paradigm* (Focus Series), CRC Press, Taylor & Francis Group, USA.

He is Editor-in-Chief of the *International Journal of Machine Learning and Networked Collaborative Engineering* (IJMLNCE), ISSN 2581-3242; *International Journal of Hyperconnectivity and the Internet of Things* (IJHIoT), ISSN 2473-4365, IGI-Global, USA, Co-Editor of *Ingenieria Solidaria Journal*, ISSN 2357-6014, and Associate Editor of the *International Journal of Information Retrieval Research* (IJIRR), IGI-GLOBAL, USA, ISSN: 2155-6377/E-ISSN: 2155-6385. He has been guest editor with IGI-Global, USA, InderScience, and many more publishers. He can be contacted at spesinfo@yahoo.com or vijendersolanki@ieee.org.

Raghvendra Kumar is working as an Associate Professor in the Computer Science and Engineering Department at GIET University, India. He received a B.Tech, M.Tech, and PhD in Computer Science and Engineering, India, and Postdoc Fellow from the Institute of Information Technology, Virtual Reality and Multimedia, Vietnam. He serves as Series Editor of the *Internet of Everything (IOE): Security and Privacy Paradigm, Green Engineering and Technology: Concepts and Applications*, published by CRC Press, Taylor & Francis Group, United States, and *Bio-Medical Engineering: Techniques and Applications*, published by Apple Academic Press, CRC Press, Taylor & Francis Group, United States. He also serves as acquisition editor for *Computer Science* by Apple Academic Press, CRC Press, Taylor & Francis Group, United States. He has published a number of research papers in international journals (SCI/SCIE/ESCI/Scopus) and conferences including IEEE and Springer as well as serving as an organizing chair

(RICE-2019, 2020), volume editor (RICE-2018), keynote speaker, session chair, co-chair, publicity chair, publication chair, advisory board, and technical program committee members in many international and national conferences, and served as a guest editor in many special issues of reputed journals (indexed by: Scopus, ESCI, SCI). He also published 13 chapters in edited books published by IGI Global, Springer, and Elsevier. His researches areas are computer networks, data mining, cloud computing and secure multiparty computations, theory of computer science, and the design of algorithms. He has authored and edited 23 computer science books in the fields of the Internet of Things, data mining, biomedical engineering, big data, robotics, by IGI Global Publication, United States, IOS Press Netherland, Springer, Elsevier, and CRC Press, United States. He is Managing Editor of the *International Journal of Machine Learning and Networked Collaborative Engineering* (IJMLNCE; ISSN 2581-3242).

Preface

The purpose of this book is to inform and educate its audience about the power of the Internet of Things (IoT) in the pharma industry. The enormous growth of the IoT has caused a vast majority of companies in the manufacturing sector to make use of this technology to unlock its limitless potential. Pharmaceutical manufacturing companies are no exception to this, as the IoT has the potential to revolutionize all aspects of the pharmaceutical manufacturing process from drug discovery to manufacturing. The IoT in pharma manufacturing coupled with big data and advanced analytics can scrutinize massive amounts of data that can be harnessed to improve manufacturing efficiency. The wide variety of topics it presents offers readers multiple perspectives on a variety of disciplines, which are described in the chapters of this book.

Chapter 1 includes the delivery of medical care services, diagnosis, consultation, treatment, as well as health education and the transfer of medical data. IoT-based applications mainly include remote patient monitoring and clinical monitoring. In addition, applications based on preventive measures are also part of smart health care. All these applications require image processing-based technologies which could be integrated into medical healthcare systems. Various types of inputs taken from cameras and processing of CT and MRI images could be integrated into IoT-based medical applications.

Chapter 2 discusses various neuroimaging approaches to the brain–computer interface and its application areas.

Chapter 3 covers the basics of the IoT and its applications in logistics and the supply chain. This chapter also discuss the growth of the pharmaceutical industry with a focus on the introduction of the IoT and how it changed the face of pharma. The applications of the IoT in pharmaceutical logistics including quality control, shipment tracking, and drug development are also described. The challenges faced in the industry and how they are being conquered using the IoT are also outlined. The aim of this chapter is to make readers aware of how the pharma industry has seen a technological boom with the IoT and enabled an entire process, starting from drug development to shipment, to become extremely efficient and secure.

Chapter 4 examines drug-related problems based on IoT technologies, that is, smartphones and the Web, to support ubiquitous access, IPv6 connectivity over low power area networks technology to support ubiquitous data collection of patients, sensors, and hospitals, and radio frequency identification/near field communications to support global identification. These technologies offer a wide range of applications in health care, and improve the quality of services, reduce mistakes, and even detect health anomalies from vital signs. Drug compliance and adverse drug reactions (ADRs) are two of the most important issues regarding patient safety throughout the global healthcare sector. ADR prevalence is 6.7% throughout hospitals worldwide, with an international mortality rate of 0.32% of the total number of the patients. This rate is even higher in ambient assisted living environments, where 15% of patients suffer clinically significant interactions due to patient noncompliance with drug dosages and intake schedules, in addition to suffering from polypharmacy. These instances increase with age and cause risks of drug interactions, adverse effects, and toxicity. However, with tight follow-up of drug treatment, complications of incorrect drug use can be reduced. For that purpose, an innovative system based on the IoT for drug identification and monitoring of medication is proposed. The IoT is applied to examine drugs in order to fulfill treatment, to detect harmful side effects of pharmaceutical excipients, allergies, liver/renal contradictions, and harmful side effects during pregnancy.

Chapter 5 deals with the significant role of the IoT in the pharmaceutical industry including ethical issues and countermeasures for accessing heterogeneous data in healthcare services.

Chapter 6 describes the combination of the IoT with the growing digitization of figures in the pharma segment, leading to new directions and opening the floodgate for enormous change and industrialized overhauling of systems. While striving for ingenious inventions, there should be no risk of software lagging behind in this respect. As has been observed, artificial intelligence and machine learning are providing tremendous opportunities in the pharmaceutical industry. Similarly, the IoT offers pharma production with an assortment of innovative prospects, such as the improved drug trade mechanisms and deterrent conservation of tools.

Chapter 7 describes improved metrics in the IoT that is allowing it to become a leader in the technological world. We are witnessing the fourth revolution in the digitization world. The reasons behind this leadership and how it can be a partner in this process are discussed. Protocols have evolved that differentiate the IoT from previous methods. This also creates security concerns, with data described as the new oil, further raising data privacy challenges.

Chapter 8 discusses the various developments achieved so far in the acceleration of the transitions from the IoT to various industries.

Chapter 9 introduces the IoT technologies that can be used in pharmaceutical manufacturing and supply chains to overcome the existing challenges in the industry. Furthermore, it also presents open challenges with regards to the implementation of these technologies and the scope for future work in the field, allowing readers to understand the need to implement IoT solutions, while also realizing the struggle of integrating them with the current architecture.

Chapter 10 proposes to reduce the issues caused by cervical spondylosis and neck pain. A sensor equipped with an accelerometer and gyrometer is integrated with the neckband to monitor the position neck position and send data to the Google cloud server for maintaining records. In addition, alerts are generated and sent to the user when the wrong neck position is encountered. Machine learning algorithms are synchronized with the proposed system to increase its accuracy.

Chapter 11 discusses the main problems around the issues raised by data increases (which on a yearly basis can be from terabytes to exabytes); to overcome these problems, this chapter proposes and analyzes a solution for the pharmaceutical industry which also has high security and privacy.

Valentina Emilia Balas, Vijender Kumar Solanki and Raghvendra Kumar

About the book

The book is a unique work, containing a collection of chapters, in which the Internet of Things (IoT) and its innovative approach to the pharma industry are described in detail. In the 21st century, technology is ubiquitous, and in many industries it is widely used. This book describes the innovative use of IoT, big data, machine learning, and artificial intelligence in the pharma industry while providing real working examples. Today, researchers create approaches to illustrate and break down pharma/medical information that can help researchers to accelerate their search for new materials. The current pharma industry would not have been conceivable without the ability to carry out computations to investigate the immense quantity of measurements received from sensors and IoT-associated infrastructure. This book will help researchers to understand the computing power of and explore biomedical data in the IoT.

Chapter 1

Medical big data mining and processing in e-health care

A. Vidhyalakshmi[1] and C. Priya[2]

[1]Department of Computer Science, Vels Institute of Science, Technology and Advanced Studies (VISTAS), Chennai, India, [2]Department of Information Technology, Vels Institute of Science, Technology and Advanced Studies (VISTAS), Chennai, India

1.1 Introduction

In any industry big data can be changed by managing, analyzing, and leveraging the data. Health care is a promising area in which this could be applied to create a change in the field, which would result in many advantages. In health care, quality of life is improved by reducing treatment costs, epidemic outbreak prediction, and avoiding preventable diseases. The global population is increasing leading to new challenges for treatments and delivery methods. Health professionals are similar to business entrepreneurs, with a massive volume of data being obtained by them which could be selected and used by using the best strategies.

Any information related to human death, health conditions, quality of life, and reproductive outcomes is referred to as health data, whether for an individual or a population. Health data consist of clinical metrics along with environmental, socioeconomic, and health and wellness-based behavior information. The communication of individuals with health-care systems is collected as health data for processing and use by health-care providers. These data consist of a record of received services, service conditions of those data, and practical outcomes obtained in the clinic or information concerning those services [1]. By having this kind of framework most health data can be sourced. The advances in health information technology (IT) and eHealth care have been enhanced by the use of health data. The advancements in health care also enhance data security, data privacy, and ethical concerns. The main and major components of digital health are achieved by increasing data collection and usage.

Health data are split into two types: structured and unstructured. When the data are easily transferred between the health information system and in standardized form then the system is said to be structured [2]. The name of

An Industrial IoT Approach for Pharmaceutical Industry Growth.
DOI: https://doi.org/10.1016/B978-0-12-821326-1.00001-2

the patient, patient DOB, and output of a blood test report are stored in a structured data format whereas nonstandardized data are considered as unstructured health data [2]. Examples of unstructured data include emails, audio recordings, or physician notes about a patient. Based on the collection and use of health data, the information system is expanded, in a health-care field the data complexity is reduced with standardization [3]. In 2013, it was observed that 60% of data in the United States was unstructured [4].

The industrial internet is defined as the human services on the IoT, these terms suggest about expanding number of brilliant very quickly, then the gadgets will be moved among the connection between the gadgets and the amount of information and individuals. Some assessments have estimated that $120 billion was spent in 4 years for medicinal services in the IoT cloud, with human services IoT being the major part of this information in an unstructured format. This unstructured data information working within the Hadoop system has made Hadoop an interesting part of the progress into investigating a wide range of information. Recently, miscellaneous collections of screen gadgets are used, from glucose screens to fetal screens, fetal screens to electrocardiograms, and then to circulatory strains for each patient. Thereafter a subsequent visit from a doctor would be important with this huge amount of estimations.

When intelligent medical devices are able to interpret the results of other devices and gadgets, the requirements for direct doctor action may be replaced by a telephone call to the patient from a medical attendant. In addition patients can use devices at home and have the data transferred electronically. The possible results provided by human services IoT could result in immense improvements to patient care. In this human services substructure, utilization of the IoT can enable the installation of hardware, programming, sensors, and systems to permit these devices to communicate with one another and trade information.

The IoT has many possible cutting-edge innovations which could affect the whole range of businesses including health care. In today's framework this can be considered as the interrelation of exceptionally identifiable items and gadgets with advantages and benefits which can be achieved by going beyond the machine-to-machine environment.

1.1.1 Types of big data

The definition and types of big data and are discussed next.

1.1.1.1 Structured

Data which can be stored, processed, and retrieved in a fixed form are known as structured data. These contain well-ordered information which can be promptly and perfectly stored. These data can be acquired from a database using search engine techniques. For example, the “workers” table in a firm’s database will be designed to include details of the employees, their positions, salaries, etc., in a well-structured manner.

1.1.1.2 Unstructured

Unstructured data are data that are unavailable in any specific form or structure. These kinds of data are complex to process and the processing time for these types of data can be excessive. Email is an example of unstructured data.

1.1.1.3 Semistructured

Data which contains both structured and unstructured formats is defined as semistructured data. Specifically this indicates information which has not been put under a particular database, and which consists of tags or vital information that separates individual elements of the data.

1.1.2 Characteristics of big data

1.1.2.1 Variety

Big data that are collected from many of multiple sources may be structured, unstructured, and/or semistructured. Previously, data may have been gathered from Excel sheets and databases, however nowadays the data are also sent in the forms emails, PDFs, photos, videos, audios, SM posts, and etc.

1.1.2.2 Velocity

Velocity is defined as the rate of speed at which data are generated in real time. In a broader context, it incorporates the rate of change, interconnection of incoming data sets at various speeds, and activity bursts.

1.1.2.3 Volume

The word big data has its meaning in the words themselves. Big data is purely a large amount of data that is being created on a day-by-day basis from various resources such as human interactions, social media platforms, machines, business processes, and networks. Such a large volume of information is saved in a data repository.

1.1.3 Integration of big data with medical imaging

Medical imaging and its processing plays a vital role in medicine in the United States, where about 600 million imaging procedures are performed annually. It is difficult manually to examine and store these images and also expensive and time consuming as hospitals need to protect and store them for several years in case of the need for future analysis by radiologists.

Medical imaging distributor care streams illustrate how often images were changed while analyzing the big data in health. The algorithms developed by physicians should analyze specific patterns in the hundreds of thousands of pixels in images and convert them into a numerical format for diagnosis. Furthermore, it could be feasible that radiologists in the future will no longer need to look at the images, but instead algorithms would analyze the outcomes as they are able to produce and remember a greater number of images. This will clearly affect the work of radiologists and their education and skills.

1.1.4 Advantages of health-care data management

The advantages of using health-care data management include the following:

- Producing 360 degree views of consumers, patients, and households, and deploying personalized, guided conversations by associating data from all available sources.
- Enhanced patient engagement with predictive modeling and analysis based on health-care data.
- Improved population health outcomes in specific geographic areas by tracking current health trends and predicting upcoming ones.
- Making informed, high-impact business decisions based on data insights.
- Understanding physician activity and aligning them with the organization's goals.
- Predictive analysis is the biggest key benefit of big data. The analytics tool of big data can classify outputs correctly and, based on this, businesses and firms are able to make the correct decisions. Clearly it can help them to avoid and reduce risks and simultaneously optimize their operational efficiency.
- Analytics tools in big data can be used in social media platforms for data harnessing; businesses around the world are streamlining their digital marketing strategies to enhance the overall consumer experience. Big data provides insights into customer pain points and allows companies to improve upon their products and services.
- To be more specific, big data associates the most common information from a greater number of resources to create highly actionable insights. As much as 43% of organizations do not have the required tools to filter out noisy and unwanted data, therefore they need to spend a huge amount of money to catch the relevant data in bulk, which can be avoided by using big data tools, which save money and time.
- To develop and create more sales leads in companies, big data analytics is used to increase revenue. In businesses, big data analytics tools are used to know how effectively they are services and products are

performing in the market and how their customers are responding. From this they can get an idea of where to invest their time and money more effectively.

- In big data insights, it is a must to always remain one step ahead of competitors. The market can be monitored to understand what types of promotions and offers are provided by competitors, to enable companies to decide on better offers for their customers. Also, big data insights give an idea into customer behavior in order to gain knowledge of customer needs and to provide an improved experience.

1.1.5 Challenges of health-care data management

The amount of health-care data available was expected to reach roughly 25,000 petabytes by 2020. Planning for and managing all that data can be an overwhelming, daunting task. Health-care organizations need to transition their operations toward a data-driven mentality: administrators and physicians must be diligent about collecting patient data, marketing departments must base their programs around data insights, and patients must be prompted to provide updated data whenever possible. Making data management a priority requires involvement from all players in the health-care industry—and this can present a challenge.

1.1.6 Health care as a big data database

Health care is a multidimensional process with the goal of prediction, prevention, and treating health problems in patients. The largest part of a health-care process is the health professionals, that is physicians and nurses, health facilities like clinics and hospitals that distribute medicines and use up-to-date treatment methodologies, and a supporting financing institution. Doctors can belong to different health sectors such as physiotherapy, midwifery, dentistry, nursing, psychology, and medicine. Based on an emergency situation health care is required at many levels. Doctors are the first point of consultation, that is for primary care, acute care requiring skilled professionals comes under secondary care, advanced medical investigation and treatment are called tertiary care, and highly uncommon diagnostic or surgical procedures are known as quaternary care. At all these levels, health professionals are responsible for specific types of information, such as patient's medical history (diagnosis and prescriptions related data), medical and clinical data (like data from imaging and laboratory examinations), and other private or personal medical data. In the past, either handwritten notes or typed reports were used to store such medical records [5], even the outputs from medical tests were stored in a paper filing system. Obviously, this practice is now outdated, with the oldest case reports existing on a papyrus text from Egypt that dates back to 1600 BCE [6]. In Stanley Reiser's words, "the clinical case records freeze the episode of illness as a story in which patient, family and the doctor are a part of the plot" [7].

With the development of computer systems health-care systems have digitized all clinical exams and medical records in a standard format in a widely adopted practice. In 2003, a division of the National Academies of Sciences, Engineering, and Medicine known as the Institute of Medicine coined the term "electronic health records" to represent record maintenance for the development of the health-care sector for the benefit of patients and clinicians.

1.1.7 Benefits of medical big data

The quality of health care is improved by using medical big data in an efficient manner, with this kind of technology used to increase analytical abilities, then predict epidemics, cure disease, build efficient health profiles, and enhance patients' quality of life. Also, it plays a role in outcome improvement, preventable death avoidance, model prediction, and decreasing resource wastage. Big data is a method of understanding the biology of a disease by interlinking a huge collection of information to develop relational models with meaning. Medicinal big data is behind the drive for the process of developing models in therapy. We are able to know much more about a patient with this type of technology and share information as early as possible, by gathering the signs of severe diseases at an early stage for rapid and less expensive treatment. Even very small matters are considered in the medical field of big data analysis. The advantages of big data in the health-care domain include the following:

- Facility performance optimization;
- Reduction in energy cost;
- Ease of accessibility of information;
- Real-time updates and alerts;
- Proactive maintenance of equipment;
- Health-care delivery cost reduction;
- Reducing the costs of research and growth.

1.2 Architecture of big data in health care

A batch layer and streaming layer are useful to simultaneously receive medical data. When the data are saved in data nodes they are referred to as batch modes. In batch modes data are transferred to a semantic module which influences the meaning of the original data using an ontology store. Hereafter, data preprocessing like cleaning and filtering operations are performed to the output data before being processed. Then the extraction and selection of features is analyzed for the preprocessed data through various phases. Finally, the features of the data are used in classification models to predict future health condition of patients. This process works in an offline scenario. In the stream framework, information is collected from many other sources such as medical sensors connected to the patient's body and when measuring blood pressure. Then, data synchronization is applied for the collected data according to the time and its missing values.

To split data into blocks adaptive preprocessor techniques are used which are based on a sliding window technique. Relevant information is then retrieved by the predictor component to generate a prediction model for each row in a window. Fig. 1.1 shows layer architecture of a proposal.

1.2.1 Batch processing layer

Batch computation is applied to the obtained data from data storage through various phases.

1.2.2 Data acquisition

To monitor the health condition of a patient continuously, several types of data need to be generated. Medical data may include structured data like traditional electronic health-care records (EHRs), semistructured data such as

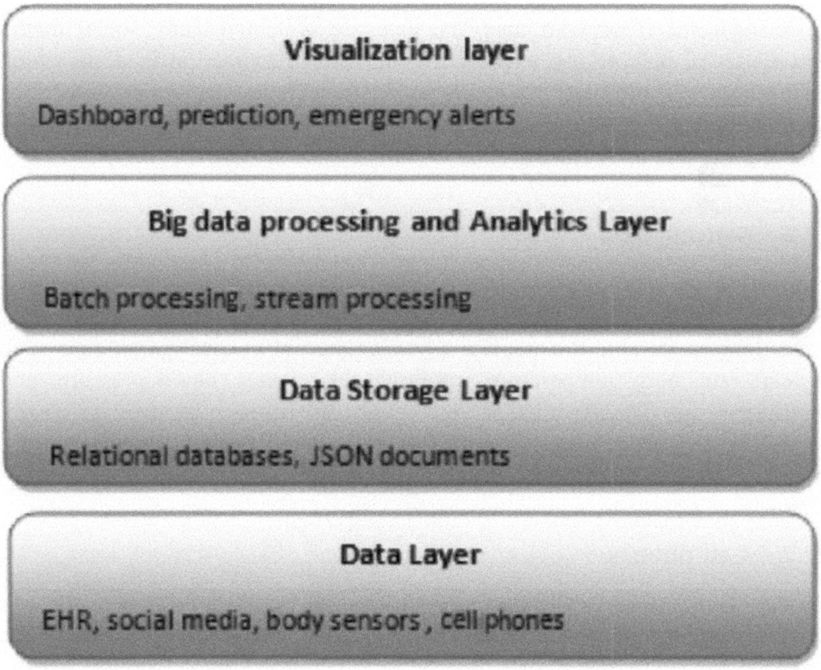

FIGURE 1.1 Layer architecture.

production of logs with medical devices, and the generation of unstructured data, such as biomedical imagery.

1.2.3 Electronic health-care records

In EHR, a digital format is followed to store a complete patient medical history. It is organized with a great deal of medical information narrating the health status of the patient, such as clinical information, demographics, medications, radiology documents, laboratory tests, diagnoses, doctor's notes, and payment notes. Therefore, EHR represents an excellent data resource for health-care analytics purposes. Moreover, EHR gives permission to exchange data between health-care professionals and the community.

1.2.4 Biomedical images

To diagnose disease and for delivery of care biomedical imaging is a powerful tool. However, processing of these kinds of images is a challenging task. Most of these images include data with more noise which needs to be removed to make the correct decisions regarding diagnosis and to aid physicians.

1.2.5 Social network analysis

To perform social network analysis, data are collected using social media such as social networking sites. Then knowledge extraction is done which can influence health-care predictive analysis such as uncovering infectious illnesses. Generally, data obtained in social networks are unreliable that can make the models untrustworthy.

1.2.6 Sensing data

In health-care monitoring systems various types of sensors have been employed. These sensors play a significant role in patient health monitoring. These devices are able to calculate a broad range of medical parameters such as heart rate, body temperature, cardiovascular status, blood pressure, and respiratory rate. To screen a patient health effectively, we need to ensure that living area of patients will be full of medical devices such as microphones, pressure sensors, and surveillance cameras.

1.2.7 Cell phones

The day-to-day use of cell phones has increased tremendously, with the cell phone becoming one of the most ubiquitous and well-known devices in the world. In the past decade, cell phones have become a basic communication device. They now offers many features and services compared to their first appearance. They are currently organized with many sensors such as accelerometers, cameras, fingerprint sensors, and satellite positioning services. They are unique devices regarding health data collection due to their wide use and multiple capabilities, allowing many health-care applications including child nutrition, heart frequency monitoring, and pregnancy monitoring.

The main goal of a data-acquiring module is to process the data collected from health-care sensors in multiple forms and then before carrying out normalization data flows.

1.2.8 Semantic module

The semantic module is depending upon ontology's, which constitute effective tools. It is used to represent practical knowledge in the biomedicine field. Ontology has the capability to gain knowledge about the biomedical field in a powerful and incremental manner which also permits automation and the ability to exchange information between various clinical systems. The major advantage of automation is to process a huge volume of patient data for medical personnel, specifically when considering that medical personnel are frequently overwhelmed by health-care tasks. The automation technology in health care is aimed at assisting medical staff, in order to improve the overall performance. It should be pointed out that automation helps people to perform their duties without replacing them. In the field of medical data interoperability is an important issue. In fact, health-care repositories lack homogeneity as they adopt different kinds of terminologies and structures. Therefore, integration and sharing information in health-care data can be complex. In this scenario, the establishment of a common structure and semantics is important in ontology, as it allows systems to share and reuse data in different ways. For example, the Web Ontology Language (OWL) denotes the standard exchange format according to the ontology data that employs XML syntax.

1.3 Preparation of data

In the big data context it is not possible to process raw data without any data preparation. Thus, it is suggested to review the data to ensure it is properly prepared to get the correct prediction and to ensure the reliability of data-mining techniques. There are two steps to data preparation: data cleaning and filtering of data.

1.3.1 Data filtering

Data filtering is defined as removing information which is not useful from health-care data in a large database on a specific criterion.

1.3.2 Data cleaning

Data cleaning is used for many functions such as noise reduction, missing data management, and normalization. To discover missing data in an entire database several methods are utilized. Noise removal is an added advantage in data cleaning because frequently medical records have noisy data which may lead to missing data. It is a tedious process to determine missing values in health-care data. If the missing data are manually corrected then there is the chance of errors being introduced which may affect the data quality. This leads to producing incorrect outputs while testing. Data-mining domains have several powerful techniques aimed at handling missing values, for example, EM algorithms and multiple imputation algorithms.

1.3.3 Noise treatment

There are two main approaches to discovering noisy data. In the first approach a data polishing methodology is used for noise correction. This methodology is implemented without difficulty and can be applied only in the case of a very amount of noise. The second technique is related to filters—these noisy filters identify and eliminate noisy content in data used for training.

Meanwhile, electronic medical software (EMR) provides incomplete information among noisy data. Electronic medical software the data sparsity search its origin when there is parameters collection are irregular over time. Hospitals record patient parameters when patients are present in the hospitals. In the case of noise reducing, several processing techniques have been utilized in biomedical imagery.

Basically, segmentation methods are used for data preparation in the biomedical of significant objects. In association with this, raw social media data preparation is a more challenging task. In addition to its large size and informal data, this type of data or information has the critical aspect of consisting of patient personal information. In social networks, data cleaning performs a major role as the processed data will be saved in the database at the end of data preparation.

1.4 Feature extraction and feature selection

In recent years multiple devices have been produced to gather medical data to monitor health care for both features and the number of instances. For a large volume of data significant features selection is crucial [8]. To handle a huge amount of features many methods have been proposed [8]. Mainly the feature extraction techniques are handled to reduce the number of attributes compared with the original data. Therefore particular tools are needed to store feature selection and extraction features. After the feature extraction process, a predictive model is generated by the selected feature subset.

1.5 Predictive model design

The main aim in the design of a predictive model is to create a design which is capable of giving predictions for new information related to previous information. Accuracy is the parameter used to predict the quality of a given predictive model. Based on the statistical and machine learning tools, the models are developed in house using the suggested architecture. A model store is used to store the batch processing outcome.

1.6 Data storage

In a big data framework storing data is one of the most challenging tasks, specifically in health-care monitoring systems which include huge volumes of information. Because of this, traditional data analysis is not suitable to manage those systems. This tool or component may be HDFS, NoSQL such as MongoDB and SQL databases, or a combination of all of them. So that it is more scalable and ensures high storage capabilities, in the proposed framework, the patient data gathered from heterogeneous sources can be separated into structured data like unstructured data such as biomedical images, or semistructured data like XML documents. These kinds of information will be stored in a raw data store in the target databases. A stream temp data store has the benefit of data streaming, such as using social media.

1.7 Stream processing layer

The stream data processing layer is defined as a combination of the data synchronization module, adaptive predictor module, and adaptive preprocessor module.

1.7.1 Data synchronization

The use of a data synchronization module in data processing is to discover whether processed data are in the correct order based on time. In addition, the process of data synchronization is avoided when inconsistent measurements and missing values are taken care of. Therefore inconsistent values are identified by fixing thresholds on the receiving data.

1.7.2 Adaptive learning

In many applications, it is assumed that the data preprocessing task is performed by learning algorithms, or at least that data have been already preprocessed before their arrival. In the majority of cases, such expectations do not match reality. This is particularly true for our proposed system which extracts streaming data from a stream temp store.

1.7.3 Adaptive preprocessor

An adaptive preprocessor initiates functions and operations by separating the data flow which arrives in time windows. In this condition, data streams are split into overlapping subsets of tuples by using a sliding window technique. The average is computed for every window and related to a predefined threshold. If the computed measure is greater than the threshold level then the value will be saved, otherwise an alarm will sounds. When the threshold values are terminated, the data collection phase proceeds by finding similar features which will be sent to the adaptive predictor component [9].

1.7.4 Adaptive predictor

By updating the predictors according to the changes made in the data there will be an increase in the accuracy level, otherwise the data will become unreliable over time due to data evolution. That is, the design of the predictor should modified based on the newly arrived data.

The connection between the adaptive predictor and adaptive processor is established by adaptive features. Depending upon this connection, the predictor transmits feedback to the preprocessor whether the updating is needed or not, hereafter the preprocessing unit uses a mapping function to provide the necessary raw data. Then the stream processing outputs store receives the outputs of the stream processing.

1.8 Query processor

The condition of patients is obtained by interlinking the query responses transferred to both the stream processing outputs store and the batch processing outputs store by the query processor.

1.9 Visualization layer

Multiple outputs are generated by the analytics layer such as, the patient health screening report and the predictive decision report. In health care, real-time visualization is based on the priorities, which is forwarded to the most complex information to make optimized decisions and avoid urgent situations. Examples of similar data include patient dashboards tracking daily health conditions, real-time alerts, and proactive messages offered by predictions. Fig. 1.2 presents the proposed big data architecture for health care.

1.10 Use of big data in biomedical research

A biological system like human cell which explores molecular and physical events of difficult interplay. To learn about the interdependencies of different devices and events in such a critical system, a biomedical experiment

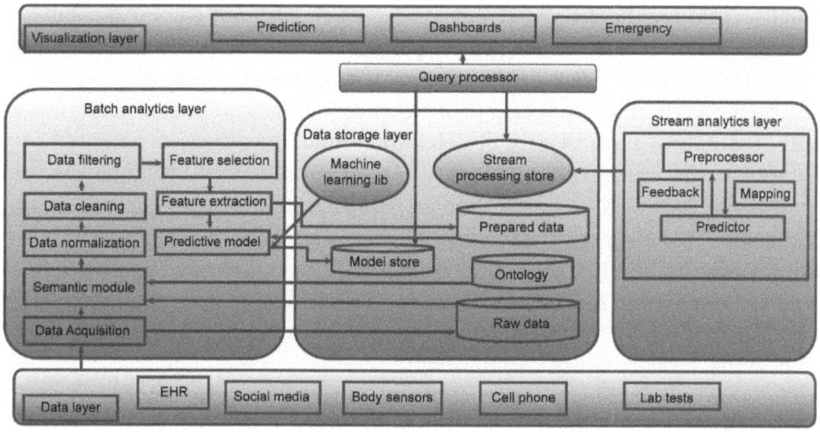

FIGURE 1.2 Big data architecture for health care.

collects data on smaller and simpler components, which needs many simplified experiments to create a large map of a given biological phenomenon. This means that data we have that are large in size help to understand the biological processes in a better way. Based on this unique idea, recent methodologies has been rapidly evaluated, for example, the quantity of generated data since the association of technologies such as genome-wide association studies and next-generation sequencing (NGS) to decode human genetics. NGS-based data give information that was not previously accessible and takes this scenario to a new level. The resolution has been increased with which we review or make recordings of biological events linked with particular diseases in a real time. The definition that huge amounts of data can tell us an efficient volume of data that frequently remains not identified or hidden in tiny experimental technologies.

1.11 Companies using big data in health care

The explanation of Topol's displays the very large potential of the big data platform for single-person medical therapy, which is not only life-saving but also improves the quality of life. Because of the developments in medical care processes wastage reduction has been achieved. There are three organizations leading in this process, as discussed below.

1.11.1 Dignity health: analytics helps prevent deadly infections

Tissue damage is a common problem, which can be caused by sepsis. Sepsis is an extremely dangerous illness which can lead to organ damage or death. In the United States, a quarter of a million people die through tissue damage annually. To identify the tissue-damaged area is very complex, especially in hectic environments like emergency rooms.

The United States has the fifth largest health system in the world and California has the largest single hospital provider. For advanced analytics in health, big data is used by Dignity Health, which can diagnose potential sepsis cases at the initial stages. Dignity Health monitors 120,000 patients per month in 34 hospitals using the Sepsis Bio Surveillance Program. From this it manages 7500 patients with potential sepsis per month. Getting the information from the EMR of all patients in its facilities, the program uses a rules engine to screening for factors including natural language processing that could identify sepsis. In the most sensitive cases, the system gives an alarm sound to the primary nurse or physician.

The implementation of the predictive analytics design in big data by Dignity Health has reduced mortality and ICU length of stay for sepsis patients. Moreover, 28 of Dignity Health's hospitals with the sepsis mortality program have shown an average 5% drop in mortality rates. This results in less time spent by patients in the ICU, and so the program has also provided benefits in the form of cost savings. Sepsis is the largest single expense for Medicare, the second highest cost for Medicaid, and the fourth most expensive cost for private insurance companies.

1.11.2 Express scripts: better decisions, healthier outcomes with big data

Express Scripts handles millions of prescriptions annually and delivers the medicines to patients at home. To crunch their big data they use high-powered analytics, and they also monitor individual patient's health effectively. This could have extremely positive outcomes for health.

- It is mandatory to know the effects of medicines before writing a prescription for painkillers, such as whether the patient is regarded as high risk or not by the health-care provider.
- The medical data could point out prescription-filling behaviors and psychosocial information that may indicate the development of a chronic illness.

1.11.3 United health care: monitoring fraud and waste, improving clinical outcomes

Big data uses advances in analytics methods for clinical developments, waste monitoring, financial analysis, and fraud detection.

"We have data that touches every aspect of the healthcare industry: member, claims, hospital, provider, clinical, operational and financial – and we use that data to get to root-causes of problems quickly, build innovative solutions and actively adapt to the changing healthcare landscape," says Ravi Shanbhag, United Healthcare's director of data science, solutions, and strategy.

1.12 Other opportunities for big data in health care

The possibilities for big data in health care are enormous. Other opportunities for big data using analytics are available for health-care organizations.

1.12.1 Episode analytics

To manage the unequal pressure on quality and cost, and also quality of care, health-care providers are required to make the best choices. To discover the performances of health-care providers, organizations use big data which gives them the opportunity to improve their quality, with the potential to redesign care delivery. To generate more versatile therapy plans for patients, episode analysis helps the health care industries' payers and providers to make the correct choices.

- *Population health analytics*. The main motive of population health analytics is to reduce the treatment costs for patients and improve the quality of service in an efficient way. We need to view how the data management and analytics is handled by the San Bernardino County and how they are taking decisions about the community.
- *Virtual care and wearable health-care technologies*. The increase in the cost of health care has been a problem for many years. Increasingly few people make the argument that the best inhibitory care is a good way of combating this problem. To give health insights into a patient streaming data with wearable devices can be combined.
- *Project data sphere*. In the field of industry-based pharmaceutical data transparency contributes a major role which also helps to protect the data which is accessed worldwide, for environmental analysis and organizations that share clinical trial information. Therefore it is important to help scientists to complete their research as rapidly as possible to improve health care for patients.

1.13 Use of health care in big data analytics

The impact and benefits of big data have been widespread and are increasing constantly. In health-care analysis big data has been very positive with lifesaving outcomes. Big data is described as a digital technology which generates a vast amount of information which is combined and analyzed using certain specific technologies. The purpose of data analysis in health care is to find or diagnose a problem before it becomes serious, which helps in providing treatment quickly and also creates an environment with greater patient involvement. When applied to health care, big data can use certain health data based on populations or individuals potentially to prevent epidemics, cure diseases, reduce costs, etc.

1.14 Unique features of big data in health care

1.14.1 Heterogeneity

Big data in health care often is in incompatible formats, which can be classified as either structured or unstructured data.

1.14.2 Incompleteness

To the extent that the data created by monitoring devices consists of continuous data streams, such as an electrocardiogram, it is difficult to consistently save it in a longitudinal record. It is also too expensive to store all big data in a health-care situation, leading to it being incomplete.

1.14.3 Data privacy

Owing to the sensitivity of health-care data, there are significant concerns regarding privacy and security. Extreme care should be taken to protect patient privacy—privacy concerns pose limitations in linking external data to individual insured data, which would improve consumer health-related experience and care.

1.14.4 Ownership

Health-care consumers legally own their health data, which may be stored and controlled by hospitals, physicians, laboratories, clinics, pharmacies, and government agencies in innumerable formats, however they may lack access to this data. To solve this problem a cooperative approach is the most effective way to meet this need.

1.15 Big data applications in health care

Applications of big data in health care include the following:

- patients thoughts and ideas for improved staffing;
- electronic health records (EHRs);
- real-time alerting;
- enhancing patient engagement;
- preventing opioid abuse in the United States;
- using health data for informed strategic planning;
- big data in the search for a cure for cancer;
- predictive analytics in health care;

- reducing fraud and enhancing security;
- telemedicine;
- integrating big data with medical imaging;
- a way to prevent/reduce unnecessary ER visits.

1.16 Internet of Things-based medical image processing

In the domain of medical information management, the IoT has wide applications. Currently, medical information management has highly significant demands in hospitals. The following terms are widely used in medical information management:

- prediction or identification;
- sample recognition;
- finding a medical record;
- diagnosis which includes patient identification;
- physician identification;
- sample identification which includes drug identification;
- medical equipment identification;
- laboratory identification;
- medical record identification which includes symptoms and identification of diseases.

These applications are more specifically described next.

1.16.1 Patient information management

Patient information management includes the medical history of the patient's family, the patient's medical history, different types of medical records, examinations, drug allergies, and other (electronic health) documents which can help physicians to improve treatment. Doctors and nurses will then take note of the patient's essential signs, including during therapies such as chemotherapy. Real-time screening is used to prevent the incorrect use of drugs/injections/dosages by automatically giving reminders, for example, to check drugs before administration.

1.16.2 Medical emergency management

In unusual circumstances, such as when there are huge numbers of casualties, family members cannot be reached, or the patient becomes, reliable and efficient storage for testing methods is required. Radio frequency identification (RFID) technologies are extremely useful, and are used t to identify relevant patient details including the emergency contact, patient's name, age, blood type, and previous medical history. This will hasten the admission formalities when the patient needs treatment urgently and provides more precious time for treatment.

3G video equipment installation in ambulances is also necessary to monitor patients in emergency situations, enabling staff at hospitals to have immediate access to details of the patient's condition on arrival. Remote medical imaging systems are also a part of rescue process, particularly when the patient location is distant from the hospital.

1.16.3 Medical waste information management

To establish a traceable medical waste tracking system, cooperation between transportation companies and hospitals is needed, so that it can be ensured that medical waste is transported properly to the treatment plant and to prevent the illegal dumping of biohazardous medical waste.

1.16.4 Drug storage

RFID technology can act as automatic storage of the entire health-care process by examining the working time and streamlining processes of that were previously managed on paper, and preventing stock shortages and helping with the recall of drugs. Overall, it will reinforce drug management and act as a safeguard to ensure medicines are prepared and provided to patients promptly.

1.16.5 Combating pharmaceutical errors

Information management in the pharmacy helps to ensure that medicines are correctly distributed to and received by patients. Pharmacy information management is currently carried out for patient use of medication, giving prescriptions, efficacy tracing, adjusting dosages, nursing administration, purchase of supplies, preservation of medicines and equipment in all climates, determination of shelf life, and inventory management.

1.16.6 Medical equipment and drug tracking

This helps us to provide accurate recording of items and patient identities, and also enables accurate accident handling, and recording of available medical devices and medicine. Basic data can be recorded accurately, including adverse events, patients involved, product use, areas with quality control problems, and locations of unused products, and faulty products, using this device.

1.16.7 Connected information sharing

Connected information sharing is used to create an improved and integrated medical network. In addition, authorized physicians can monitor the medical history of patients, including illness history, treatment, and insurance details. Patients are also able to select or replace their doctors. Also, central hospitals are connected with small town and rural hospitals for information sharing and to enable timely sharing of expert advices. Training can also be carried out and referrals arranged using this system.

1.16.8 Newborn antikidnapping system

Newborn kidnapping is a serious issue faced by general hospitals. For large general hospitals, there are restrictions to access of outsiders to hospitals' obstetrics and gynecology departments and women and children's hospitals. This is enhanced by combining mother and child identification and infant security management. Each newborn child should receive an RFID anklet with a unique ID, which has information such as the baby's unique details and the mother's information. To identify whether the family has the correct baby or not, the RFID anklet only needs to be checked by a nurse or doctor.

1.16.9 Alarm system

To track or monitor patients in hospital a real-time alarm system is used. This system will automatically produce an alarm when the patient is in distress. This system also helps to prevent patients from leaving hospital without clearance from a doctor. It also helps to prevent damage or theft of devices, and can be used to protect temperature-sensitive drugs and laboratory samples.

1.17 Internet of Things

Comparing to other industries the health-care industry has not been very quick at adapting to the movement of big data, so that the use of big data in the health-care industry is still in its infancy. For instance, to enrich health-care data, convergence of health care and biomedical big data with molecular pathology is needed. Convergence can help to untangle different instruments of action or other details of predictive biology. Therefore, to access the health status of patients, biomolecular and clinical data sets are used in a combined manner. IoT is one such source of clinical data in the health-care industry.

In health care, the IoT is a commonly used key tool which could still be improved further. The problems arising in Internet connection can lead to a lack of IoT use in objects such as cars, watches, refrigerators, and health-monitoring devices. Therefore, enhancing such objects with computer chips and sensors that enable data collection and transmission over the Internet would be welcomed. The information and communication systems used for gathering information and creating physical interactions with device technologies include near-field communication (NFC) devices and RFID tags and readers [10].

Advances in wireless sensor technology have opened up a new paradigm in the health-care industry. Ubiquitous and embedded medical monitoring is no longer a myth and will help improve the daily lives of the sick, elderly, infirm, and chronically ill people. Given the variety of medical body sensors and their applications, it is definitely of interest to implement a middleware system that can shield off these differences from the application layer. The aim of this chapter is to propose such middleware for a wireless medical body area network. The proposed middleware has the ability to support: (1) multiple sensors and applications, (2) plug and play feature, and (3) resource management. The middleware has been implemented and a prototype system has been built. The effective uses of the proposed middleware will be shown with the health-care monitoring system running on top of the middleware [11].

1.18 Use of the Internet of Things in health care

The IoT could have a great impact on the creation of more personalized and patient-oriented medicine. According to the latest calculations, the cost of creating a solution on the basis of IoT will reach up to $1 trillion by 2025. The following discusses several ways of implementing the IoT in health care.

1.18.1 Remote patient monitoring

There are many people who suffer from chronic disease and have to visit their doctors regularly. Remote patient monitoring (RTM) is a system that is suitable for people who have problems with their heart or are diabetics. The patient only needs to carry the RTM device which automatically alerts the doctor about any deterioration in the patient's condition.

1.18.2 Wearables

Nowadays, there are many devices that patients can wear every day, for example, fitness bands, blood pressure and heart rate monitoring cuffs, etc. These popular gadgets monitor not only the user's daily activity but also collect data about the number of steps taken, burnt calories, etc. These devices change patients' lives, especially elderly people, because they allow their health conditions to be constantly tracked. Wearables can send notifications to family members about changes in routine activities or any other condition variation of the user.

1.18.3 Better drug management

In order to produce and manage medicines, people spend a lot of money. IoT devices can provide an opportunity to follow all safety standards of the pharmaceutical market. One of the best examples is the smart vaccine fridge, which is able to prevent vaccines from spoiling and monitor their conditions 24/7.

1.18.4 Hospital management

There are many ways in which IoT can be used to optimize everyday hospital activities and reduce costs. Lost and stolen equipment leads to high unnecessary costs, and this problem can be solved by integrating sensors to equipment, for example, RFID or Bluetooth, allowing tracking at any time.

1.19 Health care in various countries

South Korea has one of the most advanced IT infrastructures in the world, and the application of IT in health systems is rapidly progressing from computerization to information systems, ubiquitous systems, and smart systems. However, a major problem concerning health-care resources lies in the regional disparities between medical services. Most private medical facilities are located in urban areas, and around 90% of physicians are concentrated in cities, while (only) 80% of the population lives in urban areas. South Korea is becoming an aging society faster than any other country, including Japan. With regard to the increase in the elderly population, there has been an increase in medical expenditure for chronic degenerative diseases, which has become a large social burden. In recent years, the South Korean government has attempted to reduce the financial burden through comprehensive health-care reforms encompassing the expansion of health-care facilities and the introduction of the Long-Term Care Insurance Program [12].

1.19.1 Simultaneous reporting and monitoring

To save lives in emergency situations such as diabetes, asthma attacks, and heart failure, real-time monitoring with connected devices is needed.

When a smart medical device is interfaced with a smart phone application it is able to gather the required medical and health-based information. The physician can then transfer and collect data from a medical device into the smart phone.

A review which was conducted by the Center of Connected Health Policy intimated that RTM provided for heart failure patients had a 50% reduction in the 30-day readmission rate. We can transfer and collect most private data like oxygen, blood pressure, weight, ECG results, and blood sugar levels using IOT devices, and this can be stored in the cloud and the data used when needed; the data can also easily be shared with authorized persons. IoT devices allow physicians, insurance companies, participating health firms and/or external consultants to view patient data wherever and whenever required.

1.19.2 End-to-end connectivity and affordability

Patient care can be monitored using automated systems. IoT can act as an automatic system with the help of health-care mobility solutions and many new technologies, including next-generation health-care facilities.

IoT in health care introduces machine-to-machine communication, interoperability, information exchange, and data movement which make the health-care service more effective.

The protocols used for connectivity include:

- Bluetooth LE;
- Wi-Fi;
- Z-wave;
- ZigBee;
- other modern protocols.

IoT can help health-care personnel to identify illnesses and ailments remotely. In this way it provides a new revolutionary treatment. Consequently, the cost of health care is reduced by avoiding unnecessary visits. It also helps in quality resources utilization, and improvements in allocation and planning.

1.19.3 Data assortment and analysis

Transmission of a huge amount of data using a health-care device in a short time can lead to storage difficulty, however data management has become easier with the availability of the cloud.

For health-care distributors initiating data acquisition from multiple devices and sources is easier compared to analyzing the data manually; IoT devices can collect data, form reports, and analyze data in real time.

They also processes raw data more efficiently. This process can be done because of the cloud-based system, which enables providers to access their final reports with graphs.

Health-care operations allow organizations to obtain vital health-care analytics and data-driven insights to speed up decision-making and reduce errors significantly.

1.19.4 Research

To perform research work the IoT can be used. This is because of a maximum quantity of data collected about the patient's diseases and relevant treatments collected by the IoT, but without the IOT it may not be possible to collect as much data. IoT does not only saving time but also money as it is able to perform tasks much more efficiently than if they were performed manually.

1.19.5 Data security and privacy

One of the most significant threats posed by IoT is data security and privacy, especially as some IoT devices lack data standards and protocols.

In addition, there is significant ambiguity about data ownership regulation. All of these factors make the data highly vulnerable to cybercriminals who try to hack into data systems, to illegally access doctor and patient personal health information (PHI). Cybercriminals may attempt to misuse information about patients or create dummy IDs in order to sell and buy drugs and medical equipments; they may also file fraudulent insurance claims.

1.20 Disadvantages of Internet of Things in health care

- *Privacy of patients.* Privacy is a most important issue faced by many organizations. Doctors work with patients' personal information, and so patients should be confident that their information will be kept private. Unfortunately, in IoT-based health-care systems there is an opportunity for these data to be hacked.
- *Accidental failures.* It would become a huge issue if a small error was mistakenly introduced in an IoT-based patient health monitoring system. To prevent the failure totally, none of the technology introduced yet.
- *Malware.* The frequency of malware has increased, causing infections in operating systems and IoT applications in the health-care field. However, this can be avoided by installing good antivirus software and firewalls.
- *Lack of encryption.* Another method to protect information is by implementing an encryption system. This can prevent unauthorized access to the system. Unfortunately, there is not yet a 100% fail-safe encryption standard, and the systems remain vulnerable to hackers.

1.21 Medical Internet of Things and cyber-physical systems

The health sector has been revolutionized by the data pervasiveness of wireless sensors. With the development of wireless sensor networks, a large volume of data generation, mobile health, and real-time patient monitoring from home are possible.

EMR is considered as a sole distributor of patient health-based information. However, with the introduction of real-time health systems, EMR has become just a part of the overall health IT ecosystem. Patient-based wearables, such as bio-shirts and body implants (smart capsules), feed data into the EMR, which contains the patient clinical data. A health analytic solution can then help a physician to treat an emergency or obtain an insight regarding an impending one. Different types of wearable evaluate tasks like glucose level monitoring, fall detection, position detection, and location tracking. Middleware customization should be improved to integrate different sensors with different standards. In a telehomecare-based application a middleware design includes two layers, one is a device management layer and other is a data management layer. Middleware is defined as the interface between telehomecare applications and vital signs sensors which receive vital signs data from the sensors and channels. Various other works such as those in Refs. [13] have focused on the difficulties to consider when designing middleware for various remote-sensing devices in health care.

1.22 Social network data

Traditionally, it is assumed that physicians and their measuring equipment are the only sources of a patient's data. With the advent of behavioral informatics and their applications to personalized health-care systems, it is now possible for patients to be collaborators in their treatment. For example, consider a teenager who visits hospital for treatment and the physician performs various measurements such as ECG or MRI and feeds the readings into EMR to perform a diagnosis. However, when he checks the teen's Facebook page he discovers that the subject posted a suicide message the night before. Treating the patient while omitting this critical information would be inappropriate. Using his own intuition without being able to incorporate the information with other readings would also not be sufficient. Hence, there is a need for the health analytic solution to accommodate the EMR data with information on social networks to perform a complete diagnosis.

1.23 History of the Internet of Things in health care

Patient's Internet-connected devices have been introduced in different formats, with data being collected from temperature monitors, fetal monitors, electrocardiograms, health information tracking, which then needs to be communicated with a health-care professional. To deliver highly confidential and valuable data IoT devices have been used as instruments and also the medium for direct patient-to-physician communication.

In the past decade, the application of IoT devices in health care has included smart beds which detect whether the bed is occupied or not and when the patient is attempting to get up. Appropriate pressure can be adjusted by the smart bed to support the patient without the manual input of nurses.

Home medication dispensers are another smart technology in health care which upload data or information to the cloud automatically when medicine has not been taken.

1.24 Challenges of the Internet of Things in health care

The IoT is a great technology however the privacy and security of personal data are not guaranteed. As with many devices used today secure methods are provided to interact with the cloud, however they remain vulnerable to hackers.

IoT devices can cause significant harm if stolen and the personal data are misused. When they are insecure, the health-care implications can be frightening. In a 2012 episode of *Homeland*, a hacked pacemaker caused a heart attack. Dick Cheney, a former Vice President of the United States, repeatedly asked for the wireless capabilities of his pacesetter to be disabled.

1.25 Future of the Internet of Things in health care

Compared to past decades current wearable devices have been converted into products which are used by patients in order to send information directly to physicians. Therefore the use of these types of devices allows doctors to screen vital signs in real time, including statistical parameters such as glucose levels, heart rate, and even fall detection. Medical remote patient-monitoring devices which gathers real-time information effectively have been a great advance, and they can be helpful for the monitoring of patients by the physicians from their own homes.

Smart home medication dispensers are another medical device introduced not only to know whether a medicine has been taken as prescribed by health-care professionals, but also useful to ensure medicines are stored at the correct temperature.

Also, as notified in Aruba's report, IoT applications in health care help with medical inventory management, including ordering and managing resources, such as thermometers, wheelchairs, and IV poles within a hospital, and helping staff to work more effectively.

In the future, IoT devices will become increasingly used in health-care facilities and will produce more unique intelligent services for the next generation. IoT devices will be able to react to real-time data, which will be preprocessed at the device level, enabling some tasks to be executed, and then the data will be sent back to the patient and their clinical team.

1.26 Internet of Things with ThingSpeak

Based on research performed by the IoT, it is well known that due to carelessness in health-care systems every month the number deaths is increased because of inadequate attention given to patients and their health. Improper time management is one of the reasons for this problem, and another is that, due to the heavy workload given by management, staff may forget to monitor their patients' health properly. The reason behind this project is to convey to individuals their health responsibilities and the importance of health checkups. Nowadays technologies has improved so much accordingly the diseases and inability also increased because of over workload, and not proper in taking. The invention of the IoT will make will help to provide efficient treatments for patients, and the enables features like transferring patients' PHI from their cell phones to health-care providers. Based on the data provided by the patient, doctors are able to check their historical health data.

IoT-based health-care systems have recently become cheaper, with IoT becoming the backbone of health care. It is a good platform which connects patients and doctors, and IOT devices are used in measuring body parameters like temperature, moisture, heartbeat, and ECG, which are also used to measure the patient's condition giving very accurate results. ThingSpeak is a platform used to store all the information in the cloud using cloud storage. Different kinds of sensors and modules are used to perform various tasks in IoT. ThingSpeak also provides security and facilitates access to information at anytime and anywhere in the world, which is very useful for doctors at the time of treatment. In critical or emergency situations the system sounds an alarm and can send messages about medicines, location changes, conditions, etc.

IoT-based architecture and their working principles is based on a hardware system which contains the sensors to measure the parameters such as the number of controllers and modules so that further processing can be continued and data uploading to the cloud storage is performed. A cloud service provider is selected to save the collected data and process them over the website or mobile application, all the transferred data are collected on an android application which gives a friendly graphical user interface to aid access to the screened health data.

1.27 Introduction to the cloud

1.27.1 Cloud computing

Line computing is also called cloud computing, and works based on the Internet computing which produces shared processing assets and data to personal computers and also other components when there is workload pressure. It is a model that permits use in any location, with on-demand access to a pool of structured computing sources such as servers, storage, applications, services, and networks. This can be very quickly reserved and downloaded with minimal management effort. In order to save third-party information to the centers, cloud computing uses storage solution providers and enterprises. It is also very reliable for sharing of resources to achieve coherence and economy of scale, similarly to a utility network.

1.27.2 Cloud storage

Cloud storage is similar to a data storage model. In cloud storage the information is stored in logical pools, whereas the physical storage needs a span of many servers and often a number of locations. Hosting companies organize and own the physical environment storage. These cloud storage distributors are only responsible for maintaining the availability and accessibility of information and securing operation of the physical environment. Cloud storage providers sell storage to individuals and companies, which can be bought or leased from the distributors as their need.

- https://thingspeak.com;
- transfer sensor data to the cloud.

1.28 ThingSpeak channels

ThingSpeak channels are used to save data and information collected by the ThingSpeak application. Eight types of fields are included in each channel and can hold any type of data, with location data for three fields and status data for the fourth field. ThingSpeak apps are used to analyze and visualize data after its collection of data is complete.

1.28.1 Channel setting of the cloud

The channel setting process is as follows:

- Name of the Channel: Type a unique name for the ThingSpeak channel;
- Description: Type a description of the ThingSpeak channel;
- Field: Tick the squared box to enable the field, and then type the name of the field;
- Metadata: Provide information about the channel data, including XML, or CSV data;
- Tags: Type the keywords that identify the channel. Separate tags with commas;
- Latitude: Enter the position of the sensor which collects data in decimal degrees. For instance, the latitude of India is 51.5072;
- Longitude: Enter the position of the sensor which collects data in decimal degrees. For instance, the longitude of India is −0.1275;
- Elevation: Specify the position of the sensor or thing that collects data in meters. For example, the elevation of India is 35.052;

- Make Public: Tick this if you want to make the channel publicly available;
- URL: Specify the URL of your website that contains information about your ThingSpeak channel;
- Video ID: Specify the full path of the video URL, if you have a YouTube or Vimeo video that displays your channel information.

1.28.2 Using the channel

It is possible to put the data into a channel from a device, website, or by using another ThingSpeak channel. After that the data can be monitored and can sent using ThingSpeak apps.

Channel details:

Channel name: Health care system
Channel ID: 445XXX
Author: cvb@gmail.com
Access: Private, Public

1.29 Application working

This application can be used by doctors and/or patients.

1.29.1 Doctors

In this application, doctors are able to register their personal information such as address of their hospital and their work field, and also can have a unique ID for the application after successful registration. After that the app will check the status of the entire registration via the patient in the app. The meaning of status here is the information of all patients who are connected with that particular doctor. Each doctor may have any number of patients and every patient should have their unique patient ID. This unique ID will interconnect the doctor and patient so that the doctor can track or check the record of each patient.

1.29.2 Patients

When the registration is completed in the hardware device the patient will receive a unique ID and password. By using the app the patient will be able to view their personal information and all their data in the various fields in the app. The app consists of all the historical data up to date which can be filtered by patients according to their requirements. Patient can see their health status in different fields at any location or time.

1.29.3 Login credentials

The doctor will require the following two credentials for login:

Doctor ID;

Password.

The patient will also require two credentials for login:

Patient ID;

Password.

1.29.4 Registration credentials

The doctor needs to complete their personal details, including password, clinic name, address, and profile.

1.30 Summary

Big data analytics in medicine and health care is a very promising process for integrating, exploring, and analyzing a large amount of complex heterogeneous data with different natures: biomedical data, experimental data, electronic health records data, and social media data. The integration of such diverse data makes big data analytics intertwine with several fields, such as bioinformatics, medical imaging, sensor informatics, medical informatics, health informatics, and computational biomedicine. The application of IoT is becoming increasingly common in every aspect of the health-care industry. In this chapter, we have explored various applications of IoT in different areas of the medical industry. IoT changes the way facilities are delivered to the health-care industry. These technologies improve products, creating a greater cumulative effect by bringing together minor changes.

The health-care industry needs continuously to be in the forefront in the adoption and utilization of information and communication technologies (ICTs) to enable efficient health-care organization and medication. Recent improvements in ICT and the emergence of the IoT have opened up new avenues for exploration and investigation in every field, including health-care and medical industries. Hospitals have started utilizing the cell instruments for correspondence and within this plan the IoT has been used and combined with Wi-Fi sensor nodes such as NFC tags, RFID, and small sensor nodes. The use of a cellular agent in the health-care process also provides for an opportunity to investigate enhanced services to patients and staff through its enhanced mobility.

References

[1] M. Tzourakis, in: R.Y. Wang (Ed.), The Healthcare Industry and Data Quality, International Conference on Information Quality, 1996.

[2] Unstructured Data in Electronic Health Record (EHR) Systems: Challenges and Solutions. Datamark Incorporated. Oct 2001.

[3] M. Bachiri, A. Idri, J.L. Fernández-Alemán, A. Toval, Mobile personal health records for pregnancy monitoring functionalities: analysis and potential, Comput. Methods Prog. Biomed. (2016) 121–135.

[4] A. Guyon, A. Bock, L. Buback, B. Knittel, Mobile-based nutrition and child health monitoring to inform program development: an experience from Liberia, Glob. Health Sci. Pract. 4 (4) (2016) 661–674.

[5] P. Pelegris, K. Banitsas, T. Orbach, K. Marias, A novel method to detect heart beat rate using a mobile phone, in: Conference Proceedings: IEEE Engineering in Medicine and Biology Society, 2010, pp. 5488–5491.

[6] A. Jović, K. Brkić, N. Bogunović, A review of feature selection methods with applications, in: Proceedings of the 38th International Convention on Information and Communication Technology, Electronics and Microelectronics, MIPRO 2015; May 2015; Croatia, pp. 1200–1205.

[7] S. Ramírez-Gallego, B. Krawczyk, S. García, M. Woźniak, F. Herrera, A survey on data preprocessing for data stream mining: current status and future directions, Neurocomputing 239 (2017) 39–57.

[8] K. Jee, G.H. Kim, Potentiality of big data in the medical sector: focus on how to reshape the healthcare system, Healthc. Inform. Res. 19 (2) (2013) 79–85.

[9] J. Gubbi, et al., Internet of Things (IoT): a vision, architectural elements, and future directions, Future Gener. Comput. Syst. 29 (7) (2013) 1645–1660.

[10] S. Doyle-Lindrud, The evolution of the electronic health record, Clin. J. Oncol. Nurs. 19 (2) (2015) 153–154.

[11] R.F. Gillum, From papyrus to the electronic tablet: a brief history of the clinical medical record with lessons for the digital age, Am. J. Med. 126 (10) (2013) 853–857.

[12] Y. Yin, et al., The internet of things in healthcare: an overview, J. Ind. Inform. Integr. 1 (2016) 3–13.

[13] A.B. Waluyo, S. Ying, I. Pek, J.K. Wu, Middleware for wireless medical body area network, in: Proc. IEEE Biomed. Circuits Syst. Conf. (BIOCAS), November 2007, pp. 183–186.

Chapter 2

Brain–computer interfaces and their applications

Vaibhaw, Jay Sarraf and P.K. Pattnaik
School of Computer Engineering, KIIT Deemed to be University, Bhubaneswar, India

2.1 Introduction

The brain–computer interface (BCI) is a medium to record the neurological activities of the human brain, either wired or wireless, and pass it to external devices to help humans to interact with their surroundings without the use of their body. The main aim behind the developments in this area is to help people with neurological disorders. These people have full control over their brain but cannot control their body due to neuronal damage. This type of BCI can greatly improve their quality of life.

The BCI is designed in such a way that it recognizes the electromagnetic waveforms generated by the neurological activities in the brain and then decodes these to know the intention of the person, so that it can convert those patterns into commands which it can send to the computer to carry out actions on that person's behalf, such as turning a wheelchair, switching off lights, or picking up a water bottle.

The BCI generally consists of steps from signal acquisition, its preprocessing or simply a signal enhancement, then the feature extraction followed by classification [1]. The classified signals are then used as input for the control devices.

Signal acquisition (Fig. 2.1) generally captures the brainwaves or signals by means of hardware [2]. The preprocessing stage processes these signals or, in other words, it enhances the signal for further processing. The feature

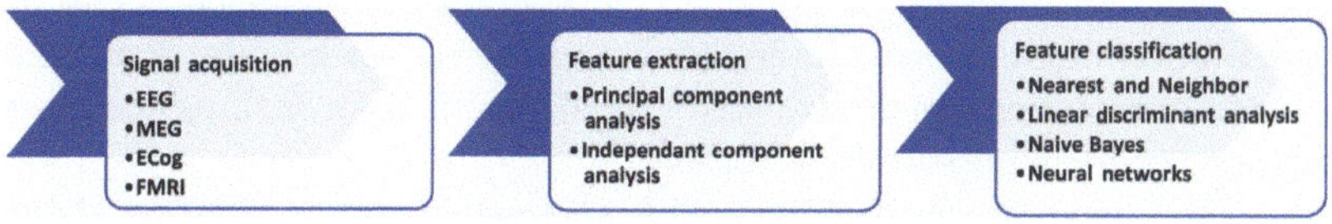

FIGURE 2.1 Summarizing the steps involved in the brain–computer interface process.

An Industrial IoT Approach for Pharmaceutical Industry Growth.
DOI: https://doi.org/10.1016/B978-0-12-821326-1.00002-4

extraction stage it responsible for identifying valuable information in the brain signal [3]. Finding and extracting these valuable features from the signal is a challenging task because they contain a lot of noise and distortion as the brainwaves are not stationary [4]. There is another problem which is that the information which is extracted from the brain may have many dimensions. These dimensions may be reduced, preserving the relevant information which also increases the complexity of this stage. The feature selection stage is optional and includes selecting relevant features according to the application. It has been found that this generally improves the classification [5–7]. The classification stage classifies those feature vectors on the basis of which features are really important for better pattern recognition. Finally, the control interface stage converts or translates those classified signals into commands that the device will understand.

BCI technology was traditionally unattractive because digging into someone's mind to extract information appeared to be strange. Also, BCI technology was limited to the laboratory as it was not portable and the technology required was not available or was extremely expensive.

The technology advances in the field of computer science over recent decades and also the development of cheap and compact computer hardware parts have greatly changed this. Also, the idea that it can serve disabled people has increased its social acceptance.

2.1.1 Neuroimaging approaches in brain–computer interfaces

The most crucial stage in the BCI is the recording stage which records the valuable brain signal. Electrophysiological and hemodynamic processes are the two main brain activities that we focus on. Electrophysiological activities are those activities which are generated when electrochemical transmitters exchange information between neurons.

Electroencephalography (EEG) is used widely as a neuroimaging modality because of its relatively low cost and portability. The BCI based on EEG includes a sensor that detects EEG signals from different areas of the brain; the skull and scalp greatly affect the standard of the signal.

Depending upon the placement of the sensor it can be noninvasive or invasive. A noninvasive procedure does not require surgery, whereas the invasive procedure is an approach that requires surgery. In simple language, in the noninvasive procedure, the sensors are placed over the brain. And in the invasive procedure, the microelectrode sensors are implanted inside the brain. The motor recovery is limited in the noninvasive approach due to the low-resolution brain signal. The invasive approach gives a brain signal with the highest resolution but it also has several health risks such as the tissue acceptance of the microelectrode sensors.

Fig. 2.2 lists the various types of approaches to record the above-discussed electrophysiological activity of the human brain.

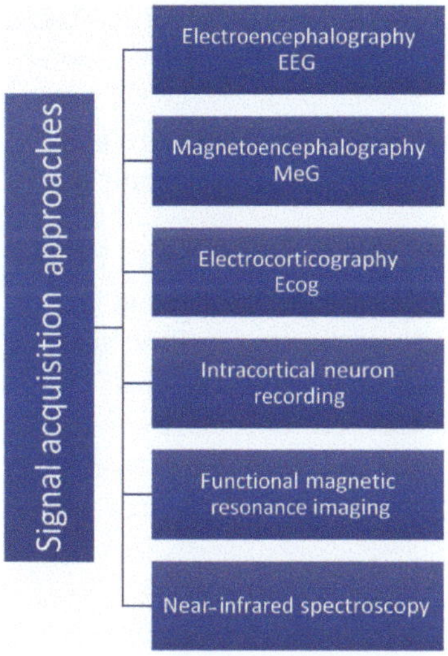

FIGURE 2.2 Various types of signal acquisition approaches.

2.1.2 Electroencephalography

Electroencephalography or EEG measures the electromagnetic brain signal-generated by neurological activities in the brain, which can be recorded using the noninvasive technique by placing the electrodes over the scalp. This is the reason why this is so widely used as a recording modality. However, the signal recorded from this type of modality is quite poor as the signal has to pass through the brain, skull, and many layers of tissues under the scalp. It is also used in brain disease detection and provides a cheap substitute for CT scanning [8].

The EEG recording device consists of electrodes attached to amplifiers, an analogue-to-digital converter, and a storage device. Electrodes are responsible for capturing those brain signals, the amplifier amplifies them, and the analogue-to-digital converter converts them into digital signals and then these signals are recorded and stored using a recording device.

These signals are very prone to external noise, which can be generated by external power lines or external electrical equipment. To reduce this noise, electromagnetic shielding can be used. These recorded signals can be further decomposed into different frequency bands on the basis of their frequency range, such as delta, theta, alpha, beta, and gamma [9].

The delta signal has a frequency below 4 Hz and is mainly found in babies and decreases as they grow. These signals are only found in adults when they are in a deep sleep state. These signals are quite abnormal in an awake adult and may suggest a neurological disease [10]. Theta waves have a frequency of 4–7 Hz. These types of waves are mostly found in young children and very low waves are found in the awake adult. These waves are associated with concentration [11,12].

Alpha waves have a frequency range 8–12 Hz and are found at the back part of the brain over the optical region [13]. These waves are associated with mental effort [14].

Beta waves have a frequency range of about 12–30 Hz. They are mainly found in the central and front brain regions. These waves are generally associated with attention and motor activity [15,16].

Gamma waves have a frequency of 30–100 Hz and are related to some motor functions. Experimental studies show that gamma waves and motor function have a relation during maximum muscle contraction and the gamma waves are replaced by beta waves during weak contractions [17]. Gamma waves are generally not used in EEG because they are affected by external noise. However, researches are on-going into their use in BCI because they provide information in higher resolution as compared to beta and alpha waves which are used traditionally.

2.1.3 Magnetoencephalography

Magnetoencephalography or MEG imaging is performed noninvasively. It works on the principle of magnetic induction. It measures the magnetic field that is produced due to the intracellular current which can be measured outside the skull [18]. The measurement of the magnetic field has an advantage over detecting electrical current as the magnetic fields are less affected by the skull and tissues [19]. MEG is highly sensitive to outer magnetic interference and so it is generally performed in a room with magnetic shielding.

It gives a high-resolution image of the brain waves as compared to EEG, however it is not generally used in BCI as it is not portable and the equipment is also very expensive.

2.1.4 Electrocorticography

Electrocorticography or ECoG is an invasive imaging technique in which the sensors are implanted inside the brain. It gives a high-resolution image as compared to EEG as the sensors are implanted inside the brain which can be responsible for various health problems. Experiments in animals have shown that the electrode implant can perform well with high accuracy over months without any recalibration. In humans, it is used to acquire alpha and beta signals [20], and also gamma signals [21], which are produced during various motor actions.

2.1.5 Intracortical neuron recording

Intracortical neuron recording is an invasive neural interface process in which the microelectrodes are implanted inside the brain for the measurement of the electrical signals. Three main signals are obtained using this technique. These signals include local field potential (LFA), single-unit activity (SUA) and multiunit activity (MUA).

LFA and SUA are obtained by low-pass and high-pass filtering of the signals from a single neuron, whereas MUA is acquired from filtering signals from multiple neurons [18]. It provides much higher resolution as compared to EEG, however the quality of the signal can decrease over the course of time due to the interaction of the tissues with the microelectrode, therefore periodic recalibration is necessary for this type of technique.

Intracortical neuron recording was first tested on monkeys and rats, and it was found that it can indicate the nature of movements of body parts and also the direction of movement, but it was unclear whether the same pattern existed when there was no physical movement. This was made clear by an experiment performed by Taylor et al. [22] on a monkey in which real and virtual arm movements were performed by the monkey, which suggested that the same pattern exists when there are no physical movements.

2.1.6 Functional magnetic resonance imaging

This technique is used to detect changes in cerebral blood volume and oxygen level in blood during the process of neural activation. These changes are detected with the help of electromagnetic fields. It is a noninvasive technique, which is performed using MRI scanners. It has high-resolution imaging capabilities but lags due to physiological delay because of the hemodynamic response [23], which makes it unsuited for rapid communication and it is also very sensitive to head motion artifacts. It is generally limited to the laboratory because of its lack of portability and also because the required equipment is relatively expensive.

2.1.7 Near-infrared spectroscopy

Near-infrared spectroscopy (NIRS) uses infrared light to acquire changes from the human cerebrum noninvasively. The infrared light can easily penetrate the scalp and skull to measure oxyhemoglobin and deoxyhemoglobin concentrations in overactive and inactive parts of the brain during neural activity.

The imaging resolution for NIRS is quite low [24] but it is cost-efficient, highly portable, and also has an acceptable resolution that is of the order of 100 ms [25].

This technique is quite sensitive to disturbance like head motions and hair obstructions. NIRS also has a limited information transfer rate because of delays in the hemodynamic response. Despite all these drawbacks, NIRS is a very good alternative to EEG, as neither a corrosive electrode nor conducting gel are required.

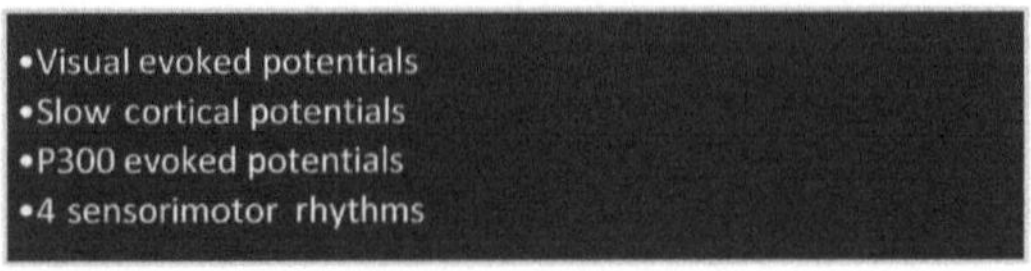

FIGURE 2.3 Types of control signals in BCI.

2.2 Control signal types in brain–computer interfaces

There are millions of brain waves at a time in the brain related to different phenomena and the origins of most remain unknown and are difficult to decode, however there are some brain waves, the physical phenomena of which can be successfully decoded so that people can learn to regulate them intentionally. These are the signals that are regarded as a control signal in BCI. Studies show that there is a large number of signals that can be successfully used as a control signal in BCI, some of them are described in Fig. 2.3.

2.2.1 Visual-evoked potentials

Visual-evoked potentials are brain activity modulations which occur after receiving visual stimuli [26]. These signals can be easily detected, as they increase greatly when the stimulus is moved in the region of the central visual field [27].

Visual-evoked potentials are classified: (1) on the basis of the morphology of the optical stimuli, they can be produced using graphic patterns or flash stimulation; (2) on the basis of frequency of visual stimulation, they can be steady-state visual-evoked potentials (SSVEPs) or transient visual-evoked potentials (TVEPs); and (3) on the basis of field stimulation, they can be half-field VEPs, part-field VEPs, or whole-field VEPs.

2.2.2 Slow cortical potentials

Slow cortical potentials (SCPs) belong to the part of EEG signals below 1 Hz [28]. They are generally related to the cortical activity; negative SCPs are associated with an increase in neuronal activity, whereas positive SCPs correlate with a decrease in individual cell activity. SCPs can be self-regulated by normal users as well as paralyzed users with the help of normal training. Thought translation devices are used to train people to generate voluntary SCP shifts. It has been discovered that SCP modulation training can be greatly affected by the user's learning capability, so self-regulation training is mostly recommended at the early stage. SCP provides acceptable accuracy but lags in providing decent information rate.

2.2.3 P300-evoked potentials

Positive peaks found in the EEG signals which are formed because of frequent auditory or visual stimuli are known as P300-evoked potentials. BCI which are based on P300-evoked potentials do not require extensive training but performance is affected as after some time the user will become used to the infrequent stimulus which will result in a decrease in the amplitude of P300-evoked potentials [29].

2.2.4 Sensorimotor rhythms

These are associated with motor movement, as the amplitude of the sensorimotor rhythm is found to vary during any motor activity. The changing amplitude of sensorimotor rhythms is not limited to actual motor movement as it was found that thinking about the motor movement generates a similar modulation pattern. BCI based on sensorimotor rhythms can be easily implemented as users can easily learn how to modulate these signals [30].

2.3 Types of brain–computer interface

The BCI is broadly classified as exogenous or endogenous on the basis of the nature of the signal used as an input or it can be grouped together in two categories as synchronous or asynchronous on the basis of processing mobility. It can also be dependent or independent.

The exogenous BCI is a type of BCI in which the neural activity is activated by an external stimulus [31]. This type of BCI requires minimal training as the control signals are easily set up. It also offers a high information transfer rate. In contrast, in an endogenous BCI the neural activity which is activated by self-regulation of brain rhythms is used. In such a case, the user must be trained to generate the specific brain patterns which the BCI can decode. The main advantage of endogenous BCI over exogenous BCI is that the user is free to control the BCI freely, whereas, in exogenous BCI the user has limited choices. Endogenous BCI can be very helpful for patients with affected sensory organs. The main drawbacks to endogenous BCI are that it requires the user to be trained for self-regulation of brain waves, which is a long process, and also the information transfer rate is low.

Synchronous BCI analyses brain signals for the chosen time period only, so the user is allowed to send a command to the BCI during the chosen time only, any brain signals sent after this time are simply ignored, which makes it a slightly more unnatural mode of interaction. The user can also avoid noise as they can blink and have other eye movements after the chosen time as these signals will not be analyzed, which makes the system simple. Asynchronous BCIs, however, continuously analyze brain signals, which makes them a more natural mode of human–computer interaction meaning they are more complex than synchronous BCI.

2.4 Features extraction and selection

The BCI is a pattern recognition device that classifies brain waves on the basis of similar patterns, with waves of a similar pattern being classified into a similar class. Different activities can result in different brain wave patterns, therefore it is necessary to group them on the basis of similarity to reduce complexity.

It is not an easy process to extract the required information from the brain waves because of the high noise content. Brain signals are nonstationary, and so the information for a certain feature occurring is very important in the analysis of brain signals. Generally, this is achieved by dividing the signals into short segments, followed by information extraction from each segment. The accuracy of the feature can be affected by the length of the signal. The technique known as stationary subspace analysis decomposes the multivariate time signal into nonstationary and stationary components and then extracts the relevant features easily. Also, the feature vector may be of a very high dimension, which is undesirable as all the information in the feature vector is not required and also it can lead to the phenomenon known as the "curse of dimensionality." Therefore, to reduce the dimension of a feature vector, dimension reduction techniques such as principal component analysis (PCA) can be applied. Some feature selection and feature extraction methods are described next.

2.4.1 Principal component analysis

PCA is a linear transformation method that converts a set of correlated data into a set of uncorrelated data. A set of components from input data is generated and sorting is done in descending order according to their variance. This is followed by projection of the input data on a k-dimension eigenspace which has k eigenvectors. Eigenvectors are calculated using the covariance matrix. Summation of the covariance matrix is obtained as:

$$\sum = \sum_{i=1}^{n} (p_i - m)(p_i - m)^t \tag{2.1}$$

where m represents the mean vector of the training sample p_i.

The number of eigenvectors and eigenvalues will be equal to the order of the matrix. With the help of eigenvalues, we know which eigenvector represents the most significant information. The dimension "k" of the matrix is selected such that it is less than the order of the covariance matrix and contains eigenvectors which have the largest eigenvalues. The projection matrix is built using eigenvectors and the feature vector is extracted using a projection matrix. These eigenvectors are arranged column-wise such that the first column is an eigenvector having the highest eigenvalue. The feature vector, v, can be calculated from the matrix, by projecting q onto the new subspace.

$$v = A^t(q - m)(2.2)$$

where m is a mean vector of the training sample p_i.

PCA does not always guarantee good classification as the discriminating feature may not be the one included in the set of the largest principal component. However, the PCA is quite efficient in reducing noise in the data as it identifies the artifacts and reconstructs the signal without the noise.

2.4.2 Independent component analysis

Independent component analysis (ICA) splits the mixed signal into its sources without knowing the nature of the signal. It works by assuming that the selected signal is a mixture of different independent signals from multiple activities. The number of sources determines the dimension of the input vector, whereas a dimension of output vectors depends on how many data channels are measured. Eq. (2.3) illustrates the resulting EEG signal $x(t)$ with corresponding source $s(t)$:

$$x(t) = f(s(t)) + n(t) \qquad (2.3)$$

where f is an unknown mixture function and $n(t)$ is the random noise.

On the basis of the solution of the unknown mixer function, the ICA can be a linear or nonlinear function. Assumptions of the linear mixing function simplify the above equation. Eq. (2.3) can also be written by considering the mixing matrix A and matrix multiplication.

$$x(t) = As(t) + n(t) \qquad (2.4)$$

Here, the noise vector $n(t)$ is removed as it is assumed that the noise in the data is too small for consideration.

2.4.3 Autoregressive components

Autoregressive (AR) spectral estimation is used for modeling EEG signals as the output random signal of a linear time-invariant filter with white noise with mean zero and some variance as input. The main aim here is to obtain different filter coefficients which can be used as the features of the EEG signals.

The EEG signal $y(t)$ can be described with the AR model of order p as:

$$y(t) = a_1y(t-1) + a_2y(t-2) + a_3y(t-3) + \ldots + a_py(t-p) + n(t) \qquad (2.5)$$

where $n(t)$ is the noise and a_i is the i-th filter coefficient.

The power spectrum of the EEG signal $y(\omega)$ can be obtained by using the coefficients as:

$$y(\omega) = \frac{1}{|1 - \sum_{k=1}^{p} a_k e^{-jk\omega}|^2} \tag{2.6}$$

where p represents the order of the AR model and a_k represents the estimated filter coefficients.

The selection of the order of the AR model is important as if its order is too low the spectrum will be too smooth, and if it is high there will be spurious peaks in the spectrum.

2.4.4 Matched filtering

Matched filtering (MF) is a feature extraction method that works on the principle of matching the signal with the predetermined signals or the templates in order to detect a specific pattern. It uses predetermined templates that represent different intentions of the user. The user's intention is revealed by comparing the EEG signals with different templates and the template with the highest correlation is selected. The sum of related components can be obtained as

$$\mathrm{MF}(n) = \sum_{k=1}^{N} a_k cos\left(\frac{2\pi k f_F}{f_s} n + \phi_k\right) \tag{2.7}$$

where n represents the template sample number, f_s and f_F represent sampling and fundamental frequency of the rhythm template, and the amplitude of individual harmonic is represented by a_t.

2.4.5 Wavelet transformation

Wavelet transformation is a widely used mathematical tool for extracting data. Nonstationary signals can be easily analyzed by wavelet transformation because of its flexibility in representing the time-frequency of a signal.

Wavelets are mathematical functions that differentiate the data into different frequency components with a limited duration, which is helpful in studying the signal in frequency as well as time domains. It has advantages over traditional the Fourier transformation method in analyzing physical situations where the signal contains sharp spikes and discontinuities. Also, FT only illustrates the different frequency contents, discarding the information on when that frequency occurs. Short-term Fourier transformation was later introduced to overcome the problem faced by FT by dividing the signal into successive time durations and then applying FT. This led to higher temporal resolution but at the same time resulted in lower frequency resolution. Wavelet transformation overcomes these problems [32].

Continuous wavelet transformation is defined as the convolution of the signal $x(t)$ as:

$$w(s,\tau) = \int_{-\infty}^{\infty} x(t)\psi^{*}_{s,\tau}(t)dt \tag{2.8}$$

where $w(S, \tau)$ represents the wavelet coefficient that corresponds to the frequency with scale S and time τ.

Continuous wavelet transformation (CWT) is like template matching for a correlation between the signal and the predetermined waveform, as in the case of a matched filter. The advantage of CWT over MF is the special property of wavelet transformation that decomposes the signal in both frequency and time domains. Continuous wavelet transformation introduces a lot of complexity and redundancy as it analyses the signal at a very high number of frequencies. To solve this problem discrete wavelet transformation was introduced to analyze and decompose the signal in certain discrete values only.

2.5 Artifacts in brain–computer interfaces

The undesirable signals obtained from the brain waves which contaminate brain activity are known as artifacts. These are basically due to noncerebral activity. They can greatly reduce the performance of the BCI. On the basis of origin, these can be classified as:

1. Physiological artifacts are those artifacts which are mainly due to the activities that are muscular, known as electromyography (EMG), ocular, known as electrooculography (EOG), and heart-related, known as electrocardiography (ECG) [33]. Electromyography or EMG are produced due to the electrical activity caused by muscle contraction, for example, during talking and chewing. Electrooculography (EOG) artifacts are caused by blinking or different eye movements whereas, electrocardiography (ECG) artifacts are produced due to the cardiac activity which introduces a pulsating signal.
2. Nonphysiological artifacts are generally due to external disturbances like noise from electrode impedance. Nonphysiological artifacts can usually be avoided by proper fitting of electrodes or by proper shielding. Therefore the main focus shifts toward reducing physiological noise. We can reduce physiological artifacts by asking the patient to avoid blinking or other movements during the procedure. However, this is not as simple as it seems, as it may be very difficult for a patient suffering from a extreme neurological disorder. Also, it is not possible for a normal patient to avoid some artifacts like heartbeats and random muscle twitching.

Artifact removal can be done manually by recognizing those patterns and removing them manually but this requires extensive manual labor and time and cannot be used in online BCI.

Artifact removal can also be done automatically by monitoring EOG and EMG and removing the brain signals when they are detected. It fails to detect them when the EOG signals are too small, and also the user loses control of the device when signals containing artifacts are rejected. Therefore, instead of discarding these signals, the approach is to remove artifacts without damaging the neurological phenomenon. Blind source separation (BSS), linear filtering, and PCA are some methods used to remove artifacts from the brain signal.

Some systems are designed such that they can acquire and process these artifacts instead of discarding them, to offer a communication path that can be used in many different tasks. However, these types of systems are not considered as BCI. Muscle–computer interfaces, man–machine interfaces, and biocontrollers are some examples of systems that employ artifact signals. These systems are more reliable than BCI but cannot be used by patients with neurological disorders in which they do not have some voluntary movements.

2.6 Classification algorithms

The classifications stage helps in recognition of user intentions by using a feature vector. This can be achieved either using a classification algorithm or a regression algorithm. Both classification and regression algorithms use the features extracted in the extraction step. Regression uses these variables to predict human intentions, however these are used to differentiate different targets in feature space by classification algorithms.

Both online and offline sessions can be used to develop a classification algorithm. The offline session includes training the machine using the data set and can be tested by using different test data so that fine-tuning can be done. In contrast, in an online session the BCI system is evaluated in a real-world environment. It is necessary to check how the algorithm performs in the real-world environment and with different users.

Traditionally the classification algorithm is trained using supervised learning which means the data set is labeled, as it was assumed that the classification algorithm will find patterns in the brain signal. However, this assumption leads to a decline in BCI performance as the brain waves are not stationary. Therefore the patterns observed during calibration in the offline session and the different patterns and correlations which are observed during the online session may be different. Also, it is trained with different users and every user has their own concentration level, mental strength, etc. An adaptive algorithm is therefore required which adapts to these changes. This can be achieved using unsupervised learning, as the labels are not present with the data set; this method attempts to find patterns and correlations between the unlabeled data in order to classify them. Reinforcement learning can also be used where the system learns from its mistakes and also prevents them from being repeated in future. This adaptation can result in excellent performance.

The curse of dimensionality and the bias–variance trade-off are a major problems in the pattern recognition task. Bias and variance should be kept low to achieve the lowest classification error. A combination of classifiers is helpful in reducing the variance. Generally a simple algorithm is preferred over a complex algorithm as they can adapt to a feature in a much simpler way. Overfitting is also a major problem in pattern recognition, as if the classifier if overfitted then it will only classify similar data and perform very badly in real-world or on-test data. By reducing the complexity of the classification procedure overfitting can be avoided.

Some classification algorithms used are described next.

2.6.1 *k*-Nearest neighbor classifier

The *k*-nearest neighbor classifiers (*k*-NNCs) work on the principle that similar features will form a different cluster in feature space with multiple data points. The classifier takes *k*-nearest neighbors in account to find similarities between the test data and the features of a different class taken one at a time. Taking a large value of *k* can be helpful in reducing the chance of errors, but can also increase the complexity, so the value of *k* is taken in between, which suits best the condition.

The nearest-*k* neighbors are ranked based on their closeness. This is done by using weights between connections. k-NNC is not generally used in BCI due to its high sensitivity to the dimension of the feature vector, however, it shows quite efficient results with low-dimension features vectors.

2.6.2 Linear discriminant analysis

Linear discriminant analysis (LDA) is generally used to classify patterns between two classes; however, it can be extended to classify multiple patterns. LDA assumes that all classes are linearly separable and according to this multiple linear discrimination function representing several hyperplanes in the feature space are created to distinguish the classes. If there are two classes then the LDA draws one hyperplane and projects the data onto this hyperplane in such a way as to maximize the separation of the two categories. This hyperplane is created according to the two criteria considered simultaneously:

- Maximizing the distance between the means of two classes;
- Minimizing the variation between each category.

It provides accepted accuracy and is widely used in BCI systems.

2.6.3 Support vector machine

Support vector machine (SVM) works in a similar fashion to linear discriminant analysis. A hyperplane or set of hyperplanes is created, in order to separate the feature vectors into several classes like LDA, but it selects the hyperplane which is at maximum distance from the nearest training samples.

SVM finds the hyperplane with the maximal margin according to Cover's theorem by mapping input data into high-dimensional space.

Cover's theorem states that when a complex classification problem is cast in high-dimensional nonlinear space it is more likely to be linearly separable as compared to when casting in low-dimensional nonlinear space. SVM also uses regularization to prevent artifacts. Nonlinear SVM can also exist with a nonlinear decision boundary with the help of kernel function, leading to greater flexibility. SVM is widely used in BCI systems as it is fast, more reliable, and also simple.

2.6.4 Artificial neural network

An artificial neural network or ANN is a nonlinear classifier that is based on how the information is processed by the brain. Neural networks once trained are capable of recognizing patterns in test data. They consist of nodes and connections which are continuously adjusted during the training phase. Training data are fed in and the results are observed, and if the results are incorrect then the weights are adjusted to give the desired output. This is done over a long period of time until the algorithm reaches stability. Mathematically, neural networks define a mapping between the input data and the output, which is defined by the vector-valued function. This mapping function is a collection of mappings which are done by each neuron. The resultant mapping is nonlinear as the information is processed nonlinearly. A single-layered neural network consists of only one hidden layer with one output and one input layer. Depending upon the need, the neural network can also be multilayered with multiple hidden layers with one input and output layer. Multilayered neural networks provide greater flexibility and can easily adapt to numerous changes. A multilayered neural network is generally used in the preprocessing stage rather than in the classification stage as it improves the separability of EEG features.

2.7 Brain–computer interface applications

The BCI can be a life-changing technology for people with several neurological disorders, including those with full control over their brain but who are unable to control their body due to neuronal damage. The BCI system greatly improves the quality of life of such patients. EEG is widely used in BCI because of its ability to offer acceptable signal quality at low cost. The target population of BCI falls into three main categories:

1. Complete locked-in state (LIS)—This includes people who suffer from crucial neural damage and have lost all motor control. For example, a patient at the terminal stage of amyotrophic lateral sclerosis (ALS).

2. LIS—This includes peoples who are paralyzed completely but have some voluntary movement, such as eye movement, eye blink, and lip twitch.
3. The third group consists of able-bodied patients who have only lost some motor control, like speech or hand control. BCI is not as critical for these patients as they can communicate more quickly and efficiently using other interfaces available.

Other types of interfaces are currently ahead of BCI as they are easier to use and also provide quite efficient results at low cost. For example, eye movements can be easily tracked using an eye tracker and it also provides greater accuracy. BCI can also be combined with other BCI or some other interface technology to form a hybrid BCI which greatly improves its performance. For now, BCI is the only choice for patients that have lost all their motor control as other interfaces are not beneficial in this regard.

Different types of BCI-based systems, including video games, brain-controlled wheelchairs, adaptive web browsers, etc. now exist. In contrast to the developments in the field of BCI there are several drawbacks:

- The information transfer rate is acceptable but does not provide quick responses.
- Artifacts and nonstationarity of brain waves complicate the interaction.
- It cannot be used directly by patients as it requires the assistant to place the electrodes.
- Midas touch—That is the patient can turn off the BCI using brain signals but is unable to turn it on without an assistant.
- It requires heavy machines and equipment to operate, which is impractical in the real world.

Despite all these drawbacks, the BCI is quite efficient in improving the lives of several patients suffering from a neurological disorder. Some applications of BCI are described next.

2.7.1 Communication

We, humans, are social animals and communication is one of our most basic needs to interact with our surroundings. BCI acts as a communication path for patients. Communication through BCI can be done using a virtual keyboard projected on the screen with the patient focusing on a symbol which is interpreted by the BCI system. By intensive training patients are able to produce negative and positive responses in SCP changes to move the virtual cursor over the letters.

P300 spellers which were designed by Farwell and Donchin [34] in 1988 are still considered among the best P300 spellers available. They contain 26 letters with other symbols displayed in 6×6 matrix which has flashing columns and rows. The user focuses his or her attention on the letter to be written. P300 evoked signals are elicited for the desired letter when the column and row containing that letter flashes. In this way, the system recognizes the symbol. The spelling rate with this method is about two characters per minute. It provides a relatively high accuracy rate but can have errors because of an evoked P300 response due to other flashing rows or columns which are adjacent to the letter which is improved by changing the distribution of the letters.

BCI can also be used in web browsers that are adapted to the disabled user, as the Internet has become a very important parts of our daily lives. One of the drawbacks is that it provides limited choices as the number of links in a decision tree is limited. The more advanced BCI-based web browser Nessi overcomes this drawback by providing colored frames which are present around the links [35].

2.7.2 Motor restoration

BCI can be used in motor restoration and can alleviate the social suffering of patients. Motion restoration is possible in quadriplegic patients through neuroprostheses by means of functional electrical stimulation. This helps in motor restoration by eliciting artificial muscle contractions.

BCI can be combined with functional electrical stimulation (FES) as BCI to provide the control signal for FES. This can be done efficiently as EEG signals are not affected by upper-extremity muscle activation.

Pfurtscheller et al. [36] developed an application through which a patient with spinal cord injury was able to grab a cylinder using his paralyzed hands. Beta oscillations were generated by the patient by imagining foot movement, which was classified by the BCI and then used to control the FES device, resulting in muscle contraction.

Most recently, Muller-Putz and Pfurtscheller [37] demonstrated a neuroprosthetic for finger movement and for grasping objects. The device consists of an electrical hand prosthesis controlled by the brain using BCI. In this, finger movement was done using four LED lights having different frequencies which were placed on the prosthesis. This enabled the user to gage control over the movement of the fingers.

2.7.3 Environmental control

Patients with neurological disorders are generally confined to their home. Environmental control plays a major part in providing maximum independence to these patient. Environmental control includes controlling surrounding devices such as lights, Internet access, TVs, etc.

Environmental control for patients with partial paralysis can be implemented using devices like a joystick, keyboard, or mouse. Microphones or voice recognition devices can be used for patients which cannot control their motor activities. In the case of patients with several disabilities, the BCI system can be controlled by modulations in sensorimotor rhythms. This will help the patient to become independent and also gives a more natural effect.

2.7.4 Locomotion

The BCI has very important applications in locomotion. In this area, the BCI is implemented to help the patient in locomotion, for example, by controlling their wheelchair. This can greatly help in improving the patient's quality of life and independence.

This type of system uses a BCI based on EEG signals, however the EEG signal is very noisy which can greatly affect the data transfer rate of the system, meaning there is a time lag between the command and the action. Due to this, some studies suggest an invasive approach to record EEG signals, but the health risks in the invasive approach cannot be ignored, which has led research toward finding noninvasive techniques.

The first application of wheelchair control was demonstrated by Tanaka et al. [38] using an EEG-based BCI in which the user controls the wheelchair by imagining right or left limb movement.

Synchronous P300-based BCIs are also introduced to provide better accuracy and fast information transfer. Rebsamen et al. [39] presented a simpler approach in which guidelines are placed on the floor with the help of an assistant, these guidelines are barcodes which are continuously read by the device. The user then has only to select the destination and the wheelchair follows the guideline and stops at the destination. This can also be modified by providing a set of scanners that will detect potential obstacles before executing the user commands. This is done to ensure the safety of the user and the surroundings.

2.7.5 Entertainment

BCI can also be used in the field of entertainment. Interest in entertainment applications has increased with the technological advancements in BCI. It is used for making video games more challenging and interesting with the help of brain activity.

Entertainment-based BCI has adapted well-known games like Pacman, pinball, etc. Middendorf et al. [40] demonstrated a flight simulator which was controlled with the help of steady-state visual-evoked responses (SSVERs)-based BCI in which the left or right control command was determined by the strength and frequency of the SSVER.

BCI can also serve as a control system where the user can easily control objects in a game and another simulator by analysis of their brain waves.

TABLE 2.1 A comparison table related to works in the application areas of the BCI.

Application areas	Paper 1	Paper 2	Paper 3	Paper 4	Paper 5
Communication	Farwell and Donchin [34] presented a P300 speller of the ERP for locked-in patients	Bensch et al. [35] demonstrated a method to control web browser-based SCPs	Brumberg et al. [41] presented an intracortical microelectrode brain–computer interface for predicting intended speech information	Li et al. [42] introduced a P300-based BCI messenger for locked-in patients	Mugler et al. [43] presented an Internet browser based on P300 signal
Motor restoration	Muller-Putz and Pfurtscheller [37] demonstrated a neuroprosthetic for restoration of finger movement	Pfurtscheller et al. [36] presented a noninvasive restoration of hand movement using EEG and FES	Pfurtscheller et al. [44] presented an electrically driven hand orthosis to restore grasp function	Buch et al. [45] presented an MEG-based BCI to control a paralyzed hand with an orthosis attached	Broetz et al. [46] proposed a combination of MEG- and EEG-based BCI for boosting rehabilitation
Environmental control	Aloise et al. [47] introduced a P300-based asynchronous BCI for environmental control	Ou et al. [48] presented a BCI-based smart environmental control system using EEG	Gao et al. [49] presented an environmental controller using BCI based on steady-state visual-evoked potential	Cincotti et al. [50] presented a detailed study on integration of BCI in the domestic environment	Hochberg et al. [51] used neuromotor prostheses to control a cursor

Locomotion	Tanaka et al. [38] presented EEG-based control of an electric chair	Rebsamen et al. [52] presented a brain-controlled wheelchair using P300 EEG signals	Zhang et al. [53] proposed an FES rehabilitation system based on a central pattern generator for locomotion	Jia et al. [54] presented a strategy to control a quadruped locomotion robot using EEG signals	Iturrate et al. [55] described a noninvasive BCI based on P300 for wheelchair control
Entertainment	Middendorf et al. [40] presented a flight simulator using steady-state visual-evoked response-based BCI	Lalor et al. [56] presented the "MindBalance" game based on SSVEP-generated responses	Finke et al. [57] implemented the "MindGame" using P300 events-based BCI	Bonnet et al. [58] presented a multiuser football video game, "BrainArena," using two BCIs	Bordoloi et al. [59] developed a BCI maze game using noninvasive BCI based on EEG

2.7.6 Other brain–computer interface applications

The BCI also finds applications in some fields other than those discussed above. The main benefit of BCI is that it can be used efficiently to acquire selective control over certain regions of the brain with the help of neurofeedback, with the aim of achieving behavioral changes. It can help in improving several qualities, like speech skills, affection, pain management, etc. It is also very helpful in the treatment of various mental disorders, such as depression and alcohol dependence.

Neuromarketing is a new field which applies neuroscientific methods to marketing research. It can greatly help in an efficient trade-off between cost and benefit (Table 2.1).

2.8 Conclusion

This chapter has presented a simple view of the topic "brain–computer interface." It started by defining the topic and the main reasons for the developments in this field. From acquiring the brain signals to converting them into commands for the BCI system, different stages involved have been discussed. Several neuroimaging approaches were discussed with the EEG being the most commonly used neuroimaging approach in BCI as it enables a noninvasive approach and is also price efficient. However, other neuroimaging approaches like fMRI, MEG, and NIRS are also very effective. The invasive approach is presented as a valuable method in providing high-quality signals, however the use of a noninvasive approach was encouraged because it is a less harmful approach.

A variety of feature extraction and classification algorithms have been considered and modified on the basis of their performance, which has significantly reduced the user training time, with a faster data transfer rate and an asynchronous approach providing a more natural mode of interaction. The faster data transfer rate and reduced user training time let to the development of more BCI-based applications such as web browsers, wheelchair control, and word processing. These developments have greatly helped in improving the lives of disabled people.

References

[1] M.B. Khalid, N.I. Rao, I. Rizwan-i-Haque, S. Munir, F. Tahir, Towards a brain computer interface using wavelet transform with averaged and time segmented adapted wavelets, in: 2009 2nd International Conference on Computer, Control and Communication, February 2009, pp. 1–4, IEEE.

[2] E.J. Candès, M.B. Wakin, An introduction to compressive sampling [a sensing/sampling paradigm that goes against the common knowledge in data acquisition], IEEE Signal Process. Mag. 25 (2) (2008) 21–30.

[3] M.A. Mazurowski, Radiogenomics: what it is and why it is important, J. Am. Coll. Radiol. 12 (8) (2015) 862–866.

[4] T. Sato, R. Kajiwara, I. Takashima, T. Iijima, Wavelet correlation analysis for quantifying similarities and real-time estimates of information encoded or decoded in single-trial oscillatory brain waves, Wavel. Theory Its Appl. 15 (2018).

[5] M.J. Post, P. van der Putten, J.N. van Rijn, Does feature selection improve classification? A large scale experiment in OpenML, International Symposium on Intelligent Data Analysis, Springer, Cham, 2016, pp. 158–170. October.

[6] R.J. Urbanowicz, M. Meeker, W. La Cava, R.S. Olson, J.H. Moore, Reliefbased feature selection: introduction and review, J. Biomed. Inform. 85 (2018) 189203.

[7] J. Kevric, A. Subasi, Comparison of signal decomposition methods in classification of EEG signals for motor-imagery BCI system, Biomed. Signal. Process. Control 31 (2017) 398–406.

A.A. Fingelkurts, A.A. Fingelkurts, H. Rytsälä, K. Suominen, E. Isometsä, et al., Composition of brain oscillations in ongoing EEG during major depression disorder, Neurosci. Res. 56 (2) (2006) 133–144.

[8] P.K. Pattnaik, J. Sarraf, Brain Computer Interface issues on hand movement, J. King Saud. Univ.-Comput. Inf. Sci. 30 (1) (2018) 1824.

[9] D. Gajic, Z. Djurovic, S. Di Gennaro, F. Gustafsson, Classification of EEG signals for detection of epileptic seizures based on wavelets and statistical pattern recognition, Biomed. Eng.-App. Bas 26 (2) (2014) 1450021.

[10] A. Kübler, B. Kotchoubey, J. Kaiser, J.R. Wolpaw, N. Birbaumer, Brain–computer communication: unlocking the locked, Psychol. Bull. 127 (3) (2001) 358.

[11] B.K. Anand, G.S. Chhina, B. Singh, Some aspects of electroencephalographic studies in yogis, Electroencephalogr. Clin. Neurophysiol. 13 (3) (1961) 452–456.

[12] L.I. Aftanas, S.A. Golocheikine, Human anterior and frontal midline theta and lower alpha reflect emotionally positive state and internalized attention: high-resolution EEG investigation of meditation, Neurosci. Lett. 310 (1) (2001) 57–60.

[13] J.A. Pineda, The functional significance of mu rhythms: translating "seeing" and "hearing" into "doing", Brain Res. Rev. 50 (1) (2005) 57–68.

[14] L. Venables, S.H. Fairclough, The influence of performance feedback on goal-setting and mental effort regulation, Motiv. Emot. 33 (1) (2009) 63–74.

[15] P. Khanna, J.M. Carmena, Neural oscillations: beta band activity across motor networks, Curr. Opin. Neurobiol. 32 (2015) 60–67.

[16] X. Dong, M. Wang, W. Wang, Electroencephalogram alertness responses to blue light stimulus in elderly people with cataract, J. Clin. Neurosci. 57 (2018) 63–67.

[17] T. Mima, N. Simpkins, T. Oluwatimilehin, M. Hallett, Force level modulates human cortical oscillatory activities, Neurosci. Lett. 275 (2) (1999) 77–80.

[18] S. Waldert, T. Pistohl, C. Braun, T. Ball, A. Aertsen, C. Mehring, A review on directional information in neural signals for brain-machine interfaces, J. Physiol. Paris 103 (3–5) (2009) 244–254.

[19] R. Salmelin, M. Hámáaláinen, M. Kajola, R. Hari, Functional segregation of movement-related rhythmic activity in the human brain, Neuroimage 2 (4) (1995) 237–243.

[20] N.E. Crone, D.L. Miglioretti, B. Gordon, R.P. Lesser, Functional mapping of human sensorimotor cortex with electrocorticographic spectral analysis. II. Event-related synchronization in the gamma band, Brain J. Neurol. 121 (12) (1998) 2301–2315.

[21] K.J. Miller, P. Shenoy, J.W. Miller, R.P. Rao, J.G. Ojemann, Real-time functional brain mapping using electrocorticography, Neuroimage 37 (2) (2007) 504–507.

[22] D.M. Taylor, S.I.H. Tillery, A.B. Schwartz, Direct cortical control of 3D neuroprosthetic devices, Science 296 (5574) (2002) 1829–1832.

[23] N. Weiskopf, K. Mathiak, S.W. Bock, F. Scharnowski, R. Veit, W. Grodd, et al., Principles of a brain-computer interface (BCI) based on real-time functional magnetic resonance imaging (fMRI), IEEE Trans. Biomed. Eng. 51 (6) (2004) 966–970.

[24] S.M. Coyle, T.E. Ward, C.M. Markham, Brain–computer interface using a simplified functional near-infrared spectroscopy system, J. Neural Eng. 4 (3) (2007) 219.

[25] R.P. Kennan, S.G. Horovitz, A. Maki, Y. Yamashita, H. Koizumi, J.C. Gore, Simultaneous recording of event-related auditory oddball response using transcranial near infrared optical topography and surface EEG, Neuroimage 16 (3) (2002) 587–592.

[26] D. Regan, Human Brain Electrophysiology: Evoked Potentials and Evoked Magnetic Fields in Science and Medicine, Elsevier, New York, NY, USA, 1989.

[27] W. Yijun, W. Ruiping, G. Xiaorong, H. Bo, G. Shangkai, A practical VEP-based brain-computer interface, IEEE Trans. Neural Syst. Rehabil. Eng. 14 (2006) 234–240.

[28] N. Birbaumer, T. Elbert, A.G. Canavan, B. Rockstroh, Slow potentials of the cerebral cortex and behavior, Physiol. Rev. 70 (1) (1990) 1–41.

[29] D. Ravden, J. Polich, On P300 measurement stability: habituation, intra-trial block variation, and ultradian rhythms, Biol. Psychol. 51 (1) (1999) 59–76.

[30] B. Blankertz, F. Losch, M. Krauledat, G. Dornhege, G. Curio, K.R. Müller, The Berlin brain-computer interface: accurate performance from first-session in BCI-naive subjects, IEEE Trans. Biomed. Eng. 55 (10) (2008) 2452–2462.

[31] B. Kleber, N. Birbaumer, Direct brain communication: neuroelectric and metabolic approaches at Tübingen, Cognit. Process. 6 (1) (2005) 65–74.

[32] V.J. Samar, A. Bopardikar, R. Rao, K. Swartz, Wavelet analysis of neuroelectric waveforms: a conceptual tutorial, Brain Lang. 66 (1) (1999) 7–60.

[33] M. Fatourechi, A. Bashashati, R.K. Ward, G.E. Birch, EMG and EOG artifacts in brain computer interface systems: a survey, Clin. Neurophysiol. 118 (3) (2007) 480–494.

[34] L.A. Farwell, E. Donchin, Talking off the top of your head: toward a mental prosthesis utilizing event-related brain potentials, Electroencephalogr. Clin. Neurophysiol. 70 (6) (1988) 510–523.

[35] M. Bensch, A.A. Karim, J. Mellinger, T. Hinterberger, M. Tangermann, M. Bogdan, et al., Nessi: an EEG-controlled web browser for severely paralyzed patients, Comput. Intell. Neurosci. 2007 (2007).

[36] G. Pfurtscheller, G.R. Müller, J. Pfurtscheller, H.J. Gerner, R. Rupp, 'Thought'–control of functional electrical stimulation to restore hand grasp in a patient with tetraplegia, Neurosci. Lett. 351 (1) (2003) 33–36.

[37] G.R. Muller-Putz, G. Pfurtscheller, Control of an electrical prosthesis with an SSVEP-based BCI, IEEE Trans. Biomed. Eng. 55 (1) (2007) 361–364.

[38] K. Tanaka, K. Matsunaga, H.O. Wang, Electroencephalogram-based control of an electric wheelchair, IEEE Trans. Robot. 21 (4) (2005) 762–766.

[39] B. Rebsamen, E. Burdet, C. Guan, C.L. Teo, Q. Zeng, M. Ang, et al., Controlling a wheelchair using a BCI with low information transfer rate, in: 2007 IEEE 10th International Conference on Rehabilitation Robotics, 2007, June, pp. 1003–1008, IEEE.

[40] M. Middendorf, G. McMillan, G. Calhoun, K.S. Jones, Brain-computer interfaces based on the steady-state visual-evoked response, IEEE Trans. Rehabil. Eng. 8 (2) (2000) 211–214.

[41] J.S. Brumberg, A. Nieto-Castanon, P.R. Kennedy, F.H. Guenther, Brain–computer interfaces for speech communication, Speech Commun. 52 (4) (2010) 367–379.

[42] Y. Li, J. Zhang, Y. Su, W. Chen, Y. Qi, J. Zhang, et al., P300 based BCI messenger, in: 2009 ICME International Conference on Complex Medical Engineering, April 2009, pp. 1–5, IEEE.

[43] E.M. Mugler, C.A. Ruf, S. Halder, M. Bensch, A. Kubler, Design and implementation of a P300-based brain-computer interface for controlling an internet browser, IEEE Trans. Neural Syst. Rehabil. Eng. 18 (6) (2010) 599–609.

[44] G. Pfurtscheller, C. Guger, G. Müller, G. Krausz, C. Neuper, Brain oscillations control hand orthosis in a tetraplegic, Neurosci. Lett. 292 (3) (2000) 211–214.

[45] E. Buch, C. Weber, L.G. Cohen, C. Braun, M.A. Dimyan, T. Ard, et al., Think to move: a neuromagnetic brain-computer interface (BCI) system for chronic stroke, Stroke 39 (3) (2008) 910–917.

[46] D. Broetz, C. Braun, C. Weber, S.R. Soekadar, A. Caria, N. Birbaumer, Combination of brain-computer interface training and goal-directed physical therapy in chronic stroke: a case report, Neurorehabil. Neural Repair 24 (7) (2010) 674–679.

[47] F. Aloise, F. Schettini, P. Aricò, F. Leotta, S. Salinari, D. Mattia, et al., P300-based brain–computer interface for environmental control: an asynchronous approach, J. Neural Eng. 8 (2) (2011) 025025.

[48] C.Z. Ou, B.S. Lin, C.J. Chang, C.T. Lin, Brain computer interface-based smart environmental control system, in: 2012 Eighth International Conference on Intelligent Information Hiding and Multimedia Signal Processing, July 2012, pp. 281–284, IEEE.

[49] X. Gao, D. Xu, M. Cheng, S. Gao, A BCI-based environmental controller for the motion-disabled, IEEE Trans. Neural Syst. Rehabil. Eng. 11 (2) (2003) 137–140.

[50] F. Cincotti, D. Mattia, F. Aloise, S. Bufalari, G. Schalk, G. Oriolo, et al., Non-invasive brain–computer interface system: towards its application as assistive technology, Brain Res. Bull. 75 (6) (2008) 796–803.

[51] L.R. Hochberg, M.D. Serruya, G.M. Friehs, J.A. Mukand, M. Saleh, A.H. Caplan, et al., Neuronal ensemble control of prosthetic devices by a human with tetraplegia, Nature 442 (7099) (2006) 164.

[52] B. Rebsamen, E. Burdet, C. Guan, H. Zhang, C.L. Teo, Q. Zeng, et al., Controlling a wheelchair indoors using thought, IEEE Intell. Syst. 22 (2) (2007) 18–24.

[53] D. Zhang, G. Liu, G. Huan, J. Liu, X. Zhu, A hybrid FES rehabilitation system based on CPG and BCI technology for locomotion: a preliminary study, International Conference on Intelligent Robotics and Applications, Springer, Berlin, Heidelberg, 2009, pp. 1073–1084. December.

[54] W. Jia, D. Huang, X. Luo, H. Pu, X. Chen, O. Bai, Electroencephalography (EEG)-based instinctive brain-control of a quadruped locomotion robot, in: 2012 Annual International Conference of the IEEE Engineering in Medicine and Biology Society, September 2012, pp. 1777–1781, IEEE.

[55] I. Iturrate, J.M. Antelis, A. Kubler, J. Minguez, A noninvasive brain-actuated wheelchair based on a P300 neurophysiological protocol and automated navigation, IEEE Trans. Robot. 25 (3) (2009) 614–627.

[56] E.C. Lalor, S.P. Kelly, C. Finucane, R. Burke, R. Smith, R.B. Reilly, et al., Steady-state VEP-based brain-computer interface control in an immersive 3D gaming environment, EURASIP J. Adv. Signal. Process. 2005 (19) (2005) 706906.

[57] A. Finke, A. Lenhardt, H. Ritter, The MindGame: a P300-based brain–computer interface game, Neural Netw. 22 (9) (2009) 1329–1333.

[58] L. Bonnet, F. Lotte, A. Lécuyer, Two brains, one game: design and evaluation of a multi-user BCI video game based on motor imagery, IEEE Trans. Comput. Intell. AI Games 5 (2) (2013) 185–198.
[59] S. Bordoloi, U. Sharmah, S.M. Hazarika, Motor imagery based BCI for a maze game, in: 2012 4th International Conference on Intelligent Human Computer Interaction (IHCI), December 2012, pp. 1–6, IEEE.

Chapter 3

Transforming pharma logistics with the Internet of things

Anjali Jain[1] and Deepak Kumar Sharma[2]
[1]Department of Instrumentation and Control, Netaji Subhas University of Technology (formerly known as Netaji Subhas Institute of Technology), New Delhi, India, [2]Department of Information Technology, Netaji Subhas University of Technology (formerly known as Netaji Subhas Institute of Technology), New Delhi, India

3.1 Introduction

The Internet of things (IoT) has emerged as a revolutionizing IT paradigm not only in industries, but also in households and smart wearables. The IoT allows widespread connectivity, wherein objects and resources are seamlessly networked and interact with each other. The pervasiveness of the IoT has been further cemented by the advent of sensors such as radio frequency identification (RFID) tags which acquire data in real time and help in making better decisions [1]. With the advent of the IoT, the pharmaceutical industry has been revolutionized as it facilitates efficiencies in production, supply chain, logistics, and regulatory compliance. In the pharmaceutical industry there is always a risk of a change in environmental conditions which might harm the integrity of drugs, and also the need to regulate the availability of stocks to avoid shortages and ensure that packaged drugs are received by the customer in a timely fashion. To ensure that the industry is safe from such recurring challenges, the IoT is employed in manufacturing, warehouse management, and even shipment and tracking, solving these age-old problems which still exist in the industry.

In this chapter, we first discuss the basics of the IoT and its applications in logistics and the supply chain. We also discuss the growth of the pharmaceutical industry, with a focus on the introduction of IoT and how it changed the face of pharma. The applications of the IoT in pharmaceutical logistics include quality control, shipment tracking, and drug development. The challenges faced in the industry and how they are being overcome with IoT are also discussed in this chapter. Understanding how the pharma industry has

An Industrial IoT Approach for Pharmaceutical Industry Growth.
DOI: https://doi.org/10.1016/B978-0-12-821326-1.00003-6

seen a technological boom with the IoT and has developed the entire process, starting from drug development to shipment extremely efficient and secure, is the main objective of this chapter.

3.1.1 What is the Internet of things?

Things connected to the Internet make up the IoT. These things can be electrical appliances, automobiles, smart watches, and even animals with chips planted under their skin which can easily collect and share real-time information over the Internet. By making use of built-in sensors these physical objects gather real-time sensory data. The information thus collected can help boost smart decision-making, automation of manufacturing processes, and overall enhancement of customer experience. The term "Internet of things" was introduced by the founder of the Auto-ID center at the Massachusetts Institute of Technology [2]. Auto-ID specializes in defining numerous methodologies to provide enhancement in areas such as work automation, increase efficiency and error reduction. The electronic product code (EPC) identifier, which uniquely identifies every object and allows tracking of these objects as they move from one place to another, was also released by the Auto-ID center in 2003. Eventually, the IoT emerged as an alternative to EPC where microchips could be used to create a network to carry out commercial jobs [1].

The IoT is finding its applications in various sectors including manufacturing, industrial automation, transportation, farming, smart wearables, and even security, as shown in Fig. 3.1. For instance, an IoT-based smart home-monitoring system makes use of sensors to continuously monitor energy consumption and maintain the desired temperature inside the house. Also, manufacturers are using the industrial IoT (IIoT) and machine-to-machine (M2M) communication for industrial automation, prediction and prevention of equipment failures, and improvement of work safety. The fundamental characteristics of the IoT technology include the ability to provide real-time solutions anytime and anywhere, which are mainly wireless solutions applicable in both indoor and outdoor environments, and the ability to remotely monitor and track any object fitted with sensors.

3.1.2 Internet of things in logistics and the supply chain

The IoT has an extensive list of applications in almost every sector including automotive, telecommunications, intelligent systems, health care, pharmaceuticals, manufacturing processes, environmental monitoring, and agriculture. In the supply chain network, the IoT finds major applications in the design and manufacturing of medical equipment, retail, and logistics among industries and for secure transportation and traceability of goods [3]. The IoT has huge potential in supply chain management (SCM) [4]. It allows devices and

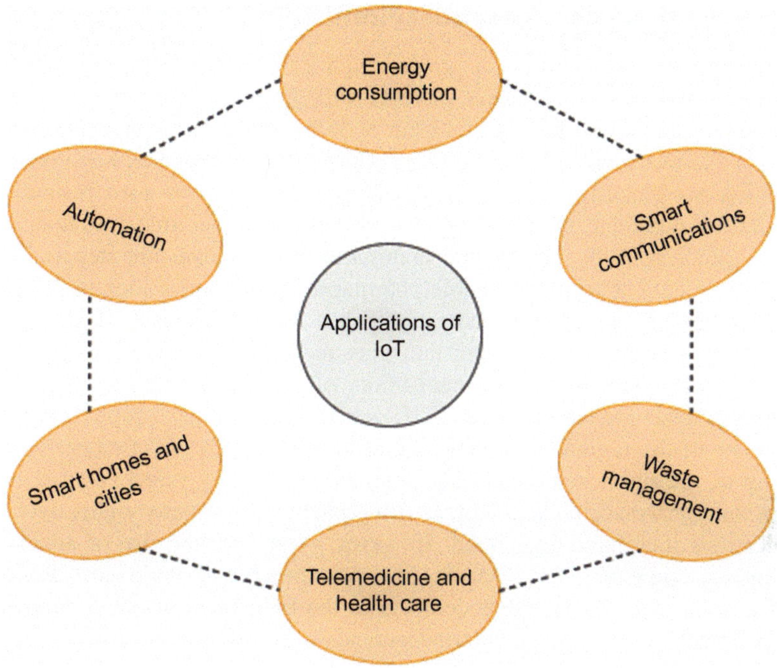

FIGURE 3.1 Applications of the IoT.

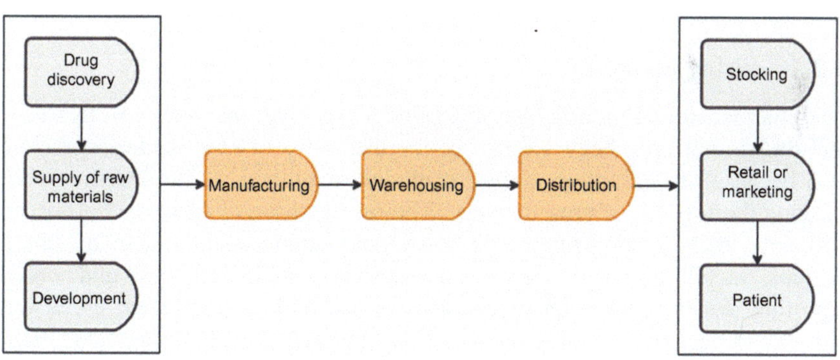

FIGURE 3.2 Role of the IoT in the pharmaceutical supply chain.

equipment in industries to communicate with ease and helps establish greater supervision at every stage of the supply chain. The IoT, through a network of computers, has the potential of not only improving process efficiency and SCM but also permitting real-time production visibility to managers, thus providing transparency in a firm. Fig. 3.2 highlights the contribution of the IoT in pharmaceutical logistics and SCM.

3.2 Growth of pharmaceutical industries

The pharmaceutical industry has grown exponentially over the past five decades and continues to empower patients with its incredible performance in drug development and supply. Johnson & Johnson, Novartis, and Roche are some of the pharmaceutical companies which have emerged as huge pharma organizations globally. A lot of research and development goes behind not just the development of drugs, but also in their manufacturing, warehousing, and distribution. Both startups and multinational companies are stepping into the pharmaceutical sector to enable pharmaceutical organizations to improve their process efficiency and come up with a standard course of action for organizations which comply with the rules and regulations laid down by the International Council for Harmonisation of Technical Requirements for Pharmaceuticals for Human Use (ICH), which is an accepted global standard [5]. Some of the regulatory practices that must be enforced for smooth functioning of the pharmaceutical sector include controlling safety, quality, and prices, and pharmaceutical scientists filing patents as property rights over the medications they have developed. However, even after the rise of advanced treatments, some chronic diseases like Alzheimer and Parkinson diseases remain incurable. Recent advancements in the field of biomedical engineering, nanotechnology, and biopharmaceuticals have provided a strong foundation to undertake these development tasks, however this does not guarantee definitive results and the patient remains subject to many life risks and uncertainties [6].

3.2.1 Rising demand

The pharmaceutical industry has changed its face over the years and this revolution has not only been a result of the discovery of new medications, but also because of technology aiding medicine manufacturers to speed up and escalate the process of drug delivery. The future of the pharmaceutical sector will have technology bringing in vivid innovations and the rise of smart devices, including smart warehousing systems, smart watches, and smart reporting systems, keeping an equal focus on finding permanent cures for diseases rather than focusing on providing immediate cures only. Other digital means of health treatments include therapeutic chatbots for treating depression by performing behavioral analysis, and blood sugar monitoring applications that help people suffering from diabetes to control their blood sugar. The pharmaceutical industry will also move toward personalized care. Researchers will shift from preparing medications for diseases to preparing medications for individual patients taking their specific needs into account. There will also be more targeted and permanent treatment measures rather than purely temporary measures. It will also become easier to predict if a patient has a particular disease or not. Pharmacogenomics, the study of how genes affect a patient's response to drugs, will redefine medications as it takes into consideration the 0.03% difference in DNA between individuals which explains why one treatment/medication works for one patient but not for another. The future of the pharmaceutical industry will be clearly reimagined. By 2050, some medical conditions will become obsolete. For instance, practitioners have come up with a cure for hepatitis C, a disease which was previously considered incurable and has affected millions of people across the globe. The arrival of the remedy has awed researchers, healthcare professionals, and patients [7].

The use of technology is gradually increasing in pharmaceutical organizations, and companies making medical instruments are constantly collaborating with software providers to provide smarter solutions. Verily Life Sciences, a subsidiary of Alphabet, invested millions of dollars in a joint enterprise to integrate electronic equipment, software, and medicines with the aim of handling the huge surge in diabetes [8]. In addition to Alphabet, the medical instruments maker Medtronic teamed up with the telecommunications giant Qualcomm to come up with a glucose-monitoring system which has the capability of providing continuous measurements of a patient's glucose levels to healthcare professionals. Technology has also enhanced the study of immunotherapy, which is a long-established area of medical research. Several immunotherapies have been developed by pharmaceutical organizations in partnership with startups and companies to provide cures for diseases. Drugs for immunotherapy provide treatments for several types of cancers and their use is also being explored in the cure and prevention of cardiovascular diseases and chronic diseases like Alzheimer and Parkinson diseases. Therefore it is clear that immunotherapies have the potential to treat these diseases and predict the risk of cardiac attacks.

3.2.2 Advent of Internet of things

The IoT, being a network of devices which can collect and transfer information across the network, has had a profound impact on the pharmaceutical industry and will perhaps be the most significant technology contribution to optimize and improve the efficiency of pharmaceutical machines and processes. Researchers have predicted that in the coming years, almost half of IoT-based technologies will be applied in health sectors making up a market worth over $120 billion dollars [9]. The combination of medicine and IoT leading to the evolution of medical informatics is bound to transform health care as it is reducing costs and inefficiencies and, most importantly, saving lives. Sensors embedded in physical devices pose multiple challenges and obstacles to IoT's acceptance in the pharmaceutical industry. One of the major challenges to implementing IoT arises in sensor/server communication because each sensor in the communication network has its own unique way of communicating with the servers. This creates an issue called the congruence of communication. Second, sensor/sensor communication poses an issue as manufacturers have their own proprietary protocols, which means that sensors by different manufacturers may not necessarily be able to communicate with each other in a smooth and uniform manner.

It was clearly evident to pharmaceutical companies that selling natural and conventional medicines in the market will neither generate the desired revenues nor sustain companies in this competitive market. This led to the introduction of the pharmaceutical IoT theory which includes digitization of both medical products and healthcare procedures using smart connected medical devices and various web and mobile applications during drug development, shipment tracking, clinical trials, and patient care. Among other possibilities which pharmaceutical IoT offers, it has the capability of offering the possibility of treatment of diseases in a totally new way. Consequently, the impact of the IoT spread over the years and the accessibility and flexibility of health-monitoring systems has increased, giving rise to process automation. Chemists have recently been more concerned with imparting knowledge of the applications and utility of these IoT-based treatments to patients. Augmented reality (AR) devices, like the HoloLens by Microsoft, allow pharmacists to access both the medical records and prescription information of patients. These technologies have the potential of solving problems before they become serious, and provide instant communication to the patient's family and real-time treatment that does not necessarily involve a visit to the doctor. Automation of pharmacy operations and the IoT have transformed the outlook of tomorrow's pharmacists. Pharmacists will be able to engage better with patients and keep digital records of all transactions and deliveries of medicines.

3.2.3 Internet of things-based pharma architecture

An IoT-based healthcare architecture is a layered architecture which has three major layers, namely a perception layer, a network transmission layer, and an application service layer [10]. However, Farooq et al. [11] discussed the existence of a fourth layer, namely the "support layer" or "middleware layer" situated between the network transmission layer and the application layer. Hence, four layers have come into existence (a pictorial representation of which is shown in Fig. 3.3), these are described below.

3.2.3.1 Perception layer

The information perception layer is the bottommost layer of the IoT-based pharma architecture and is also called the recognition layer [12]. It is primarily constituted of sensors that are used to continuously keep a check on the health of patients [13] by collecting all sorts of information from the physical world such as environmental conditions, device properties, and patient data.

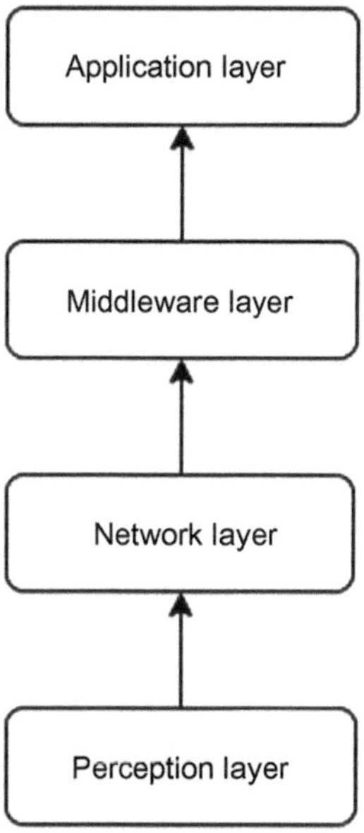

FIGURE 3.3 IoT-based healthcare architecture.

This layer includes a collection of network-enabled devices which follow the process of perceiving, detecting, collecting, and exchanging, which is applied to the system and its connected affiliates. All sorts of sensors including RFID sensors, global positioning systems (GPSs), and cameras exist at the perception layer.

3.2.3.2 Network transmission layer

The network transmission layer is responsible for transmitting data from the perception layer onto the application layer as long as the communication network, device capability, and IoT application permit. The collected information is sent over the network and stored in cloud-based data storage. IoT systems use a combination of Internet, mobile communication, satellite, and wireless networks, such as WiFi, ZigBee, and EnOcean, for transmission across these networks. Bluetooth and ZigBee technologies are employed for short-range communication, that is, from devices in the perception layer to a network gateway. In contrast, Power Line Communication, WiFi, second, third, and fourth generation (2G, 3G, and 4G, respectively) of wireless technology are employed to carry network information over long distances.

3.2.3.3 Support/middleware layer

This layer consists of information-processing modules which process the incoming data and make decisions based on that information. These systems then store the data in a database which is linked to the system itself. This layer is service-oriented and holds the capability to perform both fog computing and cloud computing. Lying just below the application layer and just above the network layer, the support layer binds them together and provides a reliable support platform for the application layer. The support platform possesses smart computing capabilities like network grid, cloud, and fog computing.

3.2.3.4 Application layer

The application service layer lies at the top of the IoT-based healthcare architecture. This layer caters to the design and management of applications to provide remote healthcare services [14]. In this layer, the incoming information is processed and analyzed to come up with optimal management strategies. These applications can also be utilized in creating smart homes, smart cities, power distribution systems, energy management, and integration of renewable energy generators. A user interface to access several IoT-based applications is also provided by the application layer. IoT applications can be utilized in various sectors such as transportation, health care, agriculture, supply chain, government, and retail.

The major challenges in providing IoT security are ensuring that each physical device in the network is provided with a unique identifier to facilitate fully secure authentication methodologies and safeguarding the connected devices in an IoT network. The confidentiality, integrity, and availability (CIA) triad, also called the security triad, is a widely accepted model in the pharmaceutical industry which implements information security within an organization's sensitive processes. Even a small violation of security rules can be hazardous to an organization, making the information vulnerable to hackers and breachers. Immediate action should be taken in the case of a data breach to prevent the company from incurring huge losses. Researchers have proposed a reliable and well-structured security architecture which can potentially abide by the CIA triad by providing data confidentiality which is paramount to an organization. The proposed architecture is displayed in Fig. 3.4.

3.3 Applications of Internet of things in pharmaceutical logistics

Although it may seem that the pharmaceutical industry revolves around the development, synthesis, and discovery of drugs and vaccines only, but in the real world manufacturing, packaging, and the supply chain form an integral part of this industry. Owing to the strict environmental conditions drugs are required to be kept under, the pharmaceutical supply chain differs from the usual supply chain networks since a minute variation could cost someone's life. Pharmaceutical companies consider SCM to be of huge significance

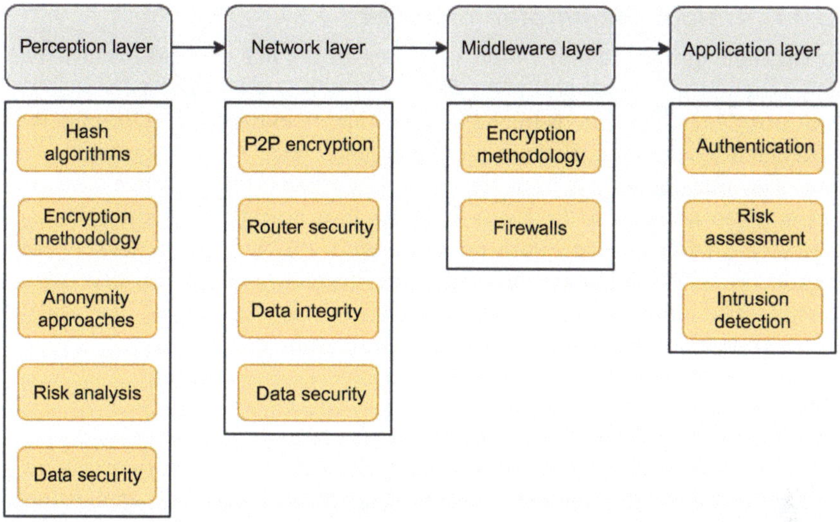

FIGURE 3.4 Security architecture of the IoT.

because any variation in the supply chain, be it at the supply, manufacturing, packaging, or even the distribution stage can lead to serious malfunctioning in the system. In order to ensure smooth transit from the manufacturing warehouse to retailers, companies are advancing from inaccurate and incomprehensive electronic data interchange reports and moving to tools like remote monitoring and network optimization analytics which are used for real-time visibility of transportation and warehouse operations which use the IoT. The IoT has been changing the pharmaceutical industry by increasing the efficiency of the transit of drugs, lowering costs, improving monitoring, and putting the focus back on better health care and medical treatments. These tools also help companies in maintaining compliance with government regulations and product quality.

3.3.1 Manufacturing of drugs and equipment

Recently the IoT has found applications in different areas of the pharmaceutical industry and in constantly transforming health care. Both patients and suppliers are gaining from the benefits of the IoT for the pharma industry as it attempts increasingly to expand its presence and utility. Some applications of medical IoT include health-tracking mobile applications and smart wearable devices that allow patients to capture almost every daily activity including exercise, food, and sleep [15]. Hospitals use the IoT to keep track of the location of medical devices, patients, and workforces by embedding wireless sensors. The IoT can be implemented in the following ways for the manufacture of drugs and equipment [1,16].

3.3.1.1 Unique identification number

To facilitate proper identification of nodes in an IoT network, which can both generate and have access to data, it is necessary to uniquely identify them using identification technologies. A unique identification (UID) number is assigned by the identification process to each node so as to provide complete data transfer to and from the node. Each individual resource in a system is given a digital UID. Every member in a hospital, including doctors, nurses, and staff members, are provided with a UID. This allows smooth transfer of information among different hospital entities in the digital domain. The universally unique identifier (UUID) is one of the widely accepted standards used for the unique identification of objects.

3.3.1.2 Real-time location system

Identifying the real-time location system (RTLS) of an entity in an IoT working model holds supreme importance in almost all the IoT applications. Location technologies like GPS and the Beidou system (BDS) are used to determine the same. However, the GPS has been identified as the most pragmatic RTLS to date. Although GPS is a satellite-dependent navigation system which is capable of locating objects under various weather conditions, both GPS and BDS are ineffective indoors because the structure of buildings causes signals emitted from satellites to attenuate and get scattered by roofs and walls. Hence, it has become necessary to develop a local positioning system (LPS) that can replace GPS and work with better accuracy indoors. An LPS is comprised of radio propagations that travel between a mobile station and an array of at least three distinct base stations or receivers aimed to locate it [17]. Smart indoor positioning network systems can also be constructed by combining a GPS or BDS with an LPS. Nonetheless, in terms of healthcare applications, the GPS can help in precisely locating ambulances, patients, and doctors, thus facilitating speedy operations.

3.3.1.3 Sensors

Sensors are an integral part of the IoT-based models as they perform the crucial task of collecting data from the surroundings, sensing changes in the physical world, and even monitoring processes. Sensors include resistance temperature detector sensors, thermocouples, wire sensors, mineral insulated sensors, etc. for controlling, monitoring, and regulating manufacturing processes. The prevalence of sensing technologies has made it possible to transform all the data received from sensors into a digital form and thereafter transmit it to a communication network. In order to ease the process of data collection, people can wear devices embedded with portable sensors which allow automated data collection and information transfer even when people are on the move.

3.3.1.4 Cloud computing

It is well known that IoT sensors and devices generate a lot of data which must be stored for further processing and operation [18]. However, the existing devices which offer network, storage, and computation facilities have certain limitations when it comes to scalability, complex analysis, and data access. In such a scenario, a technology like cloud computing comes into play [19]. To resolve this issue, the cloud data centers collect data from individual devices in an IoT system to store, perform computations, and even share it with other nodes on the communication network. With the increasing advancements of cloud in the field of IoT, it is certain that all future IoT systems will be based on the cloud. Cloud computing, being highly reliable and platform independent, is being greatly utilized in the pharmaceutical industry also. Once sent to the cloud database, the data can be formatted in a standard way through application programming interfaces (APIs). Minimal data breaches can also be ensured by adhering to security modules. Since individual IoT nodes are incapable of performing computations on data, the collected data are usually aggregated and transmitted to a more powerful node where processing can be carried out at a much faster rate. This, however, hampers the scope of a system to be scalable. The adoption of cloud computing has the power of revolutionizing how decision-making has been done so far by moving toward data-driven decisions and faster predictions at lower costs, which would provide higher revenues and reduce risks. As for the pharmaceutical and healthcare industries, the setup of a public cloud in the hospital promotes resource sharing among patients, cost reduction, and building an even more reliable medical monitoring and management system with greater efficiency [20].

3.3.1.5 Communication technologies

An IoT-based architecture requires networking between its nodes for data transfer. Communication technologies enable networking between the nodes of an IoT-based medical monitoring system's infrastructure. These technologies are of two types, namely short-distance and long-distance communication technologies. As the names suggest, short-distance communication technologies allow data transfer over short distances and long-distance technologies allow data transfer over large distances in the IoT model. While long-distance communication technologies work around regular means of communication like the Internet and mobile phones, short-distance communication revolves around wireless technologies which includes WiFi, Bluetooth, Ultra-wideband, Infrared Data Association, ZigBee, RFID, etc. Among all the short-distance radio communication techniques, WiFi is mostly adopted for short-distance communication since it deploys a local area network (LAN) for its setup at a very low cost. Nowadays, WiFi-based LANs are available in many hospitals and find huge application in the healthcare sector.

A medical IoT solution must be designed in such a way that it ensures data integrity and that medical data stay inaccessible for unauthorized users. Only authorized nodes in the IoT network should be accessible for network services or resources and a node can never reject a message/information sent earlier, thus ensuring nonrepudiation.

3.3.2 Warehouse management

A surge in demand and sales of pharmaceutical products in the market requires an increasing number of warehouses to store them. Unfortunately, in the present scenario, logistics companies are not yet able to handle the surge in demand from the pharmaceutical industries because most do not have sufficient infrastructure. However, infrastructure is not the only barrier, huge investments are required for such specialized warehouses with temperature-controlled zones to keep the drug composition intact. Most pharmaceutical companies possess and manage a large number of storage units so that drugs and medicinal equipments do not run out of stock and a continuous supply of medicines is carried out in a cost-efficient manner [14]. By keeping the warehouse operations at their own disposal at all times, most pharmaceutical companies are strategically managing their warehouse processes and storage environments internally, keeping an eye on the different cold chain conditions required for all their products. Owing to all these stringent regulations, warehousing operations turn out to be very costly and account for almost 95% of the total pharma logistics costs [14]. Also, it is imperative to have real-time visibility into warehouse operations, otherwise it can be very difficult to track products in the warehouse.

3.3.2.1 Design goals for a pharmaceutical warehouse management system

When designing a warehouse management system (WMS) for the pharmaceutical industry, a few factors, including operational factors and national legislation factors, should be kept in consideration to ensure inventory accuracy [21]. There exists a different set of expectations from all the parties involved in the pharma ecosystem, including business users, administrators, and internal warehouse operators, the design goals for whom are described below.

For business users a WMS has to accomplish the five objectives mentioned below:

- A WMS should abide by all the laws, rules, and regulations laid down by the government;
- WMS should collaborate with other WMS sharing the same enterprise system;

- Future development should be taken care of in terms of both design and operation as per the needs of the enterprise;
- The system must comply with the corporate rule book;
- In order to reduce management costs, streamlining of current business prospects can be done, which can also improve efficiency and provide an easy data-sharing process.

For administrators, a pharmaceutical WMS must accomplish the four agendas described below:

- Seamless docking should be provided among different WMS within the enterprise system for ease of data collection;
- System records should be more clear and unambiguous;
- Since data security and confidentiality are paramount for a WMS, management of the same should be carefully carried out;
- Since staff members of a WMS also have access to data, it should be ensured by all means that data are not leaked.

For internal warehouse operators, a pharmaceutical WMS aims to accomplish the four objectives mentioned below:

- Processes should be simplified in order to improve efficiency;
- The user interface of a WMS should be user-friendly and easily accessible;
- System implementation should be as simple as possible and easy to operate;
- Duplication of data should be eliminated during data collection to minimize repetitive tasks and manual operation.

3.3.2.2 Architecture of a pharmaceutical warehouse management system

Contrary to the age-old WMS design which basically stood on the client/server mechanism, the new pharmaceutical WMS uses the upgraded version of the client/server structure in order to fulfill the enterprise requirements. These stand on the browser/server mechanism for system design. Following this schematic for architecture design, it is possible to achieve a better interaction between operators and computers. This schema can also be used across different platforms, thus improving the system reliability and its applicability in other industries. Using a browser server, the operators send applications containing instructions and carry out everyday management operations. A WMS design based on the browser/server structure is comprised of three major layers which includes the web interface layer, the business logic layer, and the data access layer [21]. A system interface module is added to facilitate docking since the WMS has to collaborate with many other WMS in the enterprise management system (EMS), vendors, and customers at the same time. The architecture of a pharmaceutical WMS is shown in Fig. 3.5. As per Fig. 3.5, the system can be broken down into three major parts: a database, WMS, and users. The three segments are described in further detail below.

Database: In order to feed data to the WMS, data are obtained from three major sources, namely the RFID system from which data acquisition is done, residual data which the RFID system could not collect is manually entered by the staff, and the final chunk of data is provided by the WMS's docking system which allows collaboration among different WMS in the EMS.

An enterprise-based WMS is made up of a three-layer structure which includes the interface layer, data access layer, and the business logic layer. The three-layer structure of the WMS is introduced below.

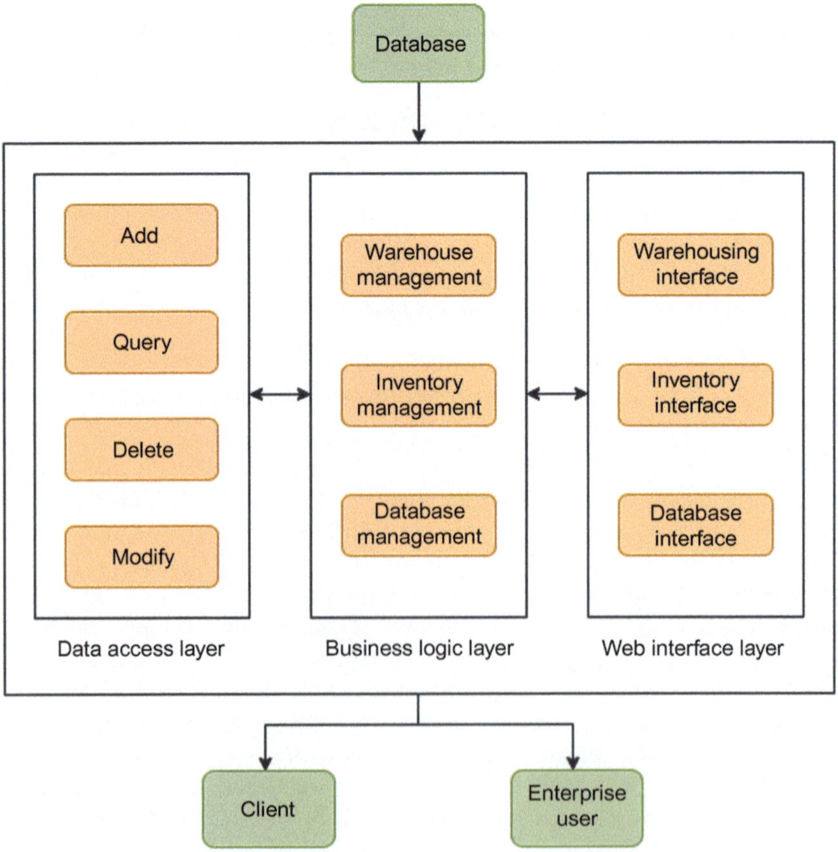

FIGURE 3.5 Pharmaceutical warehouse management system architecture design.

- Data access layer: This layer directly connects with the database and accesses the data. This layer's function is to capture the data in the database, and operate on the data which includes operations like addition, querying, deletion, and modification.
- Business logic layer: This layer works upon the orders made by the system users, which include business and process management operations like warehouse, inventory, and database management. Basically, the WMS calls in data to the business logic layer, analyzes it as per the given instructions and then the data are transferred as an interface to the web interface layer.
- Web interface layer: This layer displays the data to the users in the form of an interface. Under the browser/server-based architecture designs, the user further sends the data requirements to the browser and receives the results as per the management needs.

3.3.3 Shipment and tracking

Transportation and logistics are paramount to the pharmaceutical industry. Shipment and tracking companies have implemented IoT technologies mostly as track-and-trace applications, intending to decrease network complexity. Applications of the IoT for shipment of logistics from warehouses via trucks and planes includes capacity sensing, planning and reporting, route optimization, etc. These applications are described here:

1. Capacity sensing: Performed by systems that can detect and communicate open spaces in a warehouse, port, or parking lot.
2. Planning and reporting: Performed by systems that can detect and analyze events such as traffic accidents within a delivery network, allowing for more accurate delivery dates.
3. Route optimization: Performed by tools that can map the shortest or most fuel-efficient routes for delivery vehicles.
4. Energy management: Performed by tools that monitor and enable decision-making about the use of fuel, lighting, and heating/cooling within vehicle fleets and facilities.
5. Fault detection and resolution: Performed by systems that can monitor fleets of vehicles, aircraft, or ships for faults and maintenance needs, improving uptime for the fleet.

Similarly, current IoT use cases focus on improving those factors and include environmental monitoring, threat detection and prevention, and real-time traceability down to the unit level. These applications of the IoT for shipment and tracking benefit customers for shipment of packages, containers, etc. These applications are described here:

1. Environment monitoring and management: Performed by systems that can monitor and adjust the temperature at which a package is maintained.

2. Threat detection and prevention: Performed by tools that can help detect unauthorized openings of shipping containers, helping to prevent and reduce theft.
3. Real-time traceability: Performed by systems that can track not just vehicles or shipments but individual items.

3.3.4 Quality control of drugs

Quality control ensures better drugs and pharmaceutical equipment. Its benefits include increased sales and disposal of substandard products which cannot be sold. When proper analysis of the quality of a product is done it can save disastrous recalls that can damage a company and that can cost substantial amounts of money. With the pharmaceutical industry there is little room for mistakes and substandard products can put customer's lives at risk. Additionally, the industry is highly competitive and subquality products can mean that contracts will move to competitors. There are varying regulatory standards and quality compliance guidelines for medical drugs across the world. For pharma manufacturers and supply chain networks it is an absolute necessity to adhere to and comply with the regulations of the respective markets, which brings more complexity to the supply chain networks. Under these circumstances, there is added pressure to have maximum visibility of the supply chain, optimum temperature control, and effective warehousing.

3.4 Challenges in pharma logistics

The administration and logistics of the pharmaceutical industry can be very grueling with challenges including leaks in the supply chain, partner organizations not cooperating with each other for data sharing, uncertainty in environmental conditions, huge investment requirements for rising demand in infrastructure, complying with regulatory standards, and prevention of data breaches. The following section describes the above challenges in detail.

3.4.1 Supply chain visibility

Supply chain visibility aims to improve upon the transparency of operations within the supply chain network by ensuring sharing of data between all its divisions. Delivery of supply chain visibility is termed effective only when it passes in the areas of agility, resilience, reliability, and responsiveness (source: supply-chain-visibility.com) in the supply chain industry. The applications of supply chain visibility lie in ensuring traceability of shipments, optimum inventory levels, and meeting the regulatory requirements. With the expansion of the supply chain industry and its fragmentation into different departments, sharing of data is not usually preferred so as to ensure the smooth functioning of individual departments. This, however, hampers the performance of the entire supply chain network. This loss of control and visibility over other departments results in leaks in SCM like thefts, shortage of goods, loss of shipments, and even issues with quality of drugs.

In order to ensure end-to-end supply chain visibility, the company SAP came up with a unique solution called the "supply chain control tower," proposed in the SAPinsider's supply chain conference held in 2017. Control towers consist of information units which collect and distribute information so as to take quick actions on the challenges encountered during SCM. These towers capture and organize each and every piece of information, be it of an order placed by a customer, a shipment made from a warehouse, costs encountered, or feedback provided. The three pillars managed by the control tower are people (or organizations), processes or operations, and the technology governing the entire system.

1. Process: Processes which ensure a high level of product visibility by managing control tower functions;
2. People: The workforce is responsible for outlining and running the processes by leveraging technologies;
3. Technology: For monitoring, auditing, and ensuring the overall effectiveness of the control tower concept, integration of different softwares is done with the help of a common middleware software.

After rigorous research, the following needs have been identified pertaining to supply chain visibility in the pharmaceutical industries [22]:

1. Visibility into inventories
   - Medicines stocked up in inventories are rarely fully distributed and are subjected to immediate disposal upon expiry, thus causing expenditure of a lot of money and resources. This costs inventory managers a lot of money and reducing the inventory level hence becomes a huge concern;
   - Virtual management of inventories and warehouses is required with simultaneous reduction of manual labor;
   - Real-time inventory tracking and tracing sensitive drugs is essential;
   - Maintaining a chain of custody in the pharmaceutical industry can leverage the reputation of pharmaceutical firms;
   - Preventing counterfeit drugs from reaching the warehouses is imperative.
2. Visibility into demands
   - The distributors are always keen to know the rise or fall in demand from retailers. As of now the demands are not made visible to the distributors and they are forced to carry out a prediction based on the historical data, such as the frequency of buying of certain drugs compared to others;
   - Consumption patterns need to be monitored;

- Manufacturers require distributors to provide their demands so inventory levels can be accordingly maintained.

3. Regulation and compliance visibility
   - Participation of manufacturers, packagers, wholesale distributors, and dispensers is required to abide by regulatory requirements;
   - Addition of additional layers of management and security in the supply chain becomes mandatory.

3.4.2 Security issues

Mismanagement of shipments and carelessness in tracking have also led to thefts and caused huge losses. As per the Global Cargo Theft Threat Assessment conducted in 2013, the pharmaceutical industry has incurred extremely huge losses due to cargo thefts from shipping vehicles. To prevent this, pharma manufacturers have initiated real-time monitoring of the physical conditions of cargos and the performance of vehicles transporting them in order to minimize losses and penalties because of damage to drugs inside cargos which makes them unfit for consumption. Technologies like GPS for tracking the location of cargos provide real-time visibility to the supply chain personnel when vehicles are in transit. The data collected from vehicles help in remote and intelligent fleet management. This information is collected by GPS and then transmitted to control rooms through general packet radio service (GPRS) sensors, thus enabling control-room personnel to perform analysis on the data generated out of the movement of fleet from the comfort of their centers. This can prevent drug damage when in transit and lower transportation costs. Fleet management using sensors and GPRS is diagrammatically depicted in Fig. 3.6.

3.4.3 Temperature control

Uncertainty of weather conditions can be a big obstacle to the pharmaceutical industry. For products like drugs, maintenance of temperature while in transit and warehousing holds major importance and, if not regulated, can damage the integrity of the drug by altering its physical and chemical properties, thus having serious consequences on the health of patients. Therefore

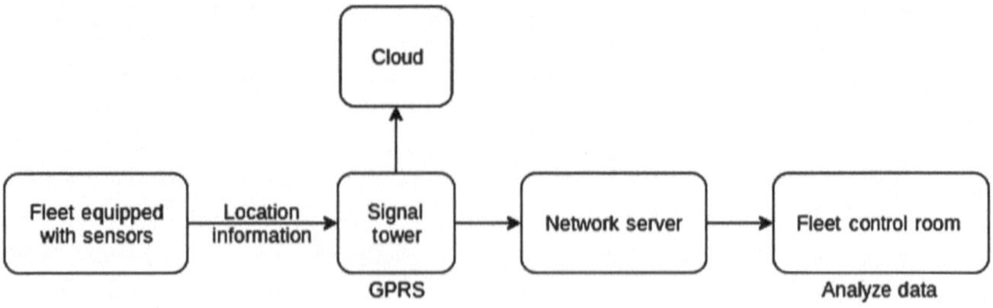

FIGURE 3.6 Fleet management using sensors and GPRS.

the temperature should be carefully regulated to ascertain the quality of the drug and its compliance with the regulatory standards.

A temperature-controlled supply chain is deployed to maintain optimal conditions and guarantee the potency of drugs. The "cold chain" mechanism is used to implement temperature-controlled logistics. Introduced in 1940 by Frederick McKinley Jones, the cofounder of Thermo King, cold chain is a mechanism which manages the storage and transportation of drugs, vaccines, and other products at a low temperature range right from the manufacturing point to the consumption point. The requirements which need to be catered to so as to successfully implement a temperature-controlled supply chain include:

1. Every stage in the pharmaceutical supply chain should have temperature-controlled storage facilities like refrigerated vehicles, and active and passing shipping systems;
2. Shipment should be successfully completed in the stipulated time to prevent potential delays. As per safety regulations, a delay could lead to an inspection which involves physical opening of packages, thus resulting in temperature deviations;
3. The cold storage unit should not be positioned near the truck's doors so that adequate air circulation can take place around the unit to maintain the optimum temperature.

To ensure secure shipment without any degradation of drugs, a quality management system (QMS) should be in place to ensure that each step during operational handling in the transportation process is executed according to the designed and qualified process. In the QMS, checks are done at seven stages for temperature-controlled fleet management:

1. Maintenance: Ensuring regulatory compliance and standard operational handling;
2. Planning: Mitigation of risk along the route;
3. Payload preparation: Ensuring air flow underneath the pallets;
4. Preinspection: To ensure appropriate insulation, set point, and temperature before loading;
5. Loading: Avoiding temperature excursions, ensuring that hot and cold spots are monitored, and that air flow is proper;
6. Transport: Ensuring temperature-controlled transport and taking action in the case of alarms;
7. Unloading: Documentation about shipment (paper work, final temperature evaluation).

3.4.4 Data security

As the pharmaceutical industry becomes more and more data intensive, lack of attention to security practices puts valuable data at risk. With the rise of big data and the need to store, analyze, and utilize large volumes of data to build powerful business models, the pharmaceutical industry has faced major cyber attacks by hackers who are constantly looking for opportunities for data breaches as the industry possesses a treasure trove of valuable information. Reports by OCISIA and UK Information Intelligence experts estimate the cost of cyber crime to the UK economy to be over £25 billion annually [23]. The same report claimed that the pharmaceutical and biotech industries are amongst the hardest hit industries. In 2017, a pharmaceutical company "Merck" fell victim to a ransomware attack known as "WannaCry" followed by "NotPetya," which led to the disruption of production of several Merck medicines and vaccines and cost them more than $300 million.

Some approaches to act on cyber attacks are described below.

1. Incident response: As the name suggests, incident response implies responding or addressing a cyber attack after the security breach takes place. This ensures that the damage is put under control and helps an organization minimize losses, mitigate exploited vulnerabilities, restore services and processes, and reduce the risks that future incidents pose. Regardless of how attackers gain access to the system, once an attacker is discovered, the basic information about the attack, called indicators of compromise (IOCs) such as IP addresses, domain names, or malware hashes, are shared across the cybersecurity community. These IOCs are then brought to use to prevent further attacks. However, there are two problems to this approach. First, to start generating IOCs and share useful information, an attack needs to take place so that it can be addressed, thus making a pharmaceutical firm suffer. Moreover, a company secures lots of its data through IOCs. In order to break into a company's valuable information, adversaries either come up with new IP addresses or they recompile their malware, thus getting a new hash value so that their attacks pass through all defenses which depend on IOCs. This proves that an incident response is not an ideal approach to deal with cyber attacks since it is highly flawed.
2. Regulatory compliance: Abiding by complex laws, regulations, and guidelines is also an effective way of securing valuable data in the pharmaceutical industry, which is so heavily regulated. Since passing a new law is a relatively time-consuming process, regulatory guidelines are preferred because they can be adopted and implemented quickly. International regulatory bodies for the pharmaceutical industry include the WHO (World Health Organization), the FDA (Food and Drug Administration) and the MHRA (Medicines and Healthcare Products Regulatory Agency). Various drug-related concerns like safety, quality, market flow, research and development incentives, and pricing can be effectively regulated through such regulatory compliances.

3. Smart SCADA: SCADA (Supervisory Control And Data Acquisition) is an architecture which provides high-level management, supervision, and control of a plant. It is a "security by design" approach wherein the software and its components are specifically designed to guarantee secure operations. "Zenon" by Copadata emerged as a powerful software solution for the pharmaceutical industry in order to provide cybersecurity. Zenon complies with FDA 21 CFR (Code of Federal Regulations) Part 11, one of the most widely accepted regulations in the pharmaceutical industry. This regulation requires organizations to implement controls, electronic audit trails, and systems validations. Zenon can be used not just for industrial automation and monitoring of existing processes but also for packaging and electronic data acquisition. Moreover, if SCADA softwares use their personal network protocols for interproduct communication, data can be transferred in digital format in binary data packages and information is never transmitted in plain text format which is easily decipherable. Further, client authentication is mandatory to set up a connection, thus preventing any unknown access to the network.

3.5 Conquering pharma logistics with Internet of things

The IoT has left its footprint in almost every working sector of pharmaceutical companies, which includes drug development, manufacturing, shipment, distribution, releasing products, and their consumption. When advanced data analytics is coupled with real-time information, pharmaceutical logistics can be conquered because of fast and much more accurate decision-making, greater efficiency, products meeting the quality standards, and regulatory compliance being ensured. Some approaches to overcoming pharma logistics with IoT are discussed next.

3.5.1 Smart warehousing

Warehouse operations are a vital part of pharmaceutical operations and cost about 90% of the total supply chain costs. Smart warehousing is beneficial because with real-time updates on inventory, floor space can be utilized to its maximum potential. Sensors can be used to track and analyze the inventory flow and asset movements. Relevant information is captured by the wireless readers and used for aggregate planning. RFID tags are utilized for tracking location and consumption rate. Any inconsistencies incurred during the processes are captured by these sensors and transmitted for analysis. These solutions altogether contribute toward higher productivity and maximum floor space utilization. The data of the inventory flow from warehouses can be used for production planning to optimize inventory levels, prevent situations where the warehouse goes out of stock, and improve demand fulfillment.

When warehouse supervisors are given a 3-D view of the warehouse along with real-time visibility into warehouse operations, they can perform cold chain temperature monitoring for temperature-sensitive materials and track the movement of pharmaceutical equipment inside the warehouse. In addition, production is also boosted by shop-floor visibility and the identification of issues which can be rectified by manpower. Refrigerators and other machines with attached active and passive temperature recorders are used for cold chain temperature monitoring by warehouse managers, thus making use of IoT-based devices for temperature logging. These temperature recorders are connected and their measurements are aggregated and compared against standard thermal stability charts and if there is any deviation from the permissible ranges, alerts are generated. IoT-based devices can also be used to monitor inventory movement. Sensors are attached to packages which generate information pertaining to their dimensions and capacity and this information is captured and aggregated by wireless readers and is forwarded to the warehouse management system for examination [14].

3.5.1.1 Business benefits of smart warehousing

3.5.1.1.1 Increasing operational efficiency in the warehouse

This can be achieved by optimizing the movement of warehouse assets. Increased operational efficiency results in a higher quality service, efficient utilization of both manpower and material resources, and certainly higher productivity [24]. Under this methodology, sensors attached to packages first transmit information on their real-time location to the warehouse managers. The managers have possession of a handheld device on which they receive all the information about inventory movement and inconsistencies, if any, in the form of a 3-D view of the warehouse. RFID tags, artificial intelligence (AI), and embedded sensor packets are attached to packages carrying material equipment which facilitate operators to have a clear look into the lifecycle of pharmaceutical equipment and products [14,25].

3.5.1.1.2 Maintain desired storage conditions

This can be achieved by tracking storage conditions of sensitive drugs. Good storage conditions result in the maintenance of high-quality drugs and prevents degradation due to temperature variations. Environmental sensors are attached to packages to govern cold chain storage conditions in real time to keep a check on the temperature conditions the drugs are being subjected to. Temperature loggers attached to refrigerators are also used to collect information and generate warnings if there is any variation from the permissible temperature ranges.

3.5.1.1.3 Orient production and demand

For every pharmaceutical company, the key to success is to provide a good-quality service to every customer while putting in less resources. This can be achieved by feeding real-time data on inventory movement as it moves through the warehouse, distribution units, and onto the retailers. Demand forecasting is employed to perform stock planning, optimize inventories and prevent them from going out of stock, and fulfill the needs of customers [24].

3.5.1.1.4 Increase manufacturing efficiency

This can be achieved by allowing supply chain visibility within the warehouse and scheduling the supplies as per the first in first out (FIFO) mechanism, that is, organizing the supplies as per their serial numbers to make sure that there are minimal errors during handling [26]. Automating the WMS to handle and act upon incoming requests can help reduce manual errors which hamper the warehouse processes [27]. This will not only reduce the incident response time but also improve production yield significantly. This results in real-time dynamic scheduling of shop-floor activities, better equipment utilization and less downtime, ability to track overall equipment effectiveness, improved productivity, efficiencies and cycle time, and certainly increased production yield. There are six methodologies to achieve this:

- Using smart devices that accumulate operational data and running status including the temperature and load of the package and checking if the manufactured goods are under maintenance or ready for use;
- Embedding sensors to pharmaceutical equipment and building a back-end platform using cloud computing services such as Amazon Web Services and Microsoft Azure;
- Performing analytics on the data collected from the sensors to predict as to when any equipment might stop functioning properly;
- Automating incident requests triggered by integration with ticket management systems;
- Using sensors that perform analysis on metadata to identify and reduce the causes of variations in processes and improve production yield;
- Movement of manpower and material should be monitored across the shop floor through tracking systems.

3.5.1.2 Basic workflow of a smart warehouse management system

The lifecycle of a WMS involves five major steps: stock planning, receiving orders and delivering them, location management, picking and packing orders, and their shipment and tracking, as demonstrated in Fig. 3.7 [24]. In this section, each warehouse operation mentioned above has been explained.

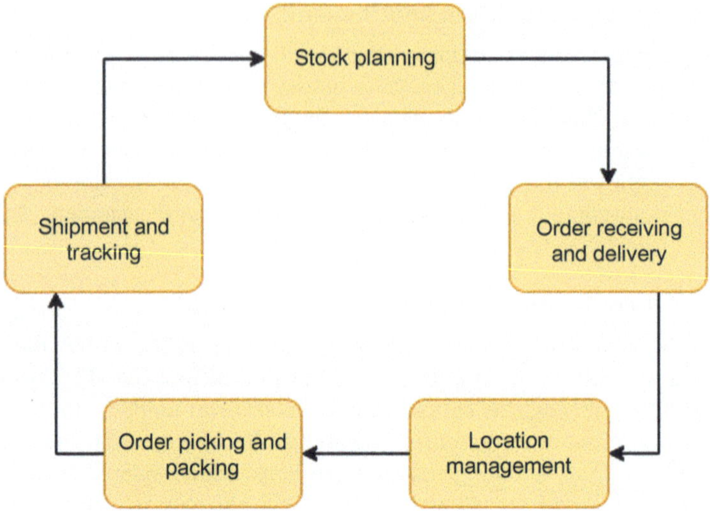

FIGURE 3.7 Basic workflow of a smart WMS.

A well-defined solution for the betterment of operations of a WMS has also been proposed.

1. Stock planning: In an attempt to fulfill the requirements of consumers, stock planning needs to be done keeping in mind the incoming demand and carrying out demand forecasting accordingly. Multiple aspects are kept in mind while doing demand forecasting including product seasonality and discounts [24].
2. Receiving and delivering orders: As soon as the products are received at the warehouse, they are required to be placed at their designated locations. If positioned properly, the products can be picked easily whenever orders arrive, thus making it easier for the warehouse staff to look after other operations in the warehouse. Clearly, receiving and positioning products is fairly easy in empty warehouses, but poses great difficulty when it comes to well-established warehouses where it is tough to keep the existing work on hold in order to rearrange products within the warehouse.
3. Location management: Technology comes to the rescue when it comes to location management, which is done using a list which determines the positions where the products should be kept within the warehouse. This list is formulated using SQL stored procedures which are SQL codes that can be saved for later use. The products are positioned in a way that those more frequently used are kept closer to the exit location of the warehouse, keeping other products further away. The algorithm used to formulate this list uses a data set which stores the information of the products and the frequency of their usage and has been described in Refs. [28,29].

4. Picking and packing orders: Millions of orders are placed by customers all over the world, and they are captured and worked upon by warehouse operators. Considering the huge amount of orders to pick and pack, their priority is set by sorting the products based on how close they are to their expiration dates [24]. Another solution has been proposed in Ref. [27] by utilizing AI as in the traveling salesman problem (TSP). The TSP calculates the distance between two points on the grid and can determine the shortest distance between two places by using dynamic programming.
5. Shipment tracking: Warehouse operations can be significantly improved if the transportation systems are embedded with sensors which facilitate real-time shipment tracking and are used to deliver drug packages to retailers. This forms an integral part of smart WMS. A three-stage algorithm runs behind the smart WMS to allow efficient shipment tracking. The first stage is the preprocessing of the data set, which sets the time taken to unload the packages from the shipping vehicle to zero for simplicity. An elementary solution is designed in the second stage. In the third stage, the Tabu Search algorithm is used to further improve the solution by using an iterative approach. Moreover, optimized transportation routes can also be suggested using the vehicle routing problem optimization technique.

3.5.2 Installing radio frequency identification tags

RFID is a technology which has gradually gained popularity as a facilitator for the IoT and its many applications [21]. Technically, RFID tags are small transponders that respond to queries from a reader by wirelessly transmitting a unique identifier which could be a serial number. They are widely used to track items in production environments and to label items in supermarkets. They are usually thought of as an advanced barcode. RFID got its first major boost when it was used for toll-tax collection on highways in Norway and a few states in the United States. RFID tags consist of the following four components: an antenna, a microchip, a case, and a battery which is needed only for active tags [30]. In most cases, the size and configuration of the microchip depend upon the antenna and the frequency at which the RFID tag will be operating. The microchip's size also depends upon the area required for the chip to meet all of its necessary components. The size ranges within a few millimeters. Some RFID tags also possess rewritable memory, where new information about identifiers and any updates between consecutive readings of the RFID tag can be stored.

Installing RFID tags into packages and equipment finds utility in the pharmaceutical industry as this can help manufacturers track the location of a package by checking where and when the attached RFID tag was scanned. To achieve this, an RFID system is constructed which has the ability to both collect and store information. The RFID system is composed of two major

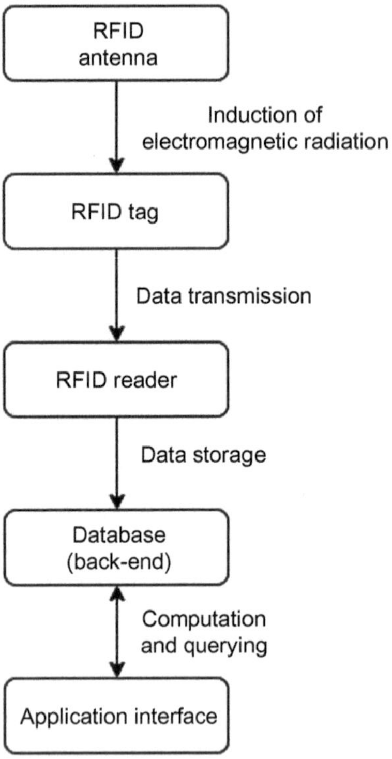

FIGURE 3.8 Structure and working of an RFID system.

systems: a tag system and a reader system. A chip and an antenna make up the tag system. The chip works as an identification label and saves the tag data. The antenna energizes the RFID tag by absorbing external electromagnetic radiation. The RFID reader system scans tags and then forwards the information to the back-end database onto the application interface where relevant computations can be performed. Fig. 3.8 further shows a block diagram of an RFID system.

3.5.3 Real-time shipment tracking

As much as the manufacturing of pharmaceutical equipment, its shipment and warehousing also hold key value in the eyes of drug manufacturers. However, monitoring the transportation of pharmaceuticals from the warehouse to the retailer never fails to pose an issue for the industry leading to product degradation and huge operational costs. With the help of IoT solutions, real-time shipment tracking can be performed. The key idea behind the utilization of equipment for real-time updates for improved process yield is connecting the equipment sensors to the cloud network so that information cannot just be analyzed on the spot but also stored for future use [3]. Pharmaceutical companies rely heavily on batch production, and this involves the activity of manpower and material on the shop floor. Moreover, current good manufacturing practice (GMP) regulations dictate that pharmaceutical equipment should be carefully calibrated to ensure that the chemical composition and effectiveness of drugs remains intact. IoT-based devices make the shop-floor activities visible to the pharmaceutical companies for better supervision. This helps in raising the production levels and assuring GMPs. This is achieved by attaching sensors to equipment in the manufacturing units. These sensors accumulate and transmit data to a cloud network which extracts the raw information from the sensors and converts it into relevant information. Sensors can be used to monitor cargo and medical equipment remotely in real time and even develop machine learning models to predict the real-time location of fleets in transit. This not only results in increased accessibility of goods but also ensures a smooth workflow in the supply chain, thus saving huge sums of money for pharmaceutical companies. Real-time shipment tracking is paramount because it not only ensures product integrity and traceability but also optimizes inventory costs, both of which have been described below.

3.5.3.1 Ensuring product integrity and traceability

To achieve this, bidirectional communication and product authentication are done at every point in the supply chain. This is done so that counterfeiting can be detected from every location in which products leave their mark as they move up the supply chain. Packages embedded with IoT-based devices continually track environmental conditions in cold chain conditions during shipment, hence ensuring that the products are of high quality and that they maintain efficiency. The IoT equipment required to achieve product integrity and traceability in the supply chain network are 2-D barcodes, RFID tags, near-field communication (NFC) sensors, and smart labels for packaging of drugs and vaccines. Both public and private cloud data centers are used to store the transmitted data as the drugs are transported from one location to another. Storage of information ensures that informed decisions are made and that the products can be traced anywhere in the supply chain.

3.5.3.2 Reduce inventory costs

To achieve reduced inventory costs visibility into the movement of products within the inventory should be increased. This results in cost-effective and optimized SCM because the quantity of products in the inventory aligns with demand. Sensors like 2-D barcodes, RFID tags, and smart labels are used here, and transmitters are also used at warehouses as they capture the movement of packages and report this to the relevant authorities. Hence, stock levels can be better optimized if informed decisions are made pertaining to the inventory by leveraging the information collected by sensors at every minute.

3.5.4 Environmental sensor installation

To maintain desired storage conditions of sensitive drugs which are highly susceptible to changes in environmental conditions, environmental sensors are used to collect data from storage areas to track the real-time conditions of the drugs. These drugs often have a large portion of costly active ingredients with shorter shelf-lives which can only be sustained under well-defined temperature conditions. Both active and passive log data are captured to generate alerts in the event of temperature changes by constant comparison with the medicinal thermal stability tables. This leads to the prevention of untimely expiration or disturbance to the chemical composition of drugs. Along with managing the temperature spikes during transportation, maintaining drug quality is also paramount. Environmental sensors are embedded inside packages kept in shipping containers to constantly monitor the temperature conditions the drugs are being subjected to. It has to be ensured that the temperature is maintained within the specified range, be this during the relocation of packages from a cooling apparatus to a shipment container or a fault in the temperature maintenance equipment. In case of discrepancies, sensors are employed to initiate alerts so that immediate action can be taken [14].

Apart from reducing the wastage of drugs and assuring efficacy, these solutions also help to ensure compliance with regional and international target market norms and regulations. Yet pharmaceutical companies have lost huge sums of money due to disintegration of drugs because of temperature deviations which might end up being an even greater threat if the damage goes unnoticed and could be life threatening for patients. For pharmaceutical shipment, personnel responsible for shipping temperature-controlled products that are handled multiple times from the manufacturer to consumers, it is paramount to assure that monitoring tools are of high quality and that all the stages in the supply chain are well coordinated. Some challenges in monitoring cold chain conditions in the pharmaceutical industry on shipping vehicles are mentioned below [31]:

1. Drug delivery: Maintaining the temperature of drugs within the specified range until they are finally delivered is always a daunting task for warehouse keepers. The temperature is generally kept between 2°C and 8°C for the pharmaceutical industry. There exist multiple factors, including the size of the package and the thickness, type and quantity of cooling agents used, which can potentially impact the duration over which a product is said to be under appropriate temperature conditions. Other traditional storage techniques which make use of frozen, water-based gel packs kept inside airtight containers have long been discarded because they can result in temperature fluctuations within the package, thus causing serious drug damage, making them unfit for consumption. Apart from temperature fluctuations, these gel packs also add to the cost of labor and shipping. Therefore, to ensure that drugs are delivered in their absolute composition,

temperature-controlled transportation has been designed which requires specialized trucks and equipment embedded with IoT-based sensors, specific driver training which again brings up the recurring challenge of shortage of drivers, and increased cost of the specialized equipment.

2. Supervision: Constant supervision is required as the goods are moved from one location to the next in the supply chain so that pharmaceutical supervisors can maintain control over their products. For companies that outsource logistics, each process runs a high level of risk, whether for their providers or carriers. In that case, the IoT assists greatly with its paramount sensor technologies which can enable pharmaceutical companies to easily monitor and manage cold chain logistics as they move from manufacturers to consumers.
3. Worldwide transportation: With globalization increasing, serialization regulations have been laid down which ensure that goods which cross national borders meet the import and export requirements of both the sender and receiver country. Regardless of being outside the country, location visibility and temperature monitoring need to be kept under control and this has led to a rise in the utility of IoT-based devices in the market for real-time tracking of goods.

References

[1] Y. Yin, Y. Zeng, X. Chen, Y. Fan, The Internet of things in healthcare: an overview, J. Ind. Inf. Integr. 1 (2016) 3–13.

[2] K. Ashton, That 'Internet of things' thing, RFiD J. 22 (2009) 97–114.

[3] A. Marathe, M. Awwad, Internet of things: opportunities and applications in pharmaceutical manufacturing and logistics, In: E-H. Ng, B. Nepal, E. Schott, and H. Keathley (eds), International Annual Conference. (2018).

[4] S.S. Kothari, S.V. Jain, A. Venkteshwar, The impact of IOT in supply chain management. Int, Res. J. Eng. Tech. (IRJET) 5 (8) (2018).

[5] H. Sharma, S. Parekh, S. Shewale, Regulatory practice in pharmaceutical industry, Pharm. Reg. Aff. 3 (2014) e125. Available from: https://doi.org/10.4172/2167-7689.1000e125.

[6] H. Grabowski, The evolution of the pharmaceutical industry over the past 50 years: a personal reflection, Int. J. Econ. Bus. 18 (2011) 161–176. Available from: https://doi.org/10.1080/13571516.2011.584421.

[7] Pharma outlook 2030: From evolution to revolution, KPMG International, 2017.[online]

[8] Sanofi, Google parent form $500 million diabetes joint venture, Reuters, 2016 [online].

[9] N. Yadav, A. Mazumder, R. Mazumder, N. Rana, Pharma-IoT: its opportunities and challenges in digital India, 70th Indian Pharmaceutical Congress, at Noida, 2018.

[10] S. Jangili, K. Bikshalu, Smart grid administration using big data and wireless sensor networks, Int. J. Adv. Res. Sci. Eng 6 (2017) 629–636.

[11] M.U. Farooq, M. Waseem, A. Khairi, S. Mazhar, A critical analysis on the security concerns of Internet of Things (IoT), Int. J. Comp. Appl 111 (7) (2015) 1–6. 0975 8887.

[12] H. Suo, J. Wan, C. Zou and J. Liu, "Security in the Internet of Things: a review," 2012 International Conference on Computer Science and Electronics Engineering, Hangzhou, 2012, pp. 648-651.

[13] S.M. Riazul Islam, D.N. Kwak, M.H. Kabir, M. Hossain, K.-S. Kwak, The Internet of things for health care: a comprehensive survey, IEEE Access 3 (2015) 678–708.

[14] A.C. Madhukar Saboo, M. Suranglikar, The Internet of things: the new Rx for pharmaceuticals manufacturing & supply chains. Cognizant 20-20 Insights, 2017, 1–12.

[15] G. Carnaz, V.B. Nogueira, "An overview of iot and healthcare", Proceedings of the 6th Informatics Days of the University of Évora. School of Science and Technology, March 2016.

[16] F. Hussain, The application of the Internet of things in healthcare, Int. J. Comp. Appl. 180 (2018) 19–23. Available from: https://doi.org/10.5120/ijca2018916454.

[17] R. Peng, M.L. Sichitiu, Angle of arrival localization for wireless sensor networks, in: Proceedings of the 3rd Annual IEEE Communications Society on Sensor and Ad Hoc Communications and Networks, Reston, VA, USA, IEEE, 2006.

[18] J. Gubbi, R. Buyya, S. Marusic, M. Palaniswami, Internet of Things (IoT): a vision, architectural elements, and future directions, Future Gener. Comp. Syst. 29 (2013) 1645–1660.

[19] M. Díaz, C. Martín, B. Rubio, State-of-the-art, challenges, and open issues in the integration of Internet of things and cloud computing, J. Netw. Comp. Appl. 67 (2016) 99–117.

[20] Y. Liu, B. Dong, B. Guo, J. Yang, W. Peng, Combination of cloud computing and Internet of Things (IOT) in medical monitoring systems, Int. J. Hybrid Inf. Technol. 8 (2015) 367–376.

[21] X. Xiang, J. Li, X. Tao, Research of a pharmaceutical enterprise warehouse management system based on RFID technology, Sci. Res. 4 (2) (2016) 43–47. Available from: https://doi.org/10.11648/j.sr.20160402.14.

[22] N.S. Zhang, W. He, P.S. Tan, Understanding local pharmaceutical supply chain visibility, Oper. Res. 25 (2008) 234–239.

[23] Regulatory compliance and cyber threats: Industrial security in the pharmaceutical sector, COPA-DATA[online].

[24] E. Žunić, S. Delalić, K. Hodžić, A. Beširević and H. Hindija, Smart warehouse management system concept with implementation, 2018 14th Symposium on Neural Networks and Applications (NEUREL), Belgrade, 2018, pp. 1-5.

[25] A. Kamali, Smart warehouse vs. traditional warehouse - review, CiiT Int, J. Autom. Auton. Syst. 11 (1) (2019).

[26] A.M. Atieh, H. Kaylani, Y. Al-abdallat, A. Qaderi, L. Ghoul, L. Jaradat, I. Hdairis, Performance improvement of inventory management system processes by an automated warehouse management system, Procedia Cirp 41 (2016) 568–572.

[27] S. Seifermann, J. Böllhoff, J. Metternich, A. Bellaghnach, Evaluation of work measurement concepts for a cellular manufacturing reference line to enable low cost automation for lean machining, Procedia CIRP 17 (2014) 588–593.

[28] Zunic, E., Besirevic, A., Skrobo, R., Hasic, H., Hodzic, K., & Djedovic, A. (2017, October). Design of optimization system for warehouse order picking in real environment. In 2017 XXVI International Conference on Information, Communication and Automation Technologies (ICAT) (pp. 1-6). IEEE.

[29] Zunic, E., Hasic, H., Hodzic, K., Delalic, S., & Besirevic, A. (2018, May). Predictive analysis based approach for optimal warehouse product positioning. In 2018 41st International Convention on Information and Communication Technology, Electronics and Microelectronics (MIPRO) (pp. 0950-0954). IEEE.

[30] Jechlitschek, C. (2006). A survey paper on Radio Frequency Identification (RFID) trends. Retreived on 29.11.12.

[31] P.S. Sedlak, Cold Chain Challenges in the Pharmaceutical Industry. https://www.jasedlak.com/blog/view/cold-chain-challenges-in-the-pharmaceutical-industry [online] (2018).

Further reading

G. Zhang, C. Li, Y. Zhang, C. Xing, J. Yang, Seman medical: a kind of semantic medical monitoring system model based on the IoT sensors, in: Proceedings of the 14th International Conference on e-Health Networking, Applications and Services (Healthcom 2012), Beijing, China: IEEE, 2012.

H.K. Nguyen, J. Song, Z. Han, Demand side management to reduce peak-to-average ratio using game theory in smart grid, in: Computer Communications Workshops (INFOCOM WKSHPS), 2012 IEEE Conference on, 2012, pp. 91–96.

S. Mondal, Inventory management techniques in pharmaceutical industry.

Shipping smarter IoT opportunities in transport and logistics, Deloitte University Press.

W. Zhang, B. Qu, Security architecture of the Internet of things oriented to perceptual layer, Int. J. Comp. Consum. Control (IJ3C) 2 (2Please provide page range details for Zhang and Qu (2013).) (2013).

Chapter 4

Drug identification and interaction checking using the Internet of Things

K. Eswari[1] and C. Priya[2]

[1]School of Pharmaceutical Sciences, Vels Institute of Science, Technology and Advanced Studies (VISTAS), Chennai, India, [2]Department of Information Technology, Vels Institute of Science, Technology and Advanced Studies (VISTAS), Chennai, India

4.1 Introduction

In recent years, the life sciences industry has seen remarkable advancements, such as the application of Internet of Things (IoT)-based devices which are benefiting patients, manufacturers, doctors, nurses, and the entire healthcare industry. Though life sciences industry has a conservative rate of adoption of new technologies, it has been among the early adopters of IoT.

In life sciences, the diagnosis, monitoring, preventive measures, and treatment of chronic conditions require proactive action and this is feasible today with the IoT by embedding sensors and smart devices into our surroundings. As per a report by Mind Commerce Publishing, the value of IoT in health care was forecast to reach $117 billion by 2020. Also, health care as a part of the life science industry is predicted to be the fastest growing segment in entire the IoT market with a compound annual growth rate of 15.1% as stated in that report.

The IoT is a network of physical objects such as smart devices, sensors, and actuators through which data are collected for analysis and action. IoT devices can be used throughout the entire pharma value chain for various purposes such as drug invention and development, manufacturing and supply chain, sales and marketing of drugs, and patient engagements. The IoT helps in assessing the overall equipment effectiveness, improving drug efficacy, detecting errors, improving regulatory compliance, improving research and development productivity, cutting costs, etc.

Patient safety is one of the most important issues in health care. Drug compliance, minimal adverse drug reactions (ADRs), and the elimination of

An Industrial IoT Approach for Pharmaceutical Industry Growth.
DOI: https://doi.org/10.1016/B978-0-12-821326-1.00004-8

medical errors are the primary goals to enhancing patient care. Noncompliance occurs when a patient neglects to take the prescribed dosage at the recommended times or decides to discontinue treatment without consulting their physician. ADR occurs when a patient is taking several drugs and the combination of their pharmaceutical excipients causes harmful effects. The major cause of ADR and drug noncompliance is polypharmacy. Polypharmacy has been defined as the long-term use of more than two drugs [1]. Therefore polypharmacy is frequent in hospitals where patients receive treatment for multiple illnesses and/or preventative drugs. The incidences which occur in hospitals due to clinical errors and negligence are responsible for disabling injuries in approximately 1 out of every 25 admissions. Most of these injuries are caused by ADR, which prolongs hospital stay, increase costs, and almost doubles the patient's risk of death. Specifically, ADR and harmful effects of pharmaceutical excipients are important clinical issues due to the high ADR rate prevalence in hospitals. Studies present an ADR ratio of 8% in Spain, 6.5% in the United Kingdom, with an average of 6.7% of severe cases and an average of 0.32% of fatalities of the total cases presented worldwide. A recent study conducted by the Royal Liverpool University Hospital (United Kingdom) shows an example of the ADR consequences: 80% of cases required hospital admission, with an average bed stay of 8 days and a total cost of $847m [2].

The main technologies utilized in IoT for drug identification range from existing technologies such as barcodes to new technologies such as radio frequency identification (RFID), together with its version for smartphones, near-field communications (NFCs). These technologies are ancillaries with the new communication capabilities of the future Internet with IPv6 connectivity over low-power area networks [3]. They extend the Internet to small and smart devices from the IoT and make it feasible to identify, locate, and connect all people to technological devices.

Severe incidences take place in hospitals worldwide due to clinical errors and negligence and are responsible for disabling injuries in about 1 in 25 hospital admissions. Most of these injuries are caused by ADR events, which prolong hospital stay, increase care costs, and almost double the a patient's risk of death. About one-third of adverse drug events occur during drug administration, when interception is unlikely. The confusion caused by expressing concentrations of drug solutions in different ways is an important cause of dose errors. Converting between ratios, percentages, international units, moles, micrograms, and milligrams causes substantial difficulty, especially for less experienced physicians. For example, epinephrine, lidocaine, heparin, and potassium chloride are frequently associated with drug errors, and it may be no coincidence that the strengths of these drug solutions are typically expressed in ratios, percentages, international units, and millimoles, respectively.

4.2 Internet of Things based smart devices used in pharma and health care

Some smart devices and their implications in the life sciences industry are described next.

4.2.1 Organ on a chip

Organ on a chip is a ground-breaking invention in the life sciences field. It is a clear polymer which contains microfluidic channels lined by living cells. These microchips help to understand the functions of living human organs including the lung, brain, kidneys, skin, and intestines. It helps to improve research and development productivity, reduces data reporting time and cost, and increases drug efficacy. Organ on a chip plays a key role in accelerating the development of new drugs and advanced personalized medicine.

4.2.2 Chip in a pill

Another application of the IoT in life sciences is the *chip in a pill*. This is an ingestible pill that looks very similar to a microcamera. These pills are used to capture images of organs from inside, to better understand the inner functions and drug effects on them. These pills transmit data to a wearable device, which are then sent to healthcare professionals (HCPs) for diagnosis via the Cloud. The procedure altogether helps to improve drug efficacy and gives a prompt understanding of the problem, so that treatment time also can be reduced.

4.2.3 Google glass

In recent years, *Google glasses* or smart glasses have played an incredible role across all verticals, including health care, as patient records and history always play a vital role in treatment. These glasses enable doctors to record visual information during a conversation with patients purely by maintaining eye contact. In addition, these glasses are game changers during surgical procedures as they enable doctors to share perspectives with specialists anywhere in the world, and the recorded data also provide assistance for the future. Also, in procedural analysis and clinical training, these glasses are very helpful. Moreover, by adopting glasses, hospitals can make significant financial savings due to the differences between costs of video conferencing, which can be around $40,000, as compared to only $1500 for Google glasses.

4.2.4 iBeacons

In health care, *iBeacon* is the most commonly adopted technology among location-based technologies. iBeacons streamline hospital operations by notifying duties and assigning work to hospital staff, such as patients' medication requirements and emergency cases. iBeacons directly send patient health history and records to doctor's just before the patient arrival. Hence, it enables easy access to information and reduces uncertainty over treatment history.

4.2.5 Sensors in drug-delivery devices

This is another IoT application in health care in which sensors are attached to drug-delivery devices. It empowers healthcare providers and patients with a tool to acquire, analyze, and monitor medication usage data. Some devices such as *Amiko* are already available in the market to monitor drug dosage and to allow patients to keep track of their medications. Currently many pharmacies are using Amiko to dispense pills to minimize the chances of errors and to ensure better medication management.

4.2.6 Smart wheelchairs

An automatic *wheelchair* is also a good application of IoT technology. Smart wheelchairs are designed by integrating a wireless body area network with various sensors whose functions are managed via the IoT. The vital signs of the individual in the chair can be monitored and their location tracked. This tool empowers disabled people to have a greater feeling of independence and if used in hospitals can provide assistance in patient management.

4.2.7 Wristbands

Wristbands are designed to measure the user's pulse, activity, body temperature, oxygen level, and blood pressure (BP). All vital information is sent to the user's smartphone to monitor health and keep track of their daily activities. A Korean company has recently introduced Angel which is a compact and wearable BP sensor through which users can monitor their BP.

4.3 Role of the Internet of Things across various stages in the healthcare value chain

4.3.1 Stage 1: Research and development

The IoT plays a critical role in the invention and development of new drugs. Various IoT applications such as organ on a chip, wearable devices, and sensors provide real-time reporting and diagnostics, which facilitates better treatment and improves productivity. In addition, cost savings can also be attained through faster screening and analysis using sensors.

4.3.2 Stage 2: Supply chain

In recent years, the latest technological trends such as RFID and sensors have provided valuable inputs in the supply chain among all verticals including pharma. Reporting on key parameters like temperature is vital in pharma industries, with continuous logistics monitoring and reporting through RFID and sensors all parameters can be measured. Warehousing and routing of drugs through smart sensors enable better supply chain management. Hence, improved supply chain management helps to reduce costs occurring due to wastage, which in turn minimizes overall drug costs.

4.3.3 Stage 3: Marketing and sales

Sensors are very effective in the early detection of adverse drug effects. Data collected through sensors is sent to wearable devices before being forwarded to HCPs. Healthcare professionals by analyzing and monitoring drug effects, get the better capability to market their products in a better way. Also, the interactive ecosystem less opportunity for errors during research and manufacturing of drugs, which helps to improve drug efficacy. In addition, the outcome of the entire procedure ensures quality products at lower cost, which increases product sales.

4.3.4 Stage 4: End users

All technological advancements in the life science industry are ultimately made for the benefit of end users, that is, patients. Whether it is organ on a chip, wearables, smart pills, or other IoT-based tools, all have similar goals, which are to advance the life science industry with vital techniques and equipment to ensure better diagnosis and treatment procedures thus enhancing drug efficacy and minimizing errors. Though most applications are made to provide assistance in hospitals and research and development, quite a few tools are also in the area of providing independence to patients such as wearable devices, drug-delivery devices, smart wheelchairs, iBeacons, and smart wristbands. By using these IoT-based devices patients can keep track of their health and monitor their drug dosage and its effect on the body without professional interventions.

4.4 Key benefits of the Internet of Things to the life sciences industry

Some key benefits of the IoT are listed as follows.

First, the IoT provides better equipment for the diagnosis diseases in their early stages. Through continuous health monitoring by the patient or by hospitals many deadly diseases such as cancer and pulmonary diseases can be prevented and/or cured. Not only diseases but also the sources of illness can be found out through pill-shaped microcameras.

Second, IoT technology is better equipped to understand a drug's effect on the human body. After drugs are taken, continuous monitoring is required to know the medication effect and how the patient is responding. With sensors embedded in devices it is possible to understand how the medication works and to see the progress of the disease.

Third, some IoT devices are very user-friendly, with patients able to use them without any professional help. For example, patients can monitor their health by placing or wearing a wearable sensor on the wrist or elbow. These wristbands are capable of measuring body temperature, pulse, activity, and blood oxygen level. Various BP- and sugar-monitoring devices are also available in the market which indicate immediately if the BP or sugar level increases/decreases.

Fourth, IoT-based devices play a crucial role in research and development. For years, many animals' lives have been lost due to drug invention and testing, but now with the help of the organ on chip, drug effects can be monitored without sacrificing these animals.

Fifth, clinical research and testing take several years to carry out and illustrate the effectiveness of drugs. Drugs have to go through several trials which can be very expensive, and if the researcher finds the drug to be ineffective, all money invested in the project will go waste. In that case, the IoT can be very cost effective, irrespective of the result. With multistage trials and monitoring the result is achieved quicker with time and money savings.

4.5 Rule-based system

Once a patient's record has been collected the drug information, a rule engine system detects drug interaction, allergies and mapped on the ontology. This is based on Jess, and the connection between Protg and Jess is based on JessTab. The current system detects the interactions of drugs, allergies caused by drugs or excipients, and drug loops. In addition, the system detects complications from liver and kidney conditions and states such as pregnancy and breastfeeding.

PortalFarma database is a pharmaceutical intelligent information system and patient health card used to prevent ADRs [4]. In addition, this new version detects active ingredient interactions, renal impairment complications, contraindications in pregnant and lactating women, and problems related to absorption in the stomach (Fig. 4.1) [12].

The first rule for this system detects drug interactions with the properties: Active Ingredient and Active Ingredient Incompatible. Thereby, it can detect when an active ingredient appears on the list of Active Ingredient Incompatibles to the other drug. The second rule detects patient allergies, and the main properties used here are Active Ingredient and Patient Allergic Ingredients. Finally, the third rule detects drug loops, based on Side Effects and Indication properties. A light-embedded rule system is defined in the

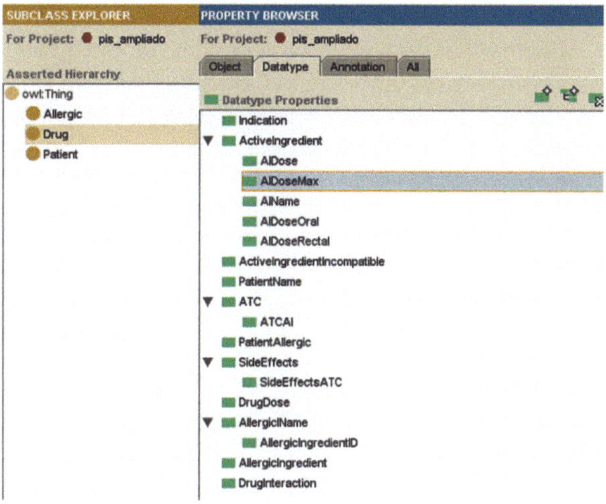

FIGURE 4.1 PortalFarma database.

Movital IrDA-based version for low-income countries, and is able to check drugs without access to the pharmaceutical industry information system (PIIS). This system functionality is focused on TB drugs with the option for a small database with physician-selected drugs in the management application. Some examples are described next [12].

4.5.1 Allergies

Drugs may generate allergic reactions in hypersensitive patients, such as skin rashes and breathing difficulties, when taking penicillin. This system warns the patient and prescriber of the risks.

4.5.2 Active ingredient interactions

This is a drug–drug interaction where one drug changes another drug's effect. Many active ingredients and excipients adversely affect patients, for example, a person with depression being treated with a selective serotonin reuptake inhibitor, such as escitalopram (an antidepressant), when also suffering from headache or joint inflammation, may take an nonsteroidal anti-inflammatory drugs (NSAID) that can be bought without a prescription or ibuprofen is prescribed. NSAID interacts with escitalopram, increasing the risk of bleeding in higher risk patients.

4.5.3 Drug loop

A frequent problem is drug loops. This occurs when a drug generates undesirable side effects which are solved by a second drug rather than changing the first. For example, augmentine is a combination of penicillins, including a beta-lactamase inhibitor, which when consumed by a patient may generate nausea. Consequently, this system detects other prescribed drugs such as bismuth subsalicylate to treat nausea. It advises the health professionals that nausea is caused by augmentine, enabling the prescriber to change the problematic drug.

4.5.4 Renal impairment

A number of drugs require dose adjustments depending on the renal and liver states, especially those that are excreted by the kidney and liver. Some doses require adjustment based on biological tests. Thereby, the prescriber is able to adjust the dose when prescribing drugs simultaneously, for example, lisinopril. The dosage needs to be calculated as a function of the glomerular filtration rate.

4.5.5 Pregnancy

Some drugs must be either avoided or controlled during pregnancy and breastfeeding due to potential damage to the fetus or mother, for example, captopril, cilazapril, and ramipril.

4.5.6 Optimization of absorption

Some ADRs or loss of drug efficacy happens when drugs are administered at the same time. In this case, the system schedules the drug administration at different times.

4.6 Social and cultural factors associated with drug abuse in adolescents

There are many factors that play a part in the initiation and maintenance of drug abuse in adolescents. Initiation of drug use is complex, with multiple factors contributing to the onset of this behavior. The social and cultural factors influencing the initiation of tobacco use vary from country to country, from developed to developing nations, region to region, and culture to culture [5].

4.6.1 Parental influence

Parents have a tremendous influence on their children and the children of smoker parents are twice as likely to become smokers as those of nonsmokers. Parental disapproval of smoking makes adolescents less likely to initiate smoking [6]. Female adolescents are more likely to be smokers if both parents are smokers. There is a strong correlation between the mother smoking and female youth becoming smokers. Raised in a home where parents smoke exposes young persons to tobacco smoke. Parents who smoke may also give easy access to cigarettes and are less likely to oppose their children's smoking. The children are also more likely to smoke when their elder siblings are smokers [7]. Cannabis, a traditional drug in Indian society, is ritualized in social and religious gatherings. It is a socially sanctioned behavior in certain cultural groups to use Bhang and Charas by adolescents and has parental approval. Parental attitudes toward alcohol play an important role in adolescents initiating alcohol consumption.

4.6.2 Family structure

Higher levels of parental education and socioeconomic variables have an inverse relationship with tobacco use and use of other psychoactive substances among adolescents. Prevalence of smoking is more common in families with low socioeconomic and educational status. Children using inhalants generally come from low socioeconomic families, and are engaged in menial work with unstable family income. Marital discord, divorce among parents, and single parenting, are associated with drug abuse among adolescents. Parents with poor monitoring of their children are more likely to have their children abusing drugs.

4.6.3 Peer influence

Friends have the greatest influence on young smokers. The initiation of tobacco smoking generally occurs in the company of a friend who is a smoker. Female adolescents with a best friend who is a smoker are nine times more at risk of becoming smokers. Smoking is a shared activity, with important socializing functions for female youth. Peer smoking also predicts continued smoking among young people who have already begun to smoke [8]. Cannabis abuse in the school-going population has been associated with poor educational performance, school dropout, and reinforcement of poor conduct. These adolescents form their own peer group.

4.6.4 Role models

Film and TV stars, pop stars, and fashion models may make smoking seem attractive and adolescents often imitate them, as they have a tremendous impact on adolescents.

4.6.5 Advertising and promotion

Advertising is an effective weapon to influence the decision-making of youth to initiate smoking. Advertising bans have been found to be very effective in reducing cigarette smoking prevalence in youth. The Indian government has banned the advertisement of cigarettes and other tobacco products [9].

4.6.6 Socioeconomic factors

Higher drug abuse rates are observed in lower income groups. Adolescents from low socioeconomic backgrounds are more likely to become smokers than their middle-class counterparts. This difference in smoking pattern may reflect divergent beliefs about tobacco use based on socioeconomic status. In India, beedi smoking is more common in adolescents due to its easy availability, low price, and convenience. Adolescents from low-income families tend to use cheap and poor-quality homemade liquor prepared illegally. The use of inhalants is also common in adolescents from poor families.

4.6.7 Availability

Availability and accessibility are important factors in the initiation and maintenance of drug abuse among adolescents. An adolescent who has easy access to drugs or alcohol because his parents or elder sibling are users is more likely to use these drugs than those whose parents or other family members do not use them. Similarly peer group members making the product available are likely to recruit new adolescents into drug use behavior.

4.6.8 Knowledge, attitudes, and beliefs

Knowledge about the detrimental health effects of drug use has a preventive effect. Some believe that moderate alcohol consumption does not have adverse effects, tobacco cessation could lead to weight gain, and cannabis is a social and religious blessing from the gods. These beliefs encourage adolescents to use drugs. A positive attitude toward drugs is likely to initiate drug use among adolescents.

4.6.9 Street children and drug abuse

According to UNICEF18, 100 million children live and work on the streets of the cities of the world: 40 million in Latin America, 25–30 million in Asia, and 10 million in Africa. India has the largest number of street children in the world. Although, India is still largely rural, urbanization is taking place rapidly, leading to the fast growth of slums and shanty towns. All these factors have resulted in an explosion of street children in the country. Eighteen million children live and work in India's slums as street children.

According to WHO estimates, up to 90% of the world's street children abuse some kind of drug. There are no systematic data on the prevalence of drug abuse in street children in India and the published literature is available from the metropolitan cities only [11]. The National Household Survey on Drug Abuse 20 surveyed 40,697 males of whom 8587 were children (aged 12–18 years). Of these, 3.8% were using alcohol, 0.6% cannabis, and 0.2% opiates. Though this survey looked into the substance use pattern of children living in houses it provided a yardstick for comparing the substance use in street children. Of those who reported for treatment of substance use in treatment centers, patients below 20 years of age constituted 5% of the total.

4.7 What happens to your brain when you take drugs?

Drugs are made of chemicals, many similar to chemicals already in our system. Drugs alter the way nerve cells normally send, receive, and process information. They do this by (1) imitating the brain's natural chemical messengers, (2) overstimulating the "reward circuit" of the brain, (3) flooding the brain with excess chemicals, and (4) binding to receptors in the brain.

For example, some drugs, such as marijuana and heroin, have a similar structure to chemical messengers, called neurotransmitters, which are naturally produced by the brain. Because of this similarity, these drugs are able to "fool" the brain's receptors and activate nerve cells to send abnormal messages. This results in the "high" you feel when you take these drugs.

Other drugs, such as cocaine or methamphetamine, can cause the nerve cells to release abnormally large amounts of natural neurotransmitters, or prevent the normal recycling of these brain chemicals, which is needed to shut off the signal between neurons. This disruption produces a greatly amplified message that results in a different type of "high."

Nearly all drugs, directly or indirectly, target the brain's reward system. The overstimulation of this system, which normally responds to natural behaviors that are linked to survival (eating, spending time with loved ones, etc.), produces euphoric effects in response to drugs. This reaction sets in motion a pattern that compels some people to repeat the behavior or abuse more drugs.

As a person continues to abuse drugs, the brain adapts to the overwhelming surges in dopamine by producing less of its own dopamine or by reducing the number of dopamine receptors in the reward circuit. As a result, dopamine's impact on the reward circuit is lessened, reducing the abuser's ability to enjoy the drugs and the things that previously brought pleasure. This decrease compels those addicted to drugs to keep abusing drugs to attempt to bring their dopamine function back to a normal level. However, they may now require larger amounts of the drug than they first did to achieve that initial high—an effect known as tolerance.

Long-term abuse also causes changes to other brain chemical systems and circuits. Drugs of abuse facilitate nonconscious (conditioned) learning, which leads the user to experience virtually uncontrollable cravings when they see a place or a person they associate with the drug experience, even when the drug itself is not available. Brain imaging studies of drug-addicted individuals show changes in areas of the brain that are critical to judgment, decision-making, learning and memory, and behavior control. Together, these changes can drive an abuser to seek out and take drugs compulsively despite adverse consequences—in other words, they become addicted to drugs.

4.7.1 Challenges for drug treatment and care

The development of drug use disorders and dependence results from a complex interaction between repeated exposure to drugs on the one hand, and biological, psychosocial, and social factors on the other. Effective treatment for such a complex, chronic condition as drug dependence requires continuing care and interaction across many disciplines, such as pharmacological, behavioral therapy, and social support [10]. Numerous challenges exist in providing treatment interventions and care for substance use that are specific to, or more pronounced for, older drug users. Owing to the possible simultaneous presence of a range of conditions, the complicated physical health needs of older drug users make drug dependence treatment more complex [11]. Historically, little attention has been paid to substance use disorders among older people, with insufficient research into and evidence on interventions for their treatment, and with limited discussion on appropriate treatment services. In combination with medical and psychiatric problems, older drug users commonly live with the negative social consequences of long-term drug use. These are important considerations in the provision of effective treatment. Older drug users are more likely to be socially and economically disadvantaged and marginalized, with a greater chance of having experienced homelessness or periods of incarceration [11].

Social exclusion and isolation from family and friends and a lack of social support are experienced more often and more acutely by older drug users than their peers or younger drug users. The absence of social support is an important predictor of relapse.

4.8 What causes drug abuse?

Drug abuse occurs when someone is unable to control their use of prescribed drugs or they are using another legal or illegal substance to the point that it interferes with their ability to function. According to the National Institute on Drug Abuse, more than 40,000 people died from accidental drug overdose in the United States in 2011 [10]. Each year, more than 22,000 people die from prescription drug abuse alone.

Drug abuse also leads to other public health problems, such as:

- drunk and drugged driving,
- violence,
- familial stress, and
- child abuse.

Intravenous drug users, who inject drugs, are also at risk of contracting and spreading infectious diseases, such as HIV and hepatitis.

Addiction involves many social and biological factors, but treatment is available. The most successful way to stop drug abuse is through prevention and education.

4.8.1 Commonly abused drugs

4.8.1.1 Alcohol

Alcohol is found in beer, wine, and liquor. It is legal for adults over the age of 21 to purchase and drink in the United States. The human body rapidly absorbs alcohol from the stomach and small intestine into the bloodstream.

One standard drink equals:

- 12 ounces of beer,
- 8 ounces of malt liquor,
- 5 ounces of wine, and
- 1.5 ounces of 80-proof distilled spirit or liquor.

When drinking alcohol, brain function and motor skills become impaired. Alcohol can damage every organ in the body. It can also damage the developing fetus if the user is pregnant.

Alcohol use increases your risks of:

- liver disease,
- stroke, and
- cancer.

Alcoholism, or alcohol use disorder, occurs when the use of alcohol affects the ability to work or maintain relationships. Alcohol abuse can threaten long-term health.

The 2014 National Survey on Drug Use and Health found that over a 30-day period:

- 52.7% of adults, age 12 and older, used alcohol at least once.
- 11.5% of young adults, between the ages of 12 and 17, used alcohol at least once.
- 16.3 million Americans reported heavy alcohol use.

4.8.1.2 Anabolic steroids

Anabolic steroids are also commonly known as:

- juice,
- gym candy,
- pumpers, and
- stackers.

Steroids are manmade substances. They mimic the male sex hormone, testosterone, and are taken orally or injected. They are illegal in the United States, however some athletes abuse them to enhance performance and build strength.

Steroids can cause serious and permanent health problems, including:

- aggressive behavior,
- liver damage,
- high BP,
- high cholesterol, and
- infertility.

Women who use steroids face additional symptoms, such as:

- facial hair growth,
- menstrual cycle changes,
- baldness, and
- deepened voice.

Teen users may:

- stunt their growth,
- have accelerated puberty, and
- experience severe acne.

4.8.1.3 Club drugs

This category of drugs refers to a wide variety of illegal drugs that young adults often use at dance parties, clubs, and bars.

They include the following:

- Gamma-hydroxybutyrate, also known as grievous bodily harm, G, and liquid ecstasy;
- Ketamine, also known as special K, K, cat valium, and vitamin K;
- Methylenedioxymethamphetamine, also known as ecstasy, XTC (Ecstasy), adam, clarity, and X;
- Lysergic acid diethylamide (LSD), also known as acid; and
- Flunitrazepam (Rohypnol), also known as roofies, rophies, roche, and forget-me pills.

Club drugs can lead to feelings of euphoria, detachment, or sedation. Roofies, in particular, have been used to commit sexual assaults on unsuspecting victims.

They can cause:

- serious short-term mental health problems, such as delirium;
- physical health issues, such as rapid heart rate, seizures, and dehydration; and
- death.

They are especially dangerous when mixed with alcohol.

4.8.1.4 Cocaine

Cocaine is also known as:

- coke,
- C,
- crack,
- snow,
- flake, and
- blow.

Cocaine is a powerful drug that leads to a strong addiction. It is sold as a fine, white powder that is injected into the veins, snorted through the nose, or smoked [12]. It can also be processed into crack cocaine, a cheaper product that is also highly addictive. In both forms, cocaine causes the user to feel energetic and euphoric.

Cocaine use increases:

- body temperature,
- BP, and
- heart rate.

Cocaine users risk:

- heart attacks,
- respiratory failure,
- strokes,
- seizures, and
- death.

The 2014 National Survey on Drug Use and Health found that 1.5 million Americans age 12 and older were current cocaine users.

4.8.1.5 Heroin

Heroin is also known as:

- smack,
- H,

- ska, and
- junk.

Heroin is an illegal opiate. Like morphine, a legal prescription drug, heroin is made from the seed of the poppy plant, or opium. It is a white or brown powder that is injected into a vein, smoked, or snorted through the nose. Users feel euphoria and experience clouded thinking followed by a drowsy state.

Heroin use leads to:

- heart problems,
- miscarriages,
- overdose, and
- death.

Regular heroin use leads to higher tolerance. Over time, users may need to take more of the drug to experience its effects. This causes addiction and severe withdrawal symptoms if the user stops taking the drug.

4.8.1.6 Inhalants

These drugs are also known as:

- huffing,
- whip-its,
- popping, and
- snappers.

Inhalants are chemical vapors that users breathe to experience mind-altering effects. They include common products, such as:

- glue,
- hair spray,
- paint, and
- lighter fluid.

The short-term effects of these drugs cause a feeling similar to alcohol use.

Inhalants are extremely dangerous. They can lead to:

- a loss of sensation,
- a loss of consciousness,
- a loss of hearing,
- spasms,
- brain damage, and
- heart failure.

The 2014 National Survey on Drug Use and Health found that about 546,000 people age 12 and older used inhalants.

4.8.1.7 Marijuana

This drug is also known as:

- ganja,
- pot,
- weed,
- grass,
- 420, and
- trees.

Marijuana is a dried mix of the cannabis plant's:

- flowers,
- stems,
- seeds, and
- leaves.

It is usually smoked, but can also be ingested in a variety of edible products. It produces feelings of euphoria, distorted perceptions, and trouble solving problems. It is the most commonly abused illegal drug in the United States. In 2014, an estimated 22.2 million Americans were users of marijuana, as reported by the Substance Abuse and Mental Health Services Administration (SAMHSA).

Research has proven and continues to explore the effectiveness of marijuana to treat certain medical conditions, such as glaucoma and the negative side effects of chemotherapy. According to the National Conference of State Legislatures, the District of Columbia and 23 states have approved marijuana for medical use, including:

- Alaska
- Arizona
- California
- Colorado
- Connecticut
- Delaware
- Hawaii
- Illinois
- Maine
- Maryland
- Massachusetts
- Michigan
- Minnesota
- Montana
- Nevada
- New Hampshire

- New Jersey
- New Mexico
- New York
- Oregon
- Rhode Island
- Vermont
- Washington

4.8.1.8 Methamphetamines

Other names for this drug include:

- chalk,
- meth,
- ice,
- crystal,
- glad,
- speed, and
- crank.

Methamphetamine is a very addictive drug that is closely related to amphetamine. It is a white or yellowish powder that is snorted, injected, or heated and smoked.

The user can experience long-term wakefulness. They may also increase their physical activity, which can lead to physical symptoms including increased:

- heart rate,
- body temperature, and
- BP.

If used for a long time, it can lead to:

- mood problems,
- violent behavior,
- anxiety,
- confusion,
- insomnia, and
- severe dental problems.

4.8.2 Prescription drugs

Many people are prescribed medication for pain and other conditions. Prescription drug abuse occurs when a medication is taken that is not prescribed for that person or they take it for reasons other than those prescribed by their doctor. Some people can become addicted, even when they are using the drug as prescribed.

These drugs may include:

- opioids for pain management, such as fentanyl, oxycodone, or hydrocodone (Vicodin);
- anxiety or sleep medicine, such as alprazolam (Xanax) or diazepam (Valium); and
- stimulants, such as methylphenidate (Ritalin) or amphetamine and dextroamphetamine (Adderall).

Their effects differ depending on the medication, but abusing prescription drugs can lead to:

- drowsiness,
- depressed breathing,
- slowed brain function,
- anxiety,
- paranoia, and
- seizures.

It may lead to long-term physical dependence and addiction. The illegal use of prescription drugs has grown over the past few decades, partially because they have become more widely available. The focus of law enforcement has been on illicit drugs.

4.8.3 Stages of drug abuse

Public health experts usually break up drug abuse into stages:

1. In the experimental use stage, the drug is used with peers or for recreation.
2. In the regular use stage, there is a change in behavior and the drug is used to fix negative feelings.
3. In the daily preoccupation, or risky use, stage, the user is preoccupied with the drug and does not care about their life outside of their drug use.
4. In the dependent stage, the user is unable to face life without using the drug. Financial and personal problems increase, and they may also take risks to obtain the drug that result in legal issues.

4.8.4 Treating drug abuse

It is important to find a program that follows the following principles of addiction treatment:

- Addiction is complex but treatable.
- There is no single treatment that works for everyone.
- Treatment is readily available.
- Treatment focuses on multiple needs.
- Treatment addresses mental health. The treatment needs are regularly evaluated to ensure the treatment is meeting them.

- It is critical to remain in treatment for an adequate amount of time. Voluntary and involuntary treatment can be effective.
- Drug use is monitored during the treatment because relapses can and do happen.

Treatment programs should check for and monitor infectious diseases while providing risk-education counseling. This encourages the patient to act responsibly and avoid contracting or spreading infectious diseases.

4.8.5 Detoxification

Depending on the drug the patient is addicted to, the first stage of treatment is often medically assisted detoxification. This process is one in which supportive care is provided as the drug is cleared from the bloodstream.

Detoxification is followed by other treatments to encourage long-term abstinence. Many treatments involve both individual and group counseling. These are given in outpatient facilities or inpatient residential recovery programs.

Medications are also helpful to reduce any withdrawal symptoms and encourage recovery. In heroin addiction, for example, the doctor may prescribe a drug called methadone. This can ease the recovery and help in coping with the intense withdrawal stage.

4.8.6 Preventing drug abuse

The best way to avoid drug abuse is to prevent initial use and addiction. Efforts usually focus on encouraging youth to avoid peer pressure. Community prevention programs work in schools, with teachers, and with community members to educate and provide information and support.

Parents play an important role in preventing their children from using drugs. They should:

- Talk openly with their children about drug-related issues.
- Find and share fact-based information about drugs and drug abuse.
- Build a strong family bond that provides a supportive environment for their children.

4.8.7 Effects of drug abuse and addiction

Drugs are chemicals that affect the body and brain. Different drugs can have different effects. Some effects of drugs include health consequences that are long-lasting and permanent. They can even continue after a person has stopped taking the substance. Unfavorable impacts characterized as the event of care hospitalization for quite a while which might be a reason for the wrong solution that is being given [10]. More information on drugs can be obtained with pharmaceutical companies providing information on use in

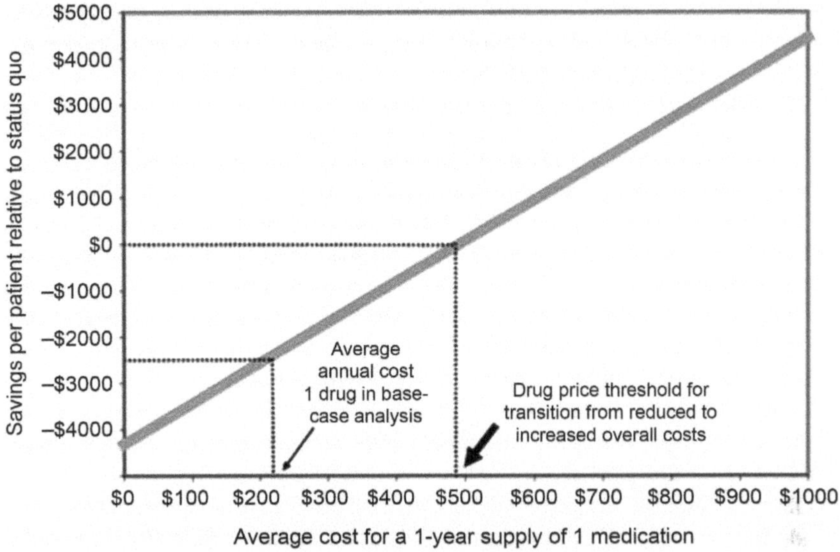

FIGURE 4.2 Cost-effectiveness of providing full drug coverage.

elderly people, children, and pregnant women, with comparisons with competitive products and types of adverse reactions, incidence, and countermeasures [13]. There are a number of ways in which a person can take drugs, including injection, inhalation, and ingestion. The effects of the drug on the body may depend on how the drug is delivered. For example, the injection of drugs directly into the bloodstream has an immediate impact, while ingestion has a delayed effect. However, all misused drugs can affect the brain.

They cause large amounts of dopamine, a neurotransmitter that helps regulate our emotions, motivation, and feelings of pleasure, to flood the brain and produce a "high." Eventually, drugs can change how the brain works and interfere with a person's ability to make choices, leading to intense cravings and compulsive drug use. Over time, this behavior can turn into a substance dependency, or drug addiction.

Today, more than 7 million people suffer from an illicit drug disorder, and one in four deaths results from illicit drug use. In fact, more deaths, illnesses, and disabilities are associated with drug abuse than any other preventable health condition. People suffering from drug and alcohol addiction also have a higher risk of unintentional injuries, accidents, and domestic violence incidents [12] (Fig. 4.2).

4.9 The effects of drug abuse on health

Substance use disorders are associated with a wide range of short- and long-term health effects. They can vary depending on the type of drug, how much and how often it is taken, and the person's general health. Overall, the effects of drug abuse and dependence can be far-reaching. They can impact almost every organ in the human body.

Side effects of drug addiction may include:

- A weakened immune system, increasing the risk of illness and infection.
- Heart conditions ranging from abnormal heart rates to heart attacks and collapsed veins and blood vessel infections from injected drugs.
- Nausea and abdominal pain, which can also lead to changes in appetite and weight loss.
- Increased strain on the liver, which puts the person at risk of significant liver damage or liver failure.
- Seizures, stroke, mental confusion, and brain damage.
- Lung disease.
- Problems with memory, attention, and decision-making, which make daily living more difficult.
- Global effects of drugs on the body, such as breast development in men and increases in body temperature, which can lead to other health problems.

4.9.1 Drug effects on behavior

Substance use disorders can lead to multiple behavioral problems, both in the short and long term, which can include:

- paranoia,
- aggressiveness,
- hallucinations,
- addiction,
- impaired judgment,
- impulsiveness, and
- loss of self-control.

These effects of drug abuse have serious consequences, including missed work, punishable offenses, accidents, and injuries. In fact, alcohol and drugs are partly to blame in an estimated 80% of offenses leading to jail time in the United States. These incidents include domestic violence, driving while intoxicated, and offenses related to damaged property. Legal and illegal drugs, excluding alcohol, are involved in about 16% of motor vehicle crashes. In 1 year, almost 12 million people drove under the influence of illicit drugs, and almost 4000 fatally injured drivers tested positive for drugs.

4.9.2 Effects of drug abuse on unborn babies

Illicit drug use poses risks for pregnant women and their babies. Drugs may contain impurities that can be harmful to an unborn baby. Pregnant women who use drugs may be more likely to harm the fetus with risky behaviors and poor nutrition. Drug use can lead to premature birth or low birth weight. It can also cause the baby to have withdrawal symptoms (sometimes in the form of neonatal abstinence syndrome), birth defects, or learning and behavioral problems later in life.

An estimated 14% of adults with an illicit substance use disorder reported receiving treatment in the previous year.

4.10 Types of drugs

One of the most devastating truths about drug abuse is that it does not just affect the user; it also affects their friends and families. Relationships strain and trust can break over the course of substance abuse.

It is difficult to pinpoint when recreational drug use triggers an outright addiction, but the physical and psychological risks of drug use are many. It is important to understand the underlying causes of addiction—from prescription drug abuse to an attempt to self-medicate mental illness.

There are seven different drug types, and each has its own set of effects and risks:

- stimulants,
- depressants,
- hallucinogens,
- dissociatives,
- opioids,
- inhalants, and
- cannabis.

4.10.1 Stimulants

Stimulants (or "uppers") impact the body's central nervous system (CNS), causing the user to feel as if they are "speeding up." These drugs increase the user's level of alertness, pumping up the heart rate, BP, breathing, and blood glucose levels.

Doctors primarily prescribe stimulants for attention deficit hyperactivity disorder (ADHD), narcolepsy, and asthma (because the drugs can open up breathing passages). The drugs can also help with weight loss, as they can decrease appetite in users. Stimulant abuse tends to occur in high school when teens wish to enhance performance in school or sports.

Stimulants often come in pill form but are also consumed via snorting or even as food and drink. For example, caffeine is found in many beverages, and cocaine is a powder that is snorted.

Examples of stimulants include:

- Adderall,
- Ritalin,

- synthetic marijuana,
- cocaine,
- methamphetamine,
- ecstasy, and
- caffeine.

4.10.1.1 Risks of stimulant abuse

When abused, stimulants can cause a variety of undesirable consequences. These effects can include:

- anxiety,
- paranoia,
- psychosis,
- high body temperature,
- depression,
- heart failure,
- stroke, and
- seizures.

4.10.2 Depressants

Like stimulants, depressants also impact the body's CNS, but with the opposite effect, making users feel as if things are "slowing down." Thus, they are often called "downers."

Doctors prescribe some depressants for anxiety, insomnia, obsessive-compulsive disorder, and other medical issues that prevent the sufferer from fully relaxing. These drugs often offer a sedative experience to users, making them a tempting choice for teens who wish to escape everyday stresses.

Examples of stimulants include:

- Rohypnol,
- barbiturates,
- Xanax,
- Valium, and
- benziodiazepines.

4.10.3 Alcohol as a depressant

Alcohol acts as a depressant, making it a popular choice for users looking to relax. Although drinking is often associated with immediate bursts of energy after intake of a small quantity, the user's vital functions inevitably slow down. Overdosing on alcohol can cause severe toxicity and even death.

4.10.4 Tobacco as a depressant

The active ingredient in tobacco is nicotine, a chemical that acts as both a stimulant and a depressant. Tobacco gives users a minor, immediate rush, followed by a feeling of relaxation. Nicotine is one of the most addictive substances known to man, and is dangerous for teens to even try.

4.10.4.1 Risks of depressant abuse

Depressants can be useful when used properly, but depressant abuse can cause a host of issues in both the long and short term:

- higher risk of high blood sugar, diabetes, and weight gain,
- increased body temperature,
- delirium,
- sluggish thinking,
- low BP,
- impaired memory,
- hallucinations, and
- death from withdrawal.

4.10.5 Hallucinogens

Hallucinogens work by disrupting communication within the brain. Users report intense, rapidly shifting emotions and perceptions of things that are not really there. For example, a hallucinogen user might believe that they see a person speaking to them—when that person does not even exist.

Hallucinogens come in many forms, and can be smoked, eaten, ingested as pills, and even mixed into beverages, and include:

- LSD,
- psilocybin,
- salvia, and
- peyote.

4.10.5.1 Risks of hallucinogen abuse

Hallucinogen abuse can have devastating effects that can last a lifetime, including:

- hallucinogen persisting perception disorder, also known as flashbacks,
- fear,
- distorted cognition,
- paranoia,
- psychosis,
- anxiety,
- increased BP, and

- nausea.

4.10.6 Dissociatives

Dissociatives distort the user's perception of reality, and cause users to "dissociate," or feel as if they are watching themselves from outside their own bodies. They may gain a false sense of invincibility, then engage in risky behavior such as driving under the influence or having unsafe sex.

These drugs work by interfering with the brain's receptors for the chemical glutamate, which plays a significant role in cognition, emotionality, and pain perception. Dissociatives can be taken as liquids, powders, solids, or gases. These drugs include:

- ketamine.
- DXM (dextromethorphan), and
- PCP (phencyclidine).

4.10.6.1 Risks of dissociative abuse

Dissociatives are very dangerous, especially when used over extended periods of time. However, their immediate impact can be quite distressing as well, including causing:

- depression,
- anxiety,
- suicidal thoughts,
- speech difficulties,
- social withdrawal,
- hallucinations,
- detachment from reality,
- numbness, and
- memory loss.

4.10.7 Opioids

Opioids are powerful painkillers that produce a sense of euphoria in users. Derived from the poppy plant, opioids are often prescribed by doctors to patients who are suffering from intense pain. They are extremely habit-forming, sometimes even causing addiction in as little as 3 days.

Opioids can be smoked, eaten, drunk, injected, or taken as pills. Examples of opioids include:

- Heroin,
- Morphine,
- hydrocodone,
- Opium,

- Vicodin,
- Oxycontin,
- Percocet, and
- codeine.

4.10.7.1 Risks of opioid abuse

Opioid abuse can devastate the life of a user. Unfortunately, when someone decides to stop using opioids, they suffer tremendously then, as well. For example, hydrocodone withdrawal can be especially nasty, causing sufferers to have flu-like symptoms for many weeks. Other effects include:

- constipation,
- liver damage,
- brain impairment,
- euphoria,
- drowsiness,
- sedation,
- pupil dilation, and
- cardiac arrest (if dose is too high).

4.10.8 Inhalants

Mostly made up of everyday household items, these drugs cause brief feelings of euphoria. As the name suggests, inhalants are always inhaled as gases or fumes. The "highs" slightly differ from inhalant to inhalant, but most abusers are willing to take any inhalant they can acquire.

Examples of inhalants include:

- fumes of markers, paint, paint thinner, gasoline, and glue,
- nitrous oxide,
- aerosol sprays, and
- room deodorizers.

4.10.8.1 Risks of inhalant abuse

Inhalant abuse can have devastating effects, both immediate and in the long term, including:

- loss of smell,
- brain damage,
- nosebleeds,
- weakness,
- euphoria,
- increased heart rate,
- loss of consciousness,
- hallucinations, and
- slurred speech.

4.10.9 Cannabis

Most commonly recognized as marijuana, cannabis acts like a hallucinogen, but also produces depressant-like effects. It is a Schedule I drug (i.e., it has a high potential for addiction) but has increasing medicinal uses in the United States. However, marijuana is often abused by those who do not medically require it.

Cannabis can be smoked, vaporized, and even eaten, if the delta9 tetrahydrocannabinol (THC) is first rendered from the plant matter. Examples of cannabis include:

- marijuana leaves,
- hashish,
- hash oil, and
- cannabis-based medicines, such as Sativex.

4.10.9.1 Risks of cannabis abuse

Cannabis abuse can destroy lives and can have both short- and long-term impacts on users:

- lowered immunity to illness,
- depression,
- chronic anxiety,
- reduced sperm count in men,
- sedation,
- slowed reaction times,
- enhanced senses, such as colors appearing to be brighter, and
- impaired sense of time.

4.11 Conclusions

IoT devices and sensors are taking the life sciences industry to the next level. These devices enable administrators to remotely manage, control, and track the production processes to enhance productivity and minimize errors. Also, by providing real-time information the quality of products can be improved and they also help to reduce the cost of production. The IoT facilitates data and message transferring from patients to home, home to HCPs, and from home to hospitals, as well serving a number of purposes across the value chain.

The system proposes an innovative system that is based on the Internet of Things for drug identification and a remote knowledge-based system (ontologies and rules-engine). Drugs are identified by means of an interchangeable mobile device and consider the incipient NFC technology and a legacy identification solution such as a barcode. Once the drug has been identified, this information is sent to the pharmaceutical intelligent information system, where decomposed active ingredients are matched with the patient's allergy profile and electronic health record, in order to detect potential reactions. The system has been designed in a flexible way. A patient with an NSAID allergy profile has been used to test the system under real conditions. The drugs database is synchronized with the Spanish Pharmaceutical Association database (PortalFarma), to embrace the entire set of current and future drugs, and keep it updated. Ongoing work is going to be, on one hand, extending the solution with additional drugs interactions; and in addition a specialized rules-engine system is going to be created to consider more specific cases from real cases.

References

[1] T. Linjakumpu, S. Hartikainen, T. Klaukka, J. Veijola, S.-K. Kivela, R. Isoaho, Use of medications and polypharmacy are increasing among the elderly, J. Clin. Epidemiol. 55 (8) (2002) 809–817.

[2] M. Pirmohamed, S. James, S. Meakin, C. Green, A.K. Scott, T.J. Walley, et al., Adverse drug reactions as cause of admission to hospital: prospective analysis of 18 820 patients, BMJ 329 (7456) (2004) 15–19.

[3] J. Hui, P. Thubert, Compression format for IPv6 datagrams over IEEE 802.15.4-based networks, RFC 6282, Internet Engineering Task Force (IETF), 2011, ISSN: 1721–2070.

[4] A.J. Jara, F.J. Belchi, A.F. Alcolea, J. Santa, M.A. Zamora, A.F.G. Skarmeta, A pharmaceutical intelligent information system to detect allergies and adverse drugs reactions based on Internet of Things, Work in progress – Porter – IEEE International Conference on Pervasive Computing and Communications, 2010.

[5] R.C. Jiloha, Social and cultural aspects of drug abuse in adolescents, Delhi Psychiatr. J. 12 (2) (2009).

[6] J.R. Eiser, M. Morgan, P. Gammage, E. Gray, Adolescent smoking: attitude, norms and parental influence, Br. J. Soc. Psychol. 28 (1989) 193–202.

[7] A.K. Elkind, The social definition of women's smoking behaviour, Soc. Sci. Med. 20 (1985) 1269–1278.

[8] S.A. McGraw, K.W. Smith, J.J. Schensul, E. Carrillo, Socio-cultural factors associated with smoking by Puerto Rican adolescents in Boston, Soc. Sci. Med. 33 (1991) 1355–1364.

[9] The Cigarette and Other Tobacco Products (production, advertisement, and regulation of trade and commerce, production supply and distribution) Act, 2003.

[10] UNODC and WHO, Principles of drug dependence treatment, discussion paper, March 2008.

[11] Ilana Crome and others, Our invisible addicts: first report of the older persons' substance misuse working group of the Royal College of Psychiatrists, Royal College of Psychiatrists, London, 2011.

[12] R. Tamblyn, et al., Adverse events associated with prescription drug cost-sharing among poor and elderly persons, J. Am. Med. Assoc. 285 (2001) 421–429.

[13] M. Yamamoto, Y. Onaka, et al., Development and utilization of a drug information system in the Japanese pharmaceutical industry, Inform. Health Soc. Care 23 (1998) 31–41.

Chapter 5

Accelerating data acquisition process in the pharmaceutical industry using Internet of Things

T. Poongodi[1], T. Lucia Agnesbeena[2], S. Janarthanan[1] and Balamurugan Balusamy[1]

[1]*School of Computing Science and Engineering, Galgotias University, Greater Noida, India,*
[2]*Department of Information Technology, St. Josephs College, Tiruchirappalli, India*

5.1 Introduction

The Internet of Things (IoT) is a newly emerging paradigm in which sensors, actuators, transceivers, and processors communicate with each other. It has become ubiquitous across the globe and affects human life in incredible ways. It is an interaction between the digital and physical worlds with sets of sensors and actuators. The network of connected things such as household objects, industrial components, vehicles, perishable products, and electrical devices, along with Internet connectivity and dynamic data analytic capabilities improve human quality of life drastically. The networking and computing capabilities are embedded with any object, and the state of the object can be identified and changed from one state to another if required. Sensors and actuators assist in communicating with the physical environment. The data gathered from sensors have to be processed efficiently to derive useful inferences from the available data [1]. A sensor may be a microwave oven or a smartphone that can be considered as such if it provides inputs about the current state and the environment. An actuator is a device that has an impact on the environmental change, such as a sprinkler that turns on automatically and puts out flames, when activated by a temperature sensor. The processing of data can be done in a remote server or in the network itself. If any preprocessing of data is required, then it is performed either at the sensor or the proximate device. Later, the processed data are transmitted to a remote server. The processing and storage capabilities of IoT devices are restricted due to resource constraints such as energy, power, size, and computational capability. Communication among IoT devices is largely wireless because the devices are located in different locations.

An Industrial IoT Approach for Pharmaceutical Industry Growth.
DOI: https://doi.org/10.1016/B978-0-12-821326-1.00005-X

Radio frequency identification (RFID) is an identification technology in which the tag (a small chip) carries data that is read by a reader. The RFID tag transmits the stored data via radio waves and works similarly to barcode technology. Line of sight communication is not required between the RFID tag and the reader and it can be identified from a long distance without human intervention. There are two types of RFID tags: active and passive. Active tags have a power source, whereas passive tags do not require a power source. Instead, it draws power from the electromagnetic waves that are emitted by the RFID reader, hence it is durable and cheap. RFID technologies are of two types, near and far. A near RFID reader has a coil that passes alternating current and creates a magnetic field. The voltage is coupled with a capacitor for accumulating a charge and the magnetic field encodes the signal to be picked up by the RFID reader. In far RFID, there is a dipole antenna in which the alternate potential difference appears and the power is used to transmit messages. Once the RFID is fixed to the object, the reader can track, detect, and record the presence of an object if it passes. It is used in several applications, such as identity authentication, access control, supply chain management, and object tracking.

RFID tags and devices, actuators, sensors, energy-harvesting technologies, software, and machine-to-machine interfaces are the basic fundamental blocks of IoT platforms. Some advanced technologies such as adjustable autonomy, telepresence, geo-sensing, geo-tagging, augmented reality, and robotics enhance the value of the basic network of things. IoT enablers can be chosen based on the requirements of an application. For instance, in logistics and maintenance services the network devices are integrated depending on the service information data availability. Depending on the requirement, RFID tag sensors, auto adaptive sensor networks with traceability enabled, and auto track tags are used. Various technologies are also integrated for configuration control, end-to-end security, predictive forecasting, etc. There are different types of sensors deployed in several IoT applications with the most generic being the smartphone. In a smartphone, many kinds of sensors are embedded including movement sensors (gyroscope, accelerometer), location sensor (GPS), microphone, light sensor, camera, magnetometer, and proximity sensor. Other types of sensors have begun to be used for measuring pressure, humidity, temperature, medical parameters of the human body, neural signals, and chemical and biomedical substances. Infrared sensors such as motion detectors, IR cameras, moisture sensors, and those for measuring the distance to nearby objects and existence of smoke and gases are now widely used in many IoT applications. The communication technologies in IoT are RFID, near-field communication (NFC), Zigbee, WiFi, Bluetooth, etc.

5.1.1 Architectural framework of the Internet of Things

The IoT connects several objects over the network by integrating heterogeneous systems or devices. Different architectures have been proposed for IoT as there is no single consensus followed in this platform. The most primitive architecture is the three-layered architecture which focuses on the basic concept of IoT but this is not adequate for many applications. Hence, multilayered architectures are proposed to meet business needs, selected technology, and technical requirements. The International Telecommunication Union (ITU) recommends the five-layered architecture consisting of a sensing layer, accessing layer, networking layer, middleware layer, and application layer [1]. The IoT consists of layered architecture that is associated with data analytics, web services, networking, communication and business structures, intelligent interface, etc. The business layer is responsible for managing the complete IoT systems including business and profit models, applications, and user's privacy. The functionalities of different layers are described below and the overall conceptual flow is depicted in Fig. 5.1.

5.1.1.1 Sensing layer

This layer is integrated with the hardware, such as RFID, sensors, and actuators in order to control/sense the physical environment and gather data. The sensors in this layer are used for sensing and collecting information about the physical environment. It senses the environment with some physical parameters and identifies smart objects. The IoT is a worldwide interconnected network in which the devices are connected and controlled remotely. Moreover, the devices are equipped with intelligent sensors that make the connectivity process much easier. Smart devices with sensors or tags are capable of automatic sensing and exchange information among heterogeneous devices. Significant advancements have enhanced the capability of IoT devices in terms of sensing, identifying things, and the environment.

5.1.1.2 Network layer

This layer facilitates data transferring over a wired or wireless network with the fundamental networking support. It connects smart objects, network devices, and servers, and is mainly responsible for transmitting and processing the sensor data that are gathered. The major role of this layer is to connect all devices together, which permits sharing information with other connected devices. In addition, it aggregates information from various infrastructures such as transportation systems, business systems, healthcare systems, Information and Communication Technologies (ICTs) systems, and power grids. The services provided by IoT devices are typically retrieved from a heterogeneous network according to the user or application requirement. Quality of Service (QoS) management and control are necessary for a dynamically changing network because of automatic discovery and mapping devices in the network. The "things" in the IoT network need to be assigned with roles automatically and the role should be able to be changed at any time whenever required. The data retrieved from the sensing layer are

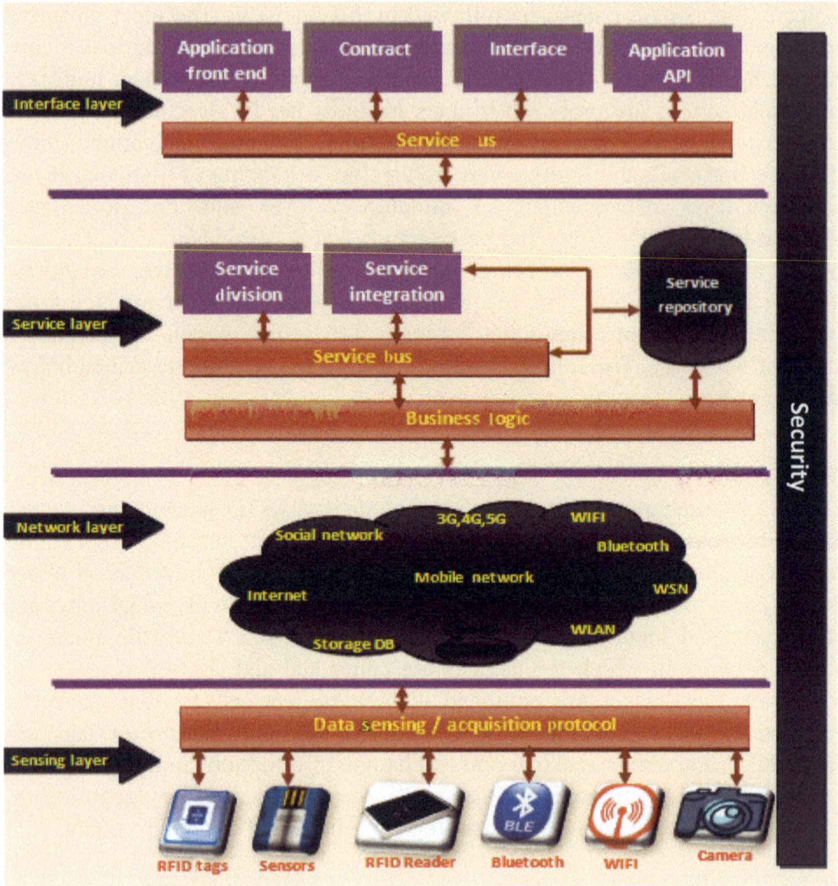

FIGURE 5.1 Layered architecture.

transferred for processing through wireless, LAN, RFID, NFC, 3G, and Bluetooth. The issues to be considered at the time of designing the network layer are network management technologies (wired, wireless, mobile), energy efficiency, QoS, data and signal processing, service discovery and retrieval, and security and privacy.

5.1.1.3 Service layer

This layer is responsible for creating, managing, and providing services in order to satisfy the user requirements. It is also known as middleware technology and provides various functionalities for the seamless integration of services and applications in the IoT. It is a cost-efficient platform since the hardware and software can be reutilized in the IoT network. A well-designed service layer identifies common application requirements and affords protocols, Application Programming Interfaces (APIs) that facilitate uninterrupted services. It processes the complete service-related issues that include information storage and exchange, communication, search engines, and data management. It is also called a processing layer which stores, analyzes, and processes the massive amount of data retrieved from the network layer. The significant role of this layer is to deliver specific services to end users. It determines the applications in which the IoT can be deployed, for example, smart cities, smart homes, smart health, smart agriculture, telemedicine systems, intelligent building, intelligent traffic, intelligent electric grid, ambient assisted living, smart environment, energy conservation, social life and entertainment, supply chain and logistics, etc. It employs several technologies such as databases, big data, cloud computing, Wireless Sensor Network (WSN), and vehicular network processing units to render a diverse range of services.

The service layer includes the following components:

1. Service discovery: Searching objects to offer the required services and information in an effective manner.
2. Service composition: Enables interaction and communication between the connected devices. It leverages the relationship between various devices to identify the desired services. It also assists in recreating or scheduling suitable services to meet the user's need.
3. Trustworthiness management: Determines trusted mechanisms to evaluate and utilize the information provided by a diverse set of services.
4. Service APIs: Leverages the interactions among the required set of services in the IoT network.

5.1.1.4 Interface layer

In the IoT, a huge number of devices are manufactured by different vendors and commonly they do not follow the same set of procedures or standards. The layer assists in enabling the interaction among the applications and users. Due to this heterogeneous nature, many interaction problems arise during the communication among things, information exchange, and cooperative event processing. Moreover, the constant increase in the number of participants in the IoT makes it harder to connect, disconnect, communicate, and operate dynamically.

5.1.2 Overview of the pharmaceutical industry

Pharma or clinical data analytics has a huge impact on the pharmaceutical industry and is a prominent area, hence it impacts with the clinical data regularly. The pharma industry is not only meant for drug development but merges many disciplinary areas such as academics, mathematics, business intelligence, and chemistry in order to develop suitable drugs and achieve good benefits. The pharma data are generated in petabytes with the help of many devices such as sensors, mobile devices, genome data collection, and image-capturing devices. The pharma data which are gathered are unstructured data that include doctors' notes, medical prescriptions, healthcare interactions, electronic health records, medical image reports, quantities dispensed, pathology reports, supplies of materials to clinics, filling medicines, discharge summaries, pharmacy reimbursements, drug barcodes, patient social communications, etc. The analytical tools are used for data processing, extraction, and visualization, to transform the massive amount of real-world unstructured data into meaningful information. The data are mainly categorized into two types: structured and unstructured data. Structured data are curves from electronic devices, statistical analysis data sets from clinical management tools, gene molecule structures, etc. Structured data are in a predefined format but will not provide meaningful information that is relatively easy to handle. Unstructured data are the information retrieved from videos, e-mails, social websites, etc. and can be about any diseases, drug descriptions, gene structures, active agents, and textual medical information. Data can be qualitative or quantitative, such as unusual symptoms taken for qualitative analysis, laboratory results, and patient treatments.

5.1.3 The Internet of Things revolution in the pharmaceutical industry

Pharmaceutical industries are striving very hard to remain competitive in terms of reducing costs in the supply chain network. Articulating IoT in the pharmaceutical industries accelerates the speed of utilization by abiding with strict regulations which often limit technological changes. Some of the challenges faced by pharmaceutical industries are highly dependent on human resources, lack of coordination among clusters, supply chain visibility, and temperature control. The IoT overcomes these challenges by acquiring accurate information, seamless collaboration, traceability in the supply chain, real-time accessibility, personalized technical care, etc. The pharmaceutical industry consists of complex processes, tasks, and organizations involved in developing and manufacturing drugs and medical products. Logistics is the significant driving factor of the pharmaceutical industry as it faces many challenges because of the complexity of dynamic business models and uncertain consumer demand. The IoT overcomes these barriers in the pharmaceutical industries by contributing more advanced promising technologies to solve those issues. Wireless sensors can be embedded with clinical devices, which are connected to patients, hospitals, and medical-related equipments. RFID tags assist in detecting counterfeit medicines and medical devices, continuous monitoring of medication storage processes, and managing patients' medical information such as efficient handling of telemedicine, medical emergency, and mobile medical care. Pharmaceuticals are closely related to health care and medical research; in the supply chain and logistics, manufacturing plays a significant role. Handling pharmaceutical drugs according to specific environmental requirements in the supply chain network differs from normal supply chain networks because of its priorities and focused points. While considering healthcare utilities and medicines, there should not be any alteration in product specifications at the time of transition from the manufacturing unit to reaching the customer. Continuous remote monitoring and optimization network analytics can be done for real-time tracking of complete transportation and warehouse operations. These tools also help in maintaining product quality and compliance with government regulations. The cold chain is a vital part in the pharmaceutical supply chain network, hence real-time monitoring and traceability of cold chains are major requirements in the network. Initially, the raw materials are sent to the manufacturer and then the medicines are manufactured and packed. Next, the drugs are distributed through a couple of steps before reaching patients.

5.2 Technical analysis of the Internet of Things-based pharmaceutical industry

IoT technologies improve QoS, reduce errors, and detect health anomalies using vital signs. IoT technology can be applied in a pharmaceutical system to identify Adverse Drug Reactions (ADRs), harmful side effects at the time of pregnancy or lactation, allergies, and complications with liver/renal defects. The physicians would feel more comfortable in drug prescribing and making clinical decisions.

5.2.1 Remote monitoring system

A remote monitoring system (RMS) uses Raspberry Pi3 to acquire data from sensors such as air flow, pulse oximeter, body temperature, and a galvanic skin and blood pressure monitor. Graphical User Interface (GUI) assists in analyzing the historical data and records the information by visualizing it. This can be accessed either locally or remotely using touch display or web devices. An IoT sensor-based platform provides solutions using historical patient data that trigger live action locally and remotely. A Remote Patient Monitoring (RPM) system explores several ways for classifying physiological data [2]. However, the sensor data vary significantly according to the physical activities engaged in by patients. In order to evade erroneous conditions, the acquisition of physiological data from body sensors is performed along with the activity using sensors. RPM enhances the medication process by enabling doctors to prescribe medication remotely and allowing caretakers to verify that the correct medication is provided to the right patient [3]. The interface with a WSN permits caregivers to monitor patient-related information continuously, where a body area network comprises of multiple sensors which are fixed on or in the body. In Refs. [4,5], Raspberry Pi 3 board and Arduino Uno boards are used for measuring the ECG signal for cardiovascular diseases. The data acquired from different types of sensors are described below and depicted in Fig. 5.2.

- Air flow checks the breathing intensity;
- Pulse oximeter (e.g., Contec CMS50D +) monitors the pulse rate and oxygen saturation level in the blood;
- Body temperature sensors track the temperature of the body;
- Galvanic skin response verifies skin conductance and resistance;
- Blood pressure monitor (e.g., Kodea KD-202F) detects the pulse rate and diastolic blood pressure.

The pulse oximeter can be directly connected to USB ports and the rest of the sensors are connected to the sensor platform which is embedded with a Raspberry Pi and Arduino.

The communication is enabled both synchronously and asynchronously to gather, save, and visualize the sensor data. Data gathered from the sensors includes air flow, pulse oximeter, and galvanic skin response, which are recorded at specific time intervals. With a body temperature sensor, the current value is compared with the previous one over a set amount of time. If temperature varies and is above the threshold level, the time interval is reset and the new value is recorded when the variation is below the threshold.

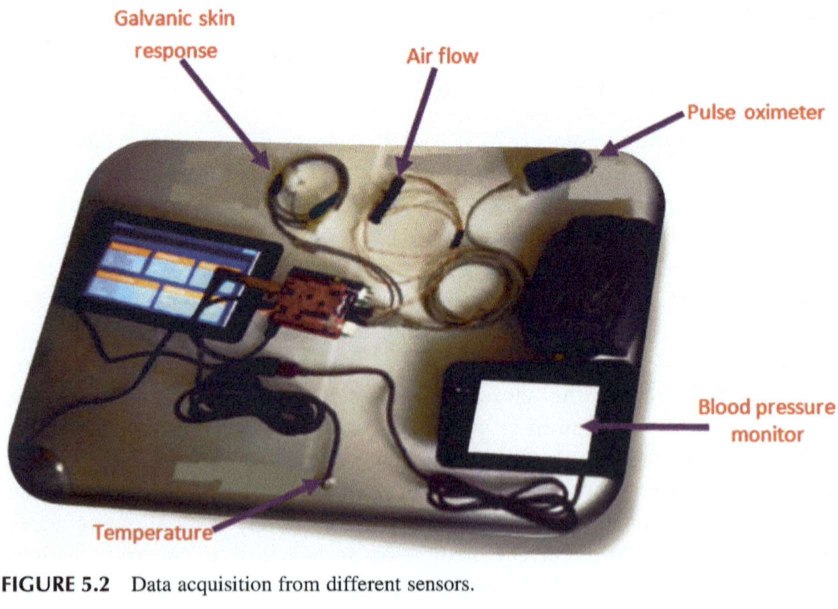

FIGURE 5.2 Data acquisition from different sensors.

5.2.2 Drug-delivery system for neurological disorders

In the unified Drug-Delivery System (DDS) framework, drug injection is provided based on the seizure detection. An Electromagnetic Actuated Valveless Micropump (EAVM) with diaphragm consists of polydimethylsiloxane and is used for delivering drugs. Epilepsy refers to the neurological disorder in which brain activity becomes abnormal, causing seizures or unusual behavior, resulting in convulsions or lack of consciousness. An average of 1% of the world's population suffers from this problem which can be controlled via surgery and antiepileptic drugs (AEDs). A very small fraction of people are willing to undergo epilepsy surgery and one-third of patients do not respond to AEDs. An effective solution for controlling seizures is vital to prevent serious consequences. The DDS is capable of detecting a seizure and injecting an AED in the respective zone of the brain at the right time. This localized and responsive injection improves the efficacy of the drug, providing an effective solution. Connectivity with other healthcare devices is established with the IoT and it enables effective analysis of health behavior and remote health monitoring. In DDS, drugs are injected into the epileptogenic zone based on seizure detection [6].

An IoT-based DDS framework provides regulated drug injection and better detection accuracy. Generally, the system consists of two units, namely seizure detection and drug-delivery units.

- Discrete wavelet transform (DWT) basically provides frequency determination and conjoint time to capture complex electroencephalography (EEG) dynamics that leads to accurate seizure detection.
- The electrostatic and piezoelectric actuation requires high voltage, while the electromagnetic actuation operates at a low voltage for membrane displacement. This reduces power consumption and paves the way for using it in low-power medical applications.
- DDS features low-power consumption, universal connectivity, and better detection accuracy. An IoT-based RMS facilitates considerable enhancement of the epilepsy patient's quality of life.

Seizure prediction, detection, and control are hot topics in on-going research. Some methodologies have been proposed using focal cooling, electrical simulation, or drug delivery for controlling seizures [7]. An electrophoretic drug-delivery device is proposed for on-demand drug delivery that functions based on the underlying principle of an organic electron ion chip. An AED can be injected directly into a particular brain region to control seizures. A seizure-detecting smartwatch detects seizures and notifies physicians about such occurrences. In DDS, initially the EEG signals are decomposed using DWT, statistical features are extracted and fed into an k-NN classifier for detection. Once the seizure is detected, AEDs are injected into the targeted area in order to end the seizure propagation. The complete processing of a DDS is shown in Fig. 5.3.

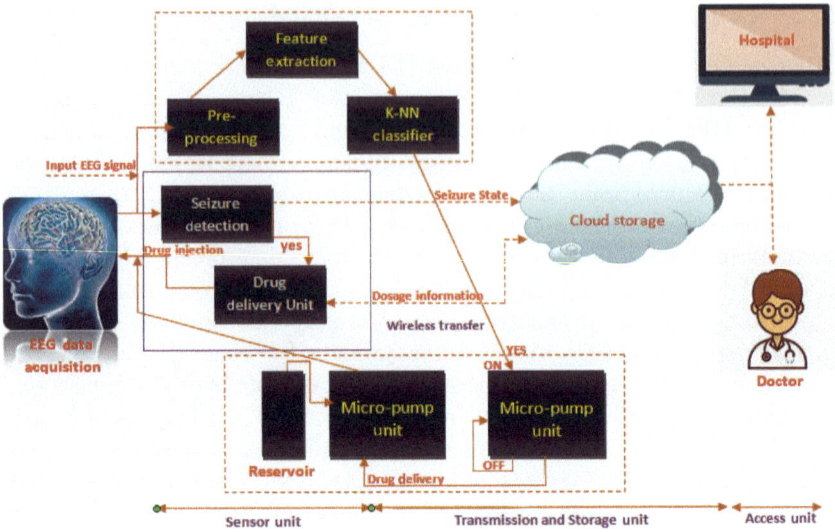

FIGURE 5.3 Drug-delivery system.

5.2.2.1 Seizure detection unit

The seizure detection unit is comprised of the following subparts: feature extraction using DWT and a k-NN classifier. EEG signals are decomposed using DWT and time frequency (TF) localization is provided. The features are extracted and given to the k-NN for classification.

DWT with TF localization is useful for analyzing nonstationary and complex EEG signals. A k-NN classifier is used in training and testing phase with a distinct data set. A Euclidean distance metric is used to compute the nearness of the data.

5.2.2.2 Drug-delivery unit

An EAVM offers advantages when compared to other micropumps including:

1. It requires lower actuation voltage;
2. It leads to faster response, higher actuation force, and deflection;
3. Elimination of valves makes it more simplified and reliable.

In the smart DDS, the k-NN classifier is followed for seizure detection and an electromagnetic micropump is used for drug delivery. This reduces the power consumption and enhances the detection accuracy. Moreover, it is most suitable for epilepsy treatment and can be preferably implemented in commercial biomedical applications.

5.2.3 Automated medicine dispenser

An online health community (OHC) platform plays a significant role and suggests ways for clinicians, caregivers, and patients to share their ideas in terms of obtaining solutions to problems [8]. The main issue prevailing in the pharmaceutical industry is medication errors caused by improper handling of dispensing medicines, transcription writing, and administration. Medication errors can be prevented by putting huge efforts into the research and development of ICTs. The steps involved in prescribing and managing a patient's medication are:

1. Ordering: The clinician must choose the appropriate dosage, medication, and frequency;
2. Transcribing: Handwritten prescriptions must be understood by the pharmacist;
3. Dispensing: The pharmacist must verify the correct dosage of medicines;
4. Administration: The medication must be provided to the appropriate patient at the right time in the right dosage.

The automatic medicine dispenser reduces patients' waiting times in hospitals and ensures error-free services. Pharmacists can gather and integrate prescriptions provided by clinicians, verify drug interactions among them, and schedule instructions for dispensers. An automatic medicine dispenser is maintained by pharmacies, and filling of medicines should be ensured, with a proper billing mechanism, and preferably for common medicines. The components of an automatic medicine dispenser are described below.

A medicine dispenser issues medicines to patients as per the prescriptions issued by the doctor and is secured by individual barcodes. The dispensers are to be maintained by the pharmacies and are to be filled with medicines whenever required. Beagle Bone is a low-power, open-source, single-board computer used to control the processing of a dispenser. Servomotor is a rotary actuator embedded with a sensor that permits precise control of linear or angular velocity and acceleration. It is connected with the Beagle Bone and controls the dispenser unit which in turn controls the number of pills dispensed. Pill containers are used to store pills ensuring cost efficiency.

The cloud data store stores the history and sequence of medications of OHC users.

The user interface assists in creating prescriptions for patients registered in the OHC.

5.2.4 Smart rehabilitation system

IoT-based smart rehabilitation systems are becoming better at mitigating problems associated with the shortage of health experts and aging populations. Although they have already started to be used in real time, some critical issues still exist in the design automation and reconfiguration of the system that respond to patients' specific requirements promptly. A smart rehabilitation system assists in understanding the medical resources and symptoms, which reconfigures medical resources and creates a rehabilitation strategy automatically according to the patient's requirements. Meanwhile, the IoT is an efficient platform which interconnects all resources and enables uninterrupted interaction. Compared to traditional rehabilitation, smart rehabilitation provides more convenience, adequate interaction, rapid configuration, and effective treatment [9]. The IoT is recognized as the next revolutionary and leading technology for associating all medical resources in a smart rehabilitation system. The IoT enables communication among a wide range of devices, objects, and appliances using wireless networking technologies.

In IoT-based healthcare systems, patients are monitored continuously and situations are detected automatically whenever medical interventions are required. Generally, the system gathers information from various sensing devices via middleware that provides security and interoperability in the context of IoT-based systems. The monitoring application is completely responsible for storing, aggregating, and consolidating data and providing a security alarm such as a network manager. The IoT establishes an information network that connects healthcare devices, hospitals, homes, communities, and other terminals. With the huge amount of available data and system complexity, there are some key challenges to IoT-based rehabilitation: (1) rapid diagnosis, (2) resource reconfiguration, and (3) strategy creation. The smart rehabilitation system assists in finding the solution rapidly for large-scale systems also.

An IoT-based rehabilitation system is established using RFID and WiFi for short-distance radio communication, GPS, unique identifier (UID), and service-oriented architecture (SOA). The SOA framework is designed to design, implement, deploy, invoke, and manage healthcare services. The topological structure of a smart rehabilitation system includes the medical resources (e.g., doctors, nurses, hospitals, ambulances, rehabilitation centers, and medical devices), database server, intermediary processing proxy for data analysis, communities and patients, critical event detection, and rehabilitation strategy creation. The components in the system are interconnected using the Ethernet with TCP/IP, and each device is assigned with a UID for easy identification. Meanwhile, the patients, doctor, nurses, and other human resources are authorized using RFID tags. A smart rehabilitation system features three main parts:

1. The master component that includes the manager (patients, doctors, nurses), end-user devices (PC, tablet, smartphone), and healthcare applications;
2. The server is the central coordinator that is responsible for data recording, analysis, subsystem building, rehabilitation strategy creation, permission control, and devices control;

3. The "things" include human resources and smart IoT devices connected by short message service, multimedia technology, or WAN. The devices in a smart rehabilitation system are connected by assigning RFID tags.

The smart rehabilitation system promotes clinical rehabilitation by integrating hospitals, healthcare communities, rehabilitation centers, and homes.

5.3 Smart Internet of Things devices and systems

The IoT has unlocked the possibilities in the pharmaceutical industry. In particular, IoT sensors and trackers can provide better control of drug manufacturing, preventive maintenance of equipment, and improved supply chain management. In addition, smart IoT sensors and trackers guarantee flawless equipment operation and help in drug fraud prevention.

Sensors collect precise data that facilitate real-time monitoring and tracking of human activities with less power consumption. Sensors are required to measure the physical parameters such as temperature, motion, stress, acceleration, heart rate, vibration, neurological or cardiovascular diseases, chronic obstructive pulmonary diseases, and hypertension. These are of two types, namely:

1. invasive
2. noninvasive sensors

Invasive sensors use the body fluid to study health-related information. Blood is a significant body fluid that provides the required factors for different organs. An ex vivo sensor or extracorporeal sensor monitors the blood gases and pH level at the time of cardiopulmonary bypass. Invasive sensors are not preferable for continuous monitoring such as glucose monitoring, fitness level monitoring, cholesterol monitoring, and oxygen saturation monitoring. Noninvasive sensors do not require body fluids, rather they read the data using skin interstitial fluids, saliva, sweat, and tears. Some examples are listed in Table 5.1.

This section explores some of the smart sensors used in the pharmaceutical industry.

5.3.1 Medical sensors

Medical sensors are used to measure and monitor the medical parameters such as blood pressure, respiration rate, pulse rate, heart rate, body temperature, and blood glucose levels in the human body. The patient's health can be monitored either locally or remotely and the sensors provide information regarding the user's location, movement, surrounding sounds, and visual images. Subsequently, real-time feedback is provided to the caregiver, physician, doctor, patient, or relatives. Sensors such as smart watches, monitoring patches, smart textiles, and wristbands can be embedded within devices that can be worn or implantable. Smart watches are becoming more popular

TABLE 5.1 Sensor types.

Invasive sensors	Noninvasive sensors
Pacemaker	Electromyography
Retina implants	Electroencephalogram
Cochlear implants	Pulse oximeter
Wireless capsule endoscope	Temperature
Deep brain simulator	Glucose sensor
Electronic pill for drug delivery	pH value
Implantable defibrillators	Blood pressure

nowadays, and have many innovative features and some are shown in Fig. 5.4. Smart watches can be connected with smartphones, or sensors such as a heart rate monitor or an accelerometer. The data gathered by the medical sensors can be related with the contextual information like physical activities. For instance, if the heart rate increases during exercise that can be inferred based on the context. The data need to be collected and combined from different sensors to make the correct inference.

Implantable sensors are very cheap, stretchable, and disposable; these kinds of patches are worn for a few days to monitor the patient's health parameters continuously. The monitoring patches can be applied to the skin and transmit the sensed data. They are mainly used to monitor blood pressure. A sample skin patch is shown in Fig. 5.5.

5.3.2 Neural sensors

These sensors are used to understand the neural signals of the brain, enabling the current state of the brain to be inferred and it can be focused for greater detail. This process is known a neurofeedback [10]. EEG is used to read brain signals. The neurons inside the brain communicate by creating an electric field, which can be measured in terms of frequencies. These brain waves are classified into alpha, beta, gamma, theta, and delta waves based on their frequency. By reading the type of wave, it can be identified whether the brain is wandering in thoughts or in a calm state. A headband with neurosensors is shown in Fig. 5.6.

5.3.3 Temperature sensors

A temperature sensor is a device that measures the amount of heat energy that creates a physical change in temperature from a particular source. It also

FIGURE 5.4 Smart watches and fitness trackers.

converts the data to the required format of the device that is attached to it. In the manufacturing process, many machines need specific environmental temperature and device temperature. With the help of temperature sensors, the manufacturing process can constantly remain optimal. Some of the categories of temperature sensors are [11]:

- *Thermocouples*: The change in voltage is indicated by the thermocouples through the output voltage. The output voltage of the thermocouple increases, as temperature goes up.
- *Resistor temperature detectors*: The resistance of the device rises when the temperature of the device goes up.

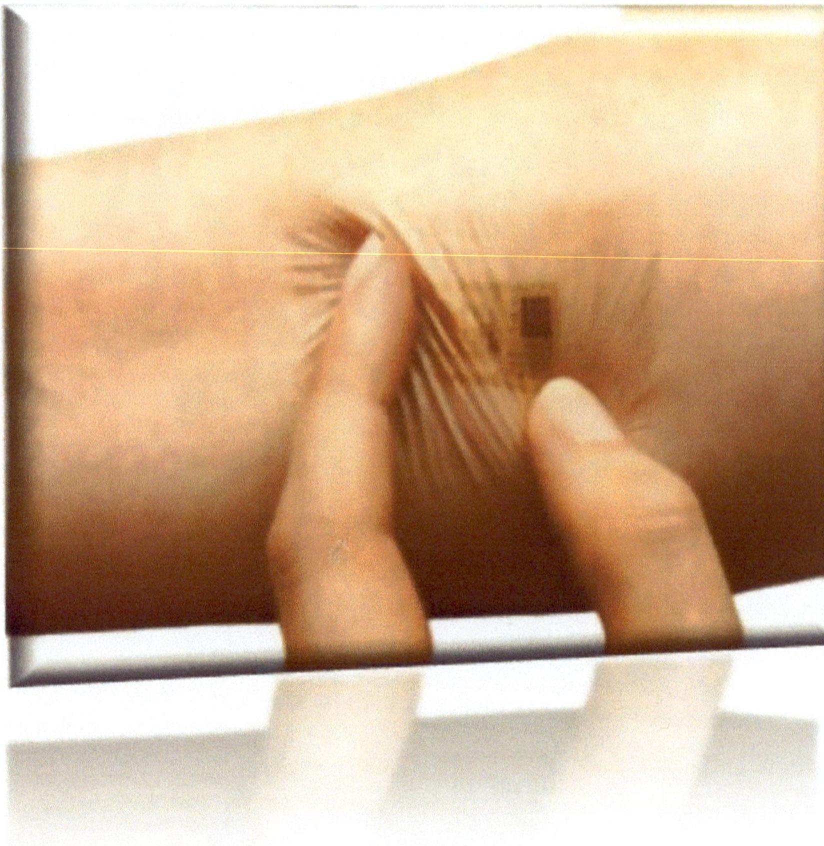

FIGURE 5.5 Implanted skin patch.

- *Thermistors*: These are temperature-sensitive resistors, that change their physical resistance once the temperature varies.
- *IC (semiconductor)*: The variable resistance property of semiconductor material is used in detecting temperature changes and the readings are denoted in digital form.
- *Infrared sensors*: The temperature of a solid or liquid is detected by capturing the intensity of the infrared energy applied to it. This is not applicable for gases as they are transparent.

5.3.4 Pressure sensors

Pressure sensors measure the pressure of any type of gas or liquid. Pressure sensors convert physical power into an electrical signal. They can also be

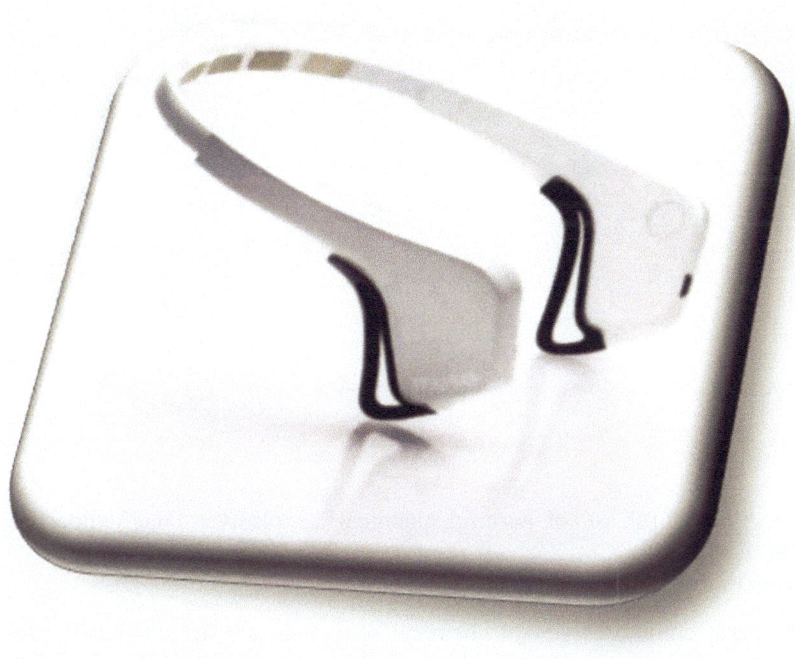

FIGURE 5.6 Headband with neurosensors.

effectively used for measuring other variables like speed and altitude [12]. *Barometers* and pressure gauges are the most common pressure sensors used for IoT systems. Barometers give an accurate measurement of ambient air that can be used in weather forecasting. *Pressure gauges* are used in closed environments of industrial sites to monitor pressure. Pressure sensors with IoT devices can be used in various areas like touchscreen devices, automotive systems, the manufacturing industry, and biomedical devices. Micropressure sensors (small in size) are also available for the measurement of pressure.

5.3.5 Water quality sensors

Water quality is detected by water quality sensors in all industries [13]. The most widespread water sensors and their purposes are listed in Table 5.2.

5.3.6 Chemical sensors

Chemical sensors are applied to find out chemical changes to liquid or air. Chemical sensors can be applied in pharma industries for the following purposes:

- Monitoring and process control;

TABLE 5.2 Different water sensors and their uses.

Sensor	Uses
Chlorine residual sensor	Measures chlorine residuals in water
Total organic carbon sensor	Measures organic elements in water
Turbidity sensor	Measures suspended solids in water
Conductivity sensor	Identifies total ionic concentrations in water solutions
pH sensor	Measure the pH level in dissolved water, to indicate acidic or basic property
Oxygen-reduction potential sensor	Measures the level of oxidation/reduction reactions occurring in the solution

- Detection of release of harmful chemicals, explosives, and radioactive chemicals;
- Recycling processes.

The most common types of chemical sensors in use are chemical field-effect transistors, chemiresistors, electrochemical gas sensors, pH glass electrodes, potentiometric sensors, fluorescent chloride sensors, hydrogen sulfide sensors, nondispersive infrared sensors, and zinc oxide nanorod sensors. These types of sensors can be used to monitor the pollution level in cities, check the food quality in kitchens, and agricultural products in the supply chain [14].

5.3.7 Gas sensors

Gas sensors are specifically used to monitor changes in air quality and detect the presence of various gases. They are used in air quality monitoring, detection of toxic or combustible gases, hazardous gas monitoring in chemical laboratory research, and manufacturing of pharmaceutical products. Some common gas sensors are air pollution sensors, breathalyzers, carbon dioxide sensors, carbon monoxide detectors, catalytic bead sensors, electrochemical gas sensors, gas detectors, hygrometers, hydrogen sensors, nitrogen oxide sensors, oxygen sensors, and ozone monitors.

5.3.8 Smoke sensors

A smoke sensor is a device that senses smoke and its level. Plugged into a system, the smoke sensor immediately notifies the user about any problems and it is used in the pharmaceutical industry. Smoke sensors are extensively used to detect fire and gas incidences. Smoke sensors detect the presence of smoke, gases, and flame surrounding their field either optically or by a physical process. A *photoelectric smoke sensor (optical)* uses the light-scattering principle trigger to residents, to notify them of any problem. An *ionization smoke sensor* uses the principle of ionization to detect molecules causing smoke and triggers an alarm.

5.3.9 Level sensors

Level sensors can determine the level or amount of fluids, liquids, or other substances that flow in an open or closed system. They are applied in recycling as well as in pharmaceutical industries. There are basically two level measurement types, these are point level sensors and continuous level sensors. *Point level sensors* usually sense the particular specific level and notify the user by an alarm or trigger, if the sensing object is above or below that level. *Continuous level sensors* assess liquid or dry material levels within a specified range and provide continuous indication of the level.

5.3.10 Humidity sensors

The amount of water vapor in an atmosphere of air or other gases is referred to as humidity, or relative humidity (RH). Humidity sensors are used in many manufacturing processes which require perfect working conditions. As sensors detect changes almost instantaneously, any sudden change is notified for immediate action. They are applied in pharmaceutical industries for heating, ventilating, and air-conditioning systems for better control to protect medicines.

The information gathered by sensors gives a full picture of the equipment utilization. Hence, planning maintenance and repair, minimizing downtime, avoiding critical issues, and ensuring workplace safety become easier. In addition, it helps to optimize performance and reduce waste. Pharma IoT facilitated with multiple sensors to monitor real-time multiple environmental indicators, like temperature, humidity, radiation, light, and CO_2 level can create greater transparency in drug production and storage environment [15]. The lab supervisor can observe environment changes that are rapidly notified by multiple sensors on a dashboard and take appropriate measures. Automatic adjustments can be triggered if these sensors are integrated with a climate control system. This automated system can also alert staff for mass evacuation in the case of a toxic substance leakage.

5.4 Implementation strategies and methodologies

5.4.1 Resource and knowledge management

The pharmaceuticals industry employs IoT applications like connected equipment, men and material tracking, smart packaging, and cold chain monitoring for efficient management. These strategies are discussed in this section.

5.4.1.1 *Equipment monitoring*

Pharmaceuticals manufacturing is generally performed in batches. Therefore the equipment has to be continually maintained and calibrated to assure the safety and efficacy of drug products. Some raw materials and finished products require very specific storage conditions. IoT technologies help the pharmaceutical industry by attaching sensors to manufacturing equipment. The sensors collect and send data to a central system that analyzes and converts the raw data into meaningful information to remotely monitor equipment. It allows technicians to develop self-learning predictive models and perform diagnostic services across the company, thereby increasing availability, reliability, and ensuring minimal disruptions in the supply chain. The role of IoT in pharma manufacturing and supply chain management is depicted in Fig. 5.7.

The real-time information about equipment availability, calibration, and utilization provided by the IoT solution can help manufacturing teams make better decisions, optimize equipment performance, reduce machine

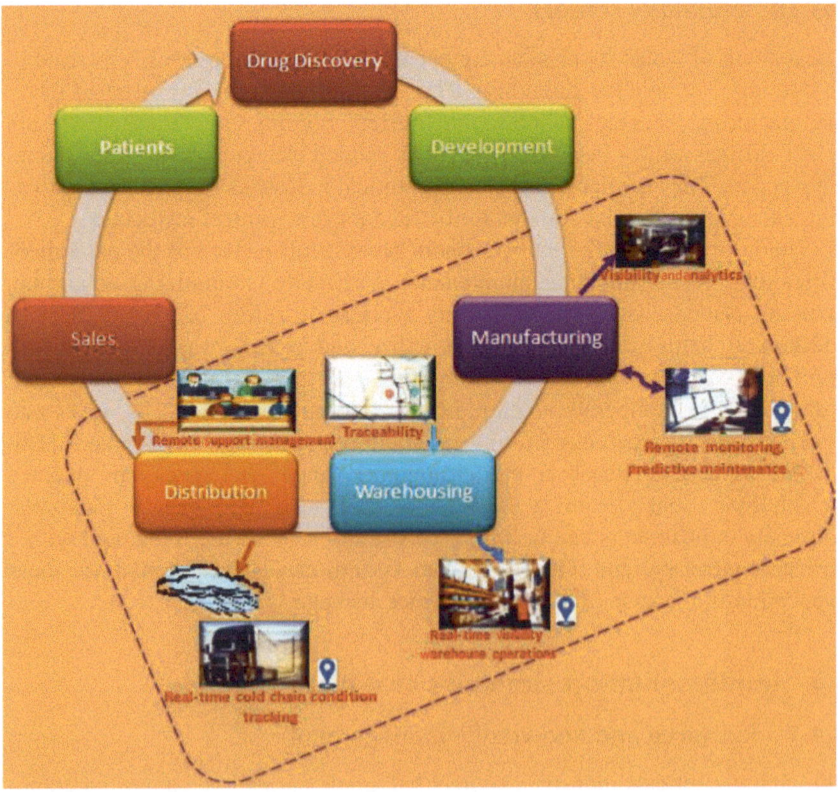

FIGURE 5.7 The IoT in pharma manufacturing and supply chain management.

downtime, and improve resource allocation. In addition, they aid in lower manufacturing costs and shorten cycle times. The major issue is connecting industrial equipment to sensors or devices to access back-end databases or cloud services. Many companies still operate heritage equipment that does not support standard communications protocols. Pharma companies have to acquire new manufacturing systems to support IoT instrumentation and the digitization of manufacturing and the supply chain. This demands open automation software that is compatible with a wide range of communication protocols for industrial IoT transition. Some industrial IoT platforms like Microsoft Azure, Amazon Web Services, or PTC's Thing Worx can be exploited.

5.4.1.2 Warehouse operations

Warehousing is a critical issue for the pharmaceuticals industry. Most companies deal with a large number of storage services across countries to ensure a continuous and timely supply of essential medicines in a cost-efficient manner. Sensors are located in storage areas and on inventory items. The data received from these sensors are used to interpret vital information like product location, inventory details, misplaced products, etc., which is then transmitted to warehouse managers' handheld devices/dashboards. Hence, it is easier for the manager to take corrective measures in real time, which significantly improves the speed, accuracy, and efficiency of the picking process. IoT-enabled smart warehouses give real-time visibility and a 3D view of warehouse operations with detailed, contextually relevant data that allows warehouse managers to:

- Examine and track the storage of sensitive medicines in controlled regions;
- Enhance warehouse floor space;
- Follow inventory of finished goods;
- Recognize problematic areas and assign resources to deal with challenges that require human assistance.

IoT technologies can also be applied in the storage of temperature-sensitive products. Active and passive temperature loggers attached to refrigerators in warehouses continuously record temperature. An IoT solution connects these devices and compares their measurements against thermostability tables. In the case of temperature variances, alerts are generated. Inventory movements can be analyzed from the sensor data. The aggregated data (volume, dimension of packages) captured from various wireless readers are investigated by the central warehouse management system.

5.4.1.3 Tracking shipments

Tracking the movement of drug inventory at every point can help the pharmaceutical industry to identify fake drugs entry and potentially save billions of dollars for the supply chain participants. 2D barcodes, RFID tags, and smart packaging labels are integrated to track each handshake in the supply chain, from manufacturing to dispensing, that keeps a record of the digital footprint. In order to ensure suitable conditions in cold chains for temperature-sensitive drugs and to have control of materials entering health-care environments, electronic chips in packaging material and automatic information data collection for smart serialization with Auto-ID are used. Thus the IoT when applied to packaging offers several advantages, including bidirectional communication, tracking, and status display mechanisms.

5.4.1.4 Managing the cold chain environment

Pharmaceuticals manufacturers increasingly deal with products such as biologics, which are highly susceptible to storage conditions. A larger proportion of high-value active ingredients of these products often have shorter shelf lives and have strict temperature requirements. Some biologics drugs require 3–46°F temperatures, and many others require temperatures lower than 77°F during cold chain transportation. Some drugs must be stored at controlled room temperatures. Sensors are embedded on tracking equipment with auto-start and shutdown mechanisms in warehouses, vehicles, or shipments. Remote monitoring of cold chain environments in real time is made by IoT solutions using an embedded sensor during transport, like reefers, to avoid the spikes that can occur in ambient containers, to protect the drugs.

5.4.1.5 Remote fleet management

Pharmaceutical manufacturers can take advantage of digital technologies such as GPS location and condition monitoring for real-time tracking of cargo conditions and vehicle performance and avoid losses and penalties from late deliveries and spoilage. By employing vehicle telematics, pharma manufacturers can unify the view of vehicles, drivers, and performance across locations. GPSs capture the vehicle details, then relay these to control rooms through a general packet radio service. This gives complete visibility of product and transport conditions and facilitates control-room personnel to analyze and compare data on fleet performance from any location. This can reduce time to market, trim down carrying costs, and prevent product damage and wastage.

5.4.2 Optimization of pharmaceutical processes

With Industry 4.0, organizations are able to get better predictability and interoperability of machines by combining the physical and digital assets used in manufacturing by employing technologies like IoT and machine learning. The pharmaceutical industry which adopts Industry 4.0 will be able to get a true predictive of the analytics capability across their supply chain from production planning to packaging and logistics. Ephlux [16] is an agile consulting and technology services company that, in collaboration with Oracle, presented its Pharma 4.0 solution. Using Pharma 4.0, the pharmaceutical industry goes beyond automation by applying the IoT and machine learning predictive analytics capabilities. The benefits of implementing Pharma 4.0 in pharmaceutical industry are:

- Improved productivity;
- Guaranteed regulatory agreement and drug safety;
- Reduced unplanned downtime;
- Monitored and predicted complex business key performance indicators.

With Pharma 4.0, a pharmaceutical company can ensure the correct amount of medicines is prepared at the right time. Fig. 5.8 illustrates the benefits of Pharma 4.0. The faults and failures can be identified before their occurrence, with real-time production monitoring with predictive ability. This helps the pharmaceutical industry to save ingredients and prepare faultless medicines. Any upcoming equipment service requests and inventory requirements can be predicted well in advance with the shop floor in sync with the enterprise resource planning.

Pharma 4.0 can help in efficient and effective production lines for the industry by reducing the potential hidden costs of service, maintenance, and additional need for manpower. This is elaborated in Fig. 5.9. It can contribute actively toward managing regulatory compliances through tracking and tracing across the supply chain, which is a key regulatory concern for the pharmaceutical industry.

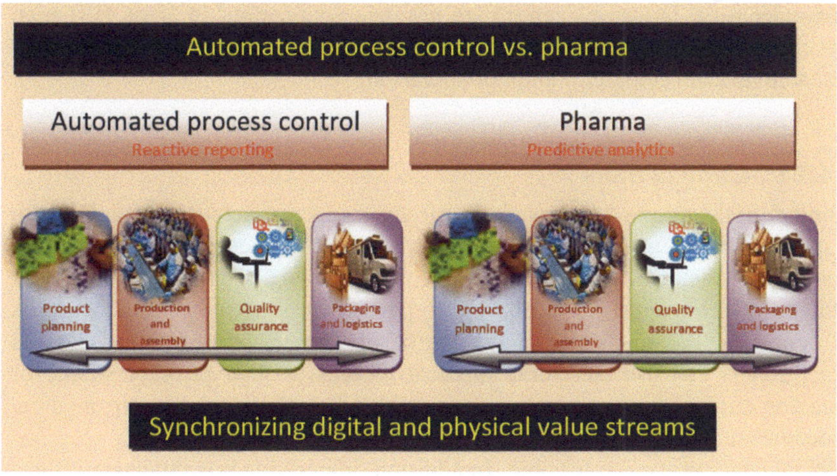

FIGURE 5.8 Pharma 4.0.

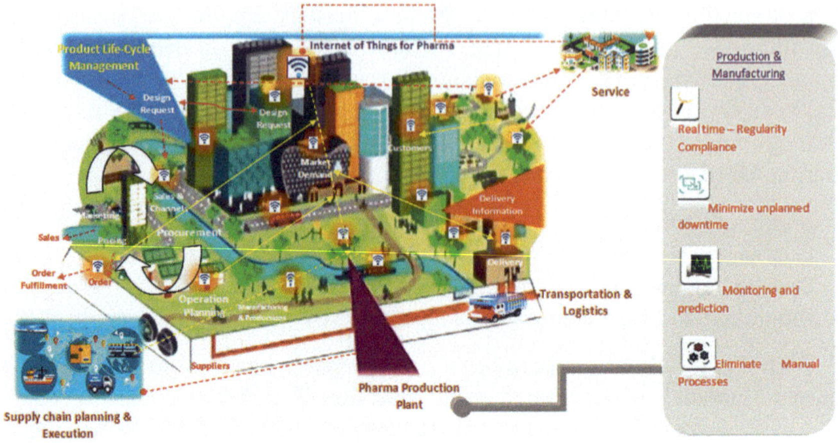

FIGURE 5.9 Production and manufacturing using Pharma 4.0.

5.4.3 Big data analytics of pharmaceutical data

The McKinsey Global Institute [17] estimated that big data strategies applied to decision-making could produce savings of up to $100 billion annually across the US healthcare system, by optimizing innovation, improving the effectiveness of research and clinical trials, and devising new tools for physicians, consumers, insurers, and regulators. In the pharmaceutical industries, data growth is generated from several sources, including the R&D process, retailers, patients, and caregivers. Successful utilization of these data will help pharmaceutical companies better identify new potential drugs and develop them into effective, approved medicines more quickly. Pharmaceutical industries can utilize big data applications in areas [18] such as clinical trials, drug discovery, drug reactions, sales and marketing, research and development, and collaboration.

5.4.3.1 Clinical trials

Clinical trials are used to ensure the effectiveness and safety of patients with a specific treatment. Clinical trials are costly and time consuming. The pharmaceutical companies have to find out the right mix of patients for a given trial. Identifying appropriate patients to participate in a trial through analysis of demographic and disease status can be performed with big data. General practitioners can use electronic medical records as their primary source of data for clinical trials, reducing data entry errors and speeding up medical procedures. With the help of big data tools, RPM, and reviewing previous clinical trial events, are feasible. In addition, identifying potential side effects before manufacturing is also achievable.

Pfizer, one of the world's premier biopharmaceutical companies, combined data from electronic medical records, clinical trial and genomic data to uncover the opportunities for "drugs for specific patient populations" [19]. The company also was able to identify that a specific genomic defect (a mutation in the *ALK* gene) caused lung cancer for a small subset of patients. Using this inference, Pfizer developed Xalkori, which was approved for use by the Food and Drug Administration (FDA) in 2011 specifically for lung cancer patients with the *ALK* gene mutation.

5.4.3.2 Drug discovery

Conventionally, researchers employed an iterative process of physically testing various plant and animal compounds to discover new drugs. Hence, drug discovery can involve a vast amount of time and resources, which may be difficult for patients with disorders such as Ebola, swine flu, and typhoid, especially during an outbreak. The cost of developing new drugs can also be extremely expensive.

By applying the power of big data in the pharmaceutical industry, researchers can exploit predictive modeling for drug discovery. Predictive modeling can facilitate researchers to forecast drug interactions, toxicity, and inhibition. Advanced mathematical models and simulations are utilized in predictive models to envisage the reaction of a particular compound with a human body. In addition, historical data collected from previous clinical studies, medical trials, and postmarketing surveillance can also be used in the predictive models to get FDA approval.

5.4.3.3 Drug reactions

In some circumstances, ADRs that have harmful effects to patients may occur during medication. This is the result of the inability to precisely replicate real-world scenarios during clinical trials. ADR reporting systems generally depend on authoritative reports released by lawyers, pharmacists, and clinicians. On the other hand, the information in ADR reports can be misinterpreted or lost. For that reason, most patients criticize ADRs or side effects on social networks such as Facebook and Twitter or on online medical forums.

Pharmaceutical companies can mine social media platforms and medical forums for ADRs and patient reviews. For this, pharmaceutical businesses can apply sentiment analysis and natural language processing. The collected data can be investigated with the aid of big data analytics to gather insights into ADRs. Hence, the process of evaluating drug reactions can be simplified by the influence of big data in the pharmaceutical industry.

5.4.3.4 Sales and marketing

Big data can influence the pharmaceutical companies to face the increasing competition in their sales and marketing operations. By analyzing information from social media, demographics, electronic medical records, customer behavior, the impact of ad campaigns, and customer retention, underserved markets may be spotted. Regularly analyzing the effectiveness of sales efforts and acquiring the feedback received by the sales force during client visits and using it effectively can help pharmaceutical companies to get an edge on their competition.

For example, Pfizer [20] has been working with Ecuador-based vendor Noux, which works with SAP, to build a more centralized method of capturing information about sales, marketing, and distribution data, for which Pfizer had previously managed the data in multiple Excel sheets spread throughout the organization. The new solution intends to give Pfizer a better understanding of doctors' prescription of a given medication, track the effectiveness of the sales fields, track competition, and appreciate the performance of individual sales representatives. This will allow them to make better decisions, faster than ever before.

5.4.3.5 Research and development

Pharma data analytics (PDA) is a relatively new research area that can be explored with the influence of big data. It also involves genome analysis, which comprises huge unstructured data. Big data analysis can lead to the prevention/cure of chronic diseases like cancer and heart-related diseases [21]. The root causes of these diseases can be identified from the genome sequence. Better analysis and early predictions from the genome sequence can provide many opportunities for the pharmaceutical industry. The patient history analyses enable the pharma company to increase its efficiency in drug discovery as well as its marketing. PDA is the process of inspecting the data set and obtaining information with the application of specialized systems and software. Through this process, patients will be given the right drug at the right time for a speedy recovery.

5.4.3.6 Collaboration

Pfizer's [19] analytics ecosystem developed a new drug, Xalkori, based on three different platforms, the genomics research repository, the life sciences hub, and the repository for electronic medical records. The genomics research repository is an open-source data management system built on tranSMART (developed in 2009 by scientists at Johnson & Johnson and Recombinant Data Corp). The platform combines genomic data sets from various internal and external sources. The life sciences hub is a cloud-based data repository to store Pfizer's own clinical trials research, along with research from its Contract Research Organization (CRO) partners. It is hosted by Oracle and assembled into Oracle's Life Sciences Hub. The repository for the electronic medical records is a storehouse of hundreds of millions of anonymized records and is built on Teradata. Tableau Software and TIBCO Spotfire were used for data analysis and visualization.

The pharma IoT concept involves the computerization of medical products and related care processes by means of smart connected medical devices and IT services during drug development, clinical trials, and patient care [22]. The effect of pharma IoT on development and clinical trials can utilize combinations of advanced technologies.

5.5 Information security and privacy protection in Internet of Things-based pharmaceutical systems

The pharmaceutical IoT infrastructure consists of a large number of devices connected in a special manner. These devices are highly resource constrained, and have limited computation capabilities, storage, and energy. IoT applications use IP-based communication protocols for connecting devices. The transition from IPV4 to IPV6 enforces rigorous modifications to the existing communication protocols, engineering designs, and standards in each layer of the IoT protocol stack [23]. As IoT uses ubiquitous computing with heterogeneous devices and networks, the standardization of the communication protocols is a big challenge. Then, standardization of IoT architecture leads to interoperability, security, and privacy issues in IoT applications. Efforts must be initiated to protect big data in a secure, private, and most effective manner.

5.5.1 Privacy Protector framework

In IoT-based health care, medical devices are more exposed to numerous security threats and attacks than other network devices. Current solutions provide protection to data during data transmission to some extent, but cannot avert some complicated threats and attacks such as collusion attacks and data leakage. Luo et al. [23] proposed a practical framework called Privacy Protector that collects the patient privacy-protected data with the aim of preventing these types of attacks. Privacy Protector applies the Slepian Wolf coding-based secret sharing (SW-SSS) for data sharing. In the framework, a distributed database consisting of multiple cloud servers, which guarantees that the privacy of patients' personal data, can remain protected as long as one of the servers remains inflexible. It includes a patient access control scheme in which patients' data are available in multiple cloud servers to offer patients' data to healthcare providers without revealing the content of the data. The Privacy Protector framework is shown in Fig. 5.10.

Privacy Protector has four modules [24]:

- The various medical sensor devices are connected to an IoT through network devices. These IoT sensor devices are used to sense the patients' bodies to acquire data.

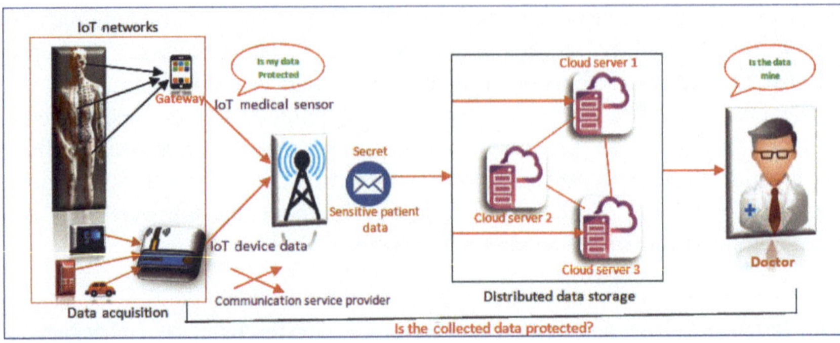

FIGURE 5.10 Privacy Protector framework.

- The patients' data collected from the medical sensor devices are then transmitted to a data storage system through the communication service provider (CSP). The CSP prepares the secret shares and distributes the secret shares to the cloud servers in the distributed data storage system.
- The patients' data in the storage system offer querying services to various healthcare providers, doctors, and health professionals.
- A patient's data access control (PDAC) system is utilized by medical users (e.g., healthcare providers) in order to gain access to the patient's data and monitor the patient's health performance.

During data collection in Privacy Protector, the sensor devices in the IoT split the patients' data into *n* parts and then transmits the data to *n* cloud servers using some secure channels. Patients' data must not be exposed to any other data server during data access. Therefore, in PDAC, any healthcare provider who needs to have access to a patient's personal data must produces a private and public key pair for an ID-based sign cryption scheme. It also generates a signature verification scheme which will be certified by the certificate authority. It also ensures confidentiality, authentication, and integrity. An attempt was made to check the performance of the Privacy Protector and it was found that the framework protected the patients' data against various attacks.

5.5.2 Data access control system

The diversity of the objects in IoT causes a heterogeneity problem in the data format in IoT applications. In the meantime, the use of IoT technology in applications has stimulated the increase of real-time data, which makes information storage and accessing more challenging. Xu et al. [24] proposed a resource-based data-accessing method to acquire and process IoT data pervasively to improve the accessibility of IoT data resources in medical services. The unified metadata model that facilitates data sharing and data interoperating is shown in Fig. 5.11.

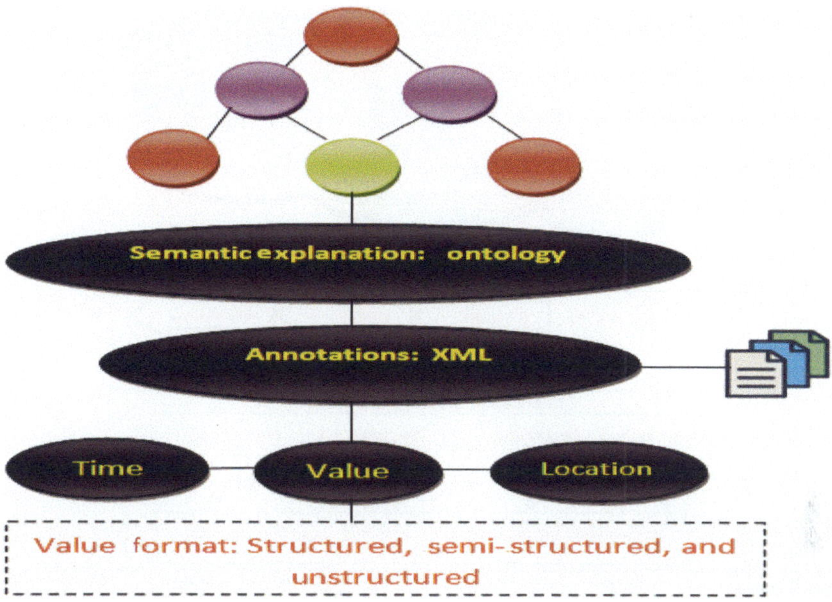

FIGURE 5.11 Unified metadata model.

This metadata model converts the traditional data structures to self-explanation of data value for accessing data ubiquitously. The XML format is used to represent the IoT data, so that it is self-described and can be accessed from the Web. To meet the requirements of big data in IoT, the data relationship is included in the ontology, so that it can be more efficiently stored using NOSQL [Non SQL] databases. For IoT applications in medical servicing, the information of time tag and location tag are added to each metadata model. The time tag is precious for real-time application system. The location tag would be notable in smart medical services for active positioning of the valuable medical apparatus and instruments. The cloud platform for ubiquitous IoT data processing is shown in Fig. 5.12.

In IoT applications, sensors or tags connected to the physical entities are represented in information systems. They are represented as entity-oriented resources (EoRs). EoR models describe the attributes of IoT physical entities in an information environment, but the business functional features of the IoT physical entities are defined as transition-oriented resources. To deal with the big data collected through sensors, a cloud platform with multitenant data management is implemented. The course of action for ubiquitous IoT data accessing can be deliberated as follows.

- The data-accessing requests are initiated by the cloud application.
- The business layer is responsible for granting the accessing right. If the request is approved, it forwards the request to the RESTful Web service.

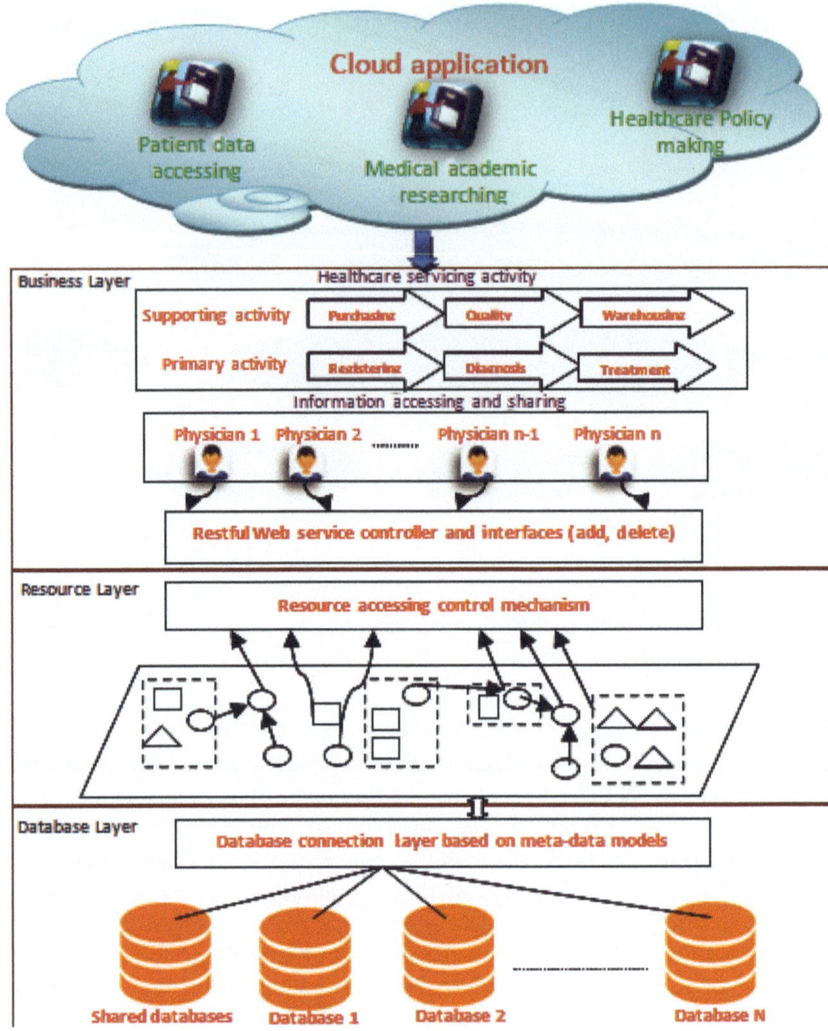

FIGURE 5.12 Cloud platform for ubiquitous IoT data processing.

- The resource layer facilitates the resource accessing request by pushing it to the appropriate database location to persuade the database processing.
- The database layer obtains data from shared and isolated databases.

In this ubiquitous IoT model, heterogeneous IoT data are represented by unique Uniform Resource Identifiers (URIs) so as to be accessed ubiquitously. This model is appropriate for information-intensive industries, such as health care. It is proved that this ubiquitous IoT data-accessing method is efficient in a distributed heterogeneous data environment for timely and ubiquitous data accessing in a cloud and mobile computing platform.

5.5.3 Privacy protection analysis of pharma data

While enjoying the benefits of the IoT-enabled pharmaceutical production facility, pharmaceutical industries are vulnerable to cyber attacks. The cost of cybercrime to the UK economy was around £27 billion annually, as reported by OCISIA in association with the UK information intelligence experts, BAE Systems Detica [23]. It also added that the pharmaceutical and biotech sectors are the hardest hit industries. In the pharmaceutical sector, starting from drug distribution companies to research and development units, most of the units hold highly sensitive data, such as personal patient data, and confidential research on drug development and testing. Thus the pharmaceutical industry becomes an attractive target for cyber attacks. An increased amount of sensors collecting production data provides more opportunities for hackers to penetrate the system. However, to protect from cyber attacks, pharmaceutical companies can use regulatory compliance, Smart supervisory control and data acquisition (SCADA) security [24], and internal security (within the organization) as a cybersecurity defense.

In 2018, the International Pharmaceutical Privacy Consortium (IPPC) and Medical Device Privacy Consortium (MDPC) combined to form the International Pharmaceutical & Medical Device Privacy Consortium (IPMPC) [25]. This consortium aspires to provide a cost-effective means for companies to be informed of and assess the recent developments in global data privacy and security laws. It offers a forum for companies to guide and share ideas about context-specific tools.

The FDA 21 CFR Part 11 is one of the most reputed regulations within the industry. The regulation insists organizations implement controls, electronic audit trails, and systems validations. The industrial security standard is established through reliable electronic documentation of the pharmaceutical manufacturing process. However, compliance with this regulation, engineering, and manufacturing data are protected against unauthorized access, modification or deletion, consistency, and completeness. Successful FDA 21 CFR Part 11 compliance will result in a more organized, efficient, and secure production process.

SCADA [24] systems like COPA-DATA's zenon are compliant with industry regulations and offer built-in cybersecurity capabilities. This software and its components are specifically designed to guarantee secure operations. Built-in software security features such as strong encryption, secure authentication, and automatic synchronization protect companies against data loss and unauthorized access of files in the network. Once the industry has assessed its system's potential vulnerabilities, it must train members of staff on industrial security. Larger organizations can appoint a chief information security officer to administer all industrial security issues and communicate the importance of cybersecurity.

Even though pharmaceutical industries follow security measures, they are exposed to challenges like integrity, confidentiality, unavailability of user data and alteration of the contents, privacy breaches, and identity disclosure during transmission. It is important for the industries to have a combination of privacy techniques to protect their data efficiently. The different privacy-preserving techniques are shown in Fig. 5.13; they include generalization, suppression, pseudonymization, bucktization, slicing, randomization, and cryptographic techniques [26].

Generalization is a widely used approach that uses predefined hierarchies or values to ensure privacy. Suppression is one type of generalization which absolutely hides original values with "*." Pseudonymization is a technique that replaces original values with pseudonyms. Bucktization stores the individual data in separate buckets, ensuring bucket protection. Slicing splits original values into segments. Randomization adds noise in the original data to confuse the attackers. Other than the above techniques, there are a variety of cryptographic approaches existing for data protection. This privacy preserving efficiently sorts out the privacy issues and preserves better utility of anonymous data.

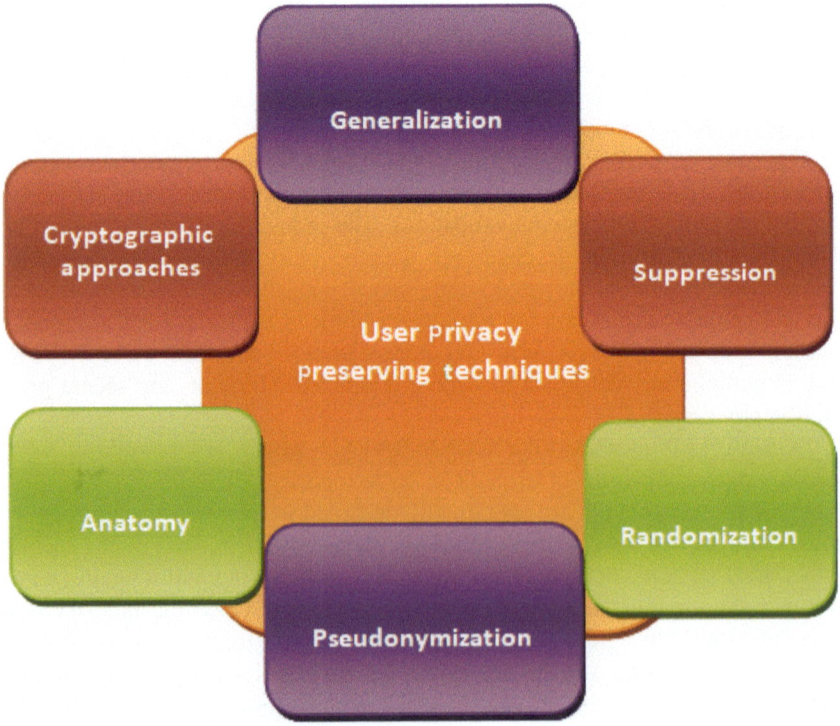

FIGURE 5.13 Privacy-preserving techniques.

5.6 Open research challenges and future directions

Increased competitiveness, growing demand, quality adherence, reduced time to market, and the rising cost of developing new drugs have increased stress on pharma companies. IoT analytics has the inherent capacity to change the whole process of pharma from research and development through manufacturing to distribution. At the same time, it is most essential in addressing existing operational challenges across the pharma supply chain. Those challenges are discussed in this section.

5.6.1 Data security and privacy integration

IoT devices are involved in real-time applications to capture and transmit data. Open data protocols and standards are necessary for IoT devices. Also, there is significant ambiguity regarding data ownership regulations. All these factors make the data highly vulnerable. Cybercriminals can exploit this situation and create fake IDs to buy drugs and medical equipment which they can later sell.

5.6.2 Need for open standards

Multiple device integration causes difficulty in the implementation of IoT in the pharmaceutical sector. The protocol for communication between this variety of devices is complicated and makes the process of data aggregation complex. It also slows down the whole process and reduces the scope of scalability of IoT.

5.6.3 Data overload and accuracy

IoT devices record voluminous data. The data collected from IoT devices are used to gain critical insights. Appropriate use of big data technology has to be employed to gain accurate insights for better decision-making. Edge/fog/cloud technology can be used to store the enormous data generated by these connected devices.

5.6.4 Future directions

The pharmaceutical sector has the responsibility of producing new and potential drugs into the market, ensuring the safety and validity of medical drugs, and in the evaluation of safe drugs, which enable quicker patient recovery. Hence, drug companies face the challenges of tracking their products in a timely fashion to avoid fake drugs entering the system. While using IoT for data acquisition, the production processes of potential drugs can be evaluated and monitored through blockchain technology [27] to prevent the production and distribution of counterfeit drugs.

Recently, Hyperledger [28], backed by the Linux Foundation, initiated a project using blockchain technology as a tool, for inspecting and fighting the production of counterfeit drugs. There is a dire need to monitor the whole pharma supply chain through the use of digital technologies worldwide. On this point, since April 2018, in one of the hospitals in the Novgorod region, Russia, a digital drug control system (DDCS), a program for monitoring the circulation of medicines based on blockchain technology, has been launched [29] for the prevention of counterfeit drugs.

Big pharmaceutical industries such as Sanofi, Pfizer, and Amgen launched a joint pilot project for the inspection and evaluation of new drugs using a blockchain-based DDCS [29]. The blockchain approach facilitates tracking the production and location of drugs at any time, improves the traceability of anecdotal drugs, ensures the security of the drug supply system, and assures the quality of drugs supplied to customers [29].

5.7 Conclusion

The IoT provides accessibility and visibility in every region of pharmaceutical companies at the time of production and operation, thus generating a massive amount of vulnerable data. Hence, strong security mechanisms must be ensured to avoid breaches that may affect the personal information of consumers and businesses. The overall scope of the IoT is found to be highly promising as it revolutionary benefits by providing the necessary infrastructure for the largest population.

References

[1] L.D. Xu, W. He, S. Li, Internet of Things in industries: a survey, IEEE Trans. Ind. Inform. 10 (4) (2014) 2233–2243.

[2] G. Sebestyen, A. Hangan, S. Oniga, Z. Gal, eHealth solutions in the context of Internet of Things, in: Proceedings of IEEE International Conference on Automation, Quality and Testing, Robotics (AQTR), Cluj-Napoca, Romania, 2014.

[3] A. Santos, J. Macedo, A. Costa, M.J. Nicolau, Internet of Things and smart objects for M-health monitoring and control, Procedia Technol. 16 (2014) 1351–1360.

[4] I. Orha, S. Oniga, Wearable sensors network for health monitoringusing e-Health platform, Carpathian J. Electron. Comp. Eng. (CJECE) 7/1 (2014).

[5] O. Yakut, S. Solak, E.D. Bolat, Measuring ECG signal using e-healthsensor platform, in: International Conference on Chemistry, Biomedical and Environment Engineering (ICCBEE), 2014.

[6] M.T. Salam, M. Mirzaei, M.S. Ly, D.K. Nguyen, M. Sawan, An implantable closed loop asynchronous drug delivery system for the treatment of refractory epilepsy, IEEE Trans. Neural Syst. Rehabil. Eng. 20 (4) (2012) 432–442.

[7] N. Verma, A. Shoeb, J. Bohorquez, J. Dawson, J. Guttag, A.P. Chandraksan, A micropower EEG acquisition SoC with integrated feature extraction processor for a chronic seizure detection system, IEEE J. Solid-State Circuit. 45 (4) (2010) 804–816.

[8] G. Suganya, M. Premalatha, A. Sharma, M. Pandya, A. Joshi, IoT based automated medicine dispenser for online health community using cloud, Int. J. Recent. Technol. Eng. (IJRTE) 7 (5S4) (2019). ISSN: 2277-3878.

[9] Y.J. Fan, Y.H. Yin, L.D. Xu, Y. Zeng, F. Wu, IoT-based smart rehabilitation system, IEEE Trans. Ind. Inform. 10 (2) (2014) 1568–1577.

[10] J.H. Gruzelier, EEG-neuro feedback for optimizing performance. I: are view of cognitive and affective outcome in healthy participants, Neurosci. Biobehav. Rev. 44 (2014) 124–141.

[11] S. Rita, Top 15 sensor types being used most by IoT application development companies. <https://www.finoit.com/blog/top-15-sensor-types-used-iot/> (accessed 11.11.19).

[12] P.H. Sureshkumar, R. Rajesh, The analysis of different types of IoT sensors and security trend as quantum chip for smart city management. <https://internetofbusiness.com/dutch-firm-improved-medicine-iot/> (accessed 11.11.19).

[13] The Internet of Things: the new Rx for pharmaceuticals manufacturing & supply chains, cognizant 20-20 insights, February 2017. Available at <https://www.cognizant.com/white-papers/the-internet-of-things-the-new-rx-for-pharmaceuticals-manufacturing-and-supply-chains-codex2437.pdf> (accessed 11.11.19).

[14] P.K. Sekhar, E.L. Brosha, R. Mukundan, F.H. Garzon, Chemical sensors for environmental monitoring and homelandsecurity, Electrochem. Soc. Interface 19 (4) (2010) 35–40.

[15] I. Shugalo, The role of IoT in pharma manufacturing and distribution. <https://www.pharma-iq.com/manufacturing/articles/the-role-of-iot-in-pharma-manufacturing-and-distribution>, 2019 (accessed 11.11.19).

[16] Industry 4.0 Solution for the Pharma Industry. <http://www.ephlux.com/industry-4-0-solution-for-the-pharma-industry-pharma4-0> (accessed 11.11.19).

[17] J. Cattell, S. Chilukuri, M. Levy, How Big Data Can Revolutionize Pharmaceutical R&D, 7, McKinsey & Company, 2013.

[18] J. Naveen, The pharmaceutical industry needs big data. Here's why, Friday 24, May 2019, <https://www.allerin.com/blog/the-pharmaceutical-industry-needs-big-data-heres-why> (accessed 13.11.19).

[19] 6 ways pharmaceutical companies are using data analytics to drive innovation & value, process excellence network. <http://www.iqpc.com/media/1001534/35903.pdf> (accessed 14.11.19).

[20] S.S. Kaddi, M.M. Patil, Pharma data analytics: an emerging trend, in: 2017 International Conference on Smart Technologies for Smart Nation (SmartTechCon), IEEE, 2017, pp. 1307–1311.

[21] D.V. Dimitrov, Medical internet of things and big data in healthcare, Healthc. Inform. Res. 22 (3) (2016) 156–163.

[22] D.V. Jose, A. Vijyalakshmi, An overview of security in Internet of things, Procedia Comput. Sci. 143 (2018) 744–748.

[23] E. Luo, M.Z.A. Bhuiyan, G. Wang, M.A. Rahman, J. Wu, M. Atiquzzaman, Privacyprotector: privacy-protected patient data collection in IoT-based healthcare systems, IEEE Commun. Mag. 56 (2) (2018) 163–168.

[24] B. Xu, L.D. Xu, H. Cai, C. Xie, J. Hu, F. Bu, Ubiquitous data accessing method in IoT-based information system for emergency medical services, IEEE Trans. Ind. Inform. 10 (2) (2014) 1578–1586.

[25] Regulatory compliance and cyber threats: industrial security in the pharmaceutical sector. <http://www.samedanltd.com/uploads/pdf/white_paper/23ccbbdb4947fb9a6587a7-fefe092aa8.pdf>.

[26] T. Lange, Optin, "Intelligent SCADA systems" <https://www.ee.co.za/wp-content/uploads/legacy/Intelligent%20SCADA%20systems.pdf> (accessed 18.11.19).
[27] International Pharmaceutical & Medical Device Privacy Consortium. <https://www.ipmpc.org> (accessed 18.11.19).
[28] A. Majeed, Attribute-centric anonymization scheme for improving user privacy and utility of publishing e-health data, J. King Saud. Univ.-Comput. Inf. Sci. 31 (4) (2019) 426–435.
[29] A. Siyal, A. Junejo, M. Zawish, K. Ahmed, A. Khalil, G. Soursou, Applications of blockchain technology in medicine and healthcare: challenges and future perspectives, Cryptography 3 (1) (2019) 3.

Further reading

S.S. Kaddi, M.M. Patil, Pharma data analytics: an emerging trend, IEEE, International Conference on Smart Technology for Smart Nation, 2017, pp. 1307–1310.
A. Marathe, Internet of Things: opportunities and applications in pharmaceutical manufacturing and logistics, in: Proceedings of the American Society for Engineering Management International Annual Conference, October 2018.
P. Taylor, Applying blockchain technology to medicine traceability, April 2016. Available online: <https://www.securingindustry.com/pharmaceuticals/applying-blockchain-technology-to-medicinetraceability/s40/a2766/> (accessed 20.11.19).
V. Plotnikov, V. Kuznetsova, The prospects for the use of digital technology "Blockchain" in the pharmaceutical market, in: MATEC Web of Conferences, vol. 193, EDP Sciences, London, 2018.

Chapter 6

Internet of Things applications in the pharmaceutical industry

Deepak Kumar Sharma[1], Saakshi Bhargava[2] and Kartik Singhal[3]
[1]Department of Information Technology, Netaji Subhas University of Technology (formerly known as Netaji Subhas Institute of Technology), New Delhi, India, [2]Department of Physical Sciences and Engineering, Banasthali Vidyapith, Tonk, India, [3]Department of Manufacturing Process and Automation Engineering, Netaji Subhas University of Technology (formerly known as Netaji Subhas Institute of Technology), New Delhi, India

6.1 Introduction to the Internet of Things

When a variety of devices are linked to one other and start communication through the Internet, the Internet of Things (IoT) becomes apparent. The factors that are responsible for the proper working of IOT is the interconnection of devices with each other, connection between them has made it possible to transfer information from one device or say point to another. The connection is not only limited to devices but has also extended to humans. The first step that initiates the process is the collection of information that takes place when we collect data from the environment via sensors then exchanging this data through the embedded [1] software, cameras, sensors, actuators, etc. The sensors that are embedded in the process of collecting data can sense different forms of energy from the environment such as temperature, pressure, light, and movement, and then can act according to the required output. The IoT is basically a model which contains physical entities known to be "things," which when combined with other parameters like software, actuators, sensors, electronics components, and the most important component called the network will allow the devices connected in the respective network to receive the data, collect the data, and then convert it into the desired form.

The main aim of the IoT is to narrow the gap between the respective user and network by providing them with the required Internet connectivity from typical devices such as smartphones, i-Pads, computers, tablets, and smart TV, to relatively inarticulate devices such as coolers, fans, microwaves, and air

An Industrial IoT Approach for Pharmaceutical Industry Growth.
DOI: https://doi.org/10.1016/B978-0-12-821326-1.00006-1

conditioners. Possessing the capability of making everything "smart," virtually, the IoT is a focused domain working on the improvement facets of human life with excellence of data collection, AI algorithms, and networks [2].

The "thing" in IoT mentioned above can be anything, it may be a person who is suffering from diabetes, thyroid, or any other health issue and wants to keep a record in order to monitor their daily health charts, or it could be an animal (a pet or based in a particular habitat) with a tracking device on it for track recording and location updates, etc. A number of surveys take place each year focusing on the new technological effects of the ongoing technology in society, one of them being the IoT, discussing the industrial advancements by working on smart devices to smart cities. A survey was carried out and it was claimed that, by the year 2020, there may be almost 20 billion IoT smart devices in the market with a worldwide expenditure on IoT technologies approaching $1.3 trillion. Large and renowned companies, namely Microsoft, Walmart, Google, Flipkart, Amazon, Cisco Systems, Nvidia, Intel, General Electric (GE), IBM, HCL, HP, Bosch, and many others are relying on IoT technology for their survival. GE established GE Digital in 2015 with the motive of incorporating the IoT also has to focus on prediction of conservational needs and sustained resource development. hence a software is build by GE that receives generated data from machines and then compare the data with the outdated data already stored in database, which then helps in making the prediction. GE also has IoT for focusing on the healthcare domain, airline diligence, and utility enterprises. Rolls Royce is another large company which uses smart sensors in military jets, civil aerospace jet engines, and other aircraft. Cisco has smart grid technologies [3], which they are using for providing customers with efficient services and energies (these energies are basically extracted in the form of oil or gas). Cisco also evaluates and gathers data related to energy use and circulation, application regulations, wireless networks, and cybersecurity.

The IoT has built an international showground which is enormous and developing exponentially. Those who were involved in the embryonic digital world and have recently observed the production and growth of IoT-enabled devices know the importance of the technology. The Iot has been a medium in making varied changes in technologies, making the technology more accessible to common groups, not only this there are number of tasks that can be focused and done through it, which is wholly responsible amplifying the level of understanding. The IoT is a system of physical devices, vehicles, and other objects entrenched with electronics, software, sensors, detectors, actuators, and proper network connectivity, which enable these things to accumulate and interchange data. Some examples are smartphones, refrigerators, cameras, and anything that is connected to the Internet. Though its definition has changed with time the main points have remained unchanged.

In 2008, the term "IoT" was used to describe a number of technologies and research areas that allow the Internet to influence the real world filled with physical objects.

In 2020, the IoT was described as "things having identities and virtual personalities operating in smart spaces using intelligent interfaces to connect and communicate within social, environmental, and user contexts." An example was stated as follows: "Smart Mother" has its own freewheel sensors known as cookies that can accumulate data related to any activity that is assigned to it by the user. The device then presents these data to the final customer with the help of a corresponding application for which the sensors were used. The applications include walking, sleeping, medication, drinking, measuring temperature, etc. This is an example on Car IQ which is a plug and play device, this device gathers records based on the present situation. which is then responsible for the electronic arrangements inside the car. These data are unraveled from innumerable electronic arrangements inside a car and exhibited to the end-user. The third example is the well-known company "Uber," which opens the question: "Is Uber a technology based on IoT technology or just a transport-based company?"

6.1.1 Internet of Things framework

The IoT is growing and is remains a state-of-the-art concept, and has become a most important part of the world of IT. Over a period of time, IoT has created interest through the visualization of a universal setup of things that are connected to each other and are properly networked. The IoT has enabled connectivity [4] at any time and at any place, for example, a person who is traveling to another city can operate appliances at their house following simple steps. The IoT can be thought of as an overall network that permits communication between humans and things, humans and other humans, and things and other things, while providing a unique identity to each device and user [5].

The IoT creates a biosphere where everything can be associated and communicated with using a smart approach, where "being connected" is not related to electronic devices such as servers, i-Pads, computers, AC, tablets, televisions, iPods, and smartphones, but the interconnection of humans with these devices or even between two humans. The framework of the IoT is basically a combination of electronic equipment, sensors (detectors), and actuators entrenched on physical things that are associated through wired and wireless networks, and thereby using a similar Internet protocol that links the user to the Internet. The chief idea behind this technology is that of a connected world. This connected world helps the devices to have a plethora of interactions with human activities, such as smart cars and smartphones. Connected devices in today's world are popularly described as smart devices. The priority here is to bring optimal, cost-effective, and feasible solutions to people in the easiest way possible. IoT technology aims to target and drive all major sectors of services and production with such systems.

"IoT upholds the idea of developing a digital relationship (exchange of data/information) between the devices and people, so that every device and person can have worldwide connectivity."

All smart devices could be things in the IoT. Any electronic device with the ability to connect and either transmit or receive data could be a "thing" in the IoT. All that is needed the combination of electronic devices, proper software, and efficient transmission and receipt of data.

6.1.1.1 *Inside the "Things" in the Internet of Things*

It is extremely important to understand the tiny components inside these devices that enable them to become an "IoT-enabled thing." IoT-enabled things mainly serve two purposes:

1. supplying any smart system with input and
2. dispersing the output.

To do so, the devices needed to be equipped with connectivity. Fig. 6.1 shows how these systems are connected through electronic devices, namely sensors and actuators, and maintain the connectivity of the entire ecosystem.

6.1.1.2 *Internet of Things technology stack*

The IoT technology stack is the classification of numerous technologies used in the IoT categorized into broad layers. The number of layers integral to an IoT ecosystem can increase over time, however there are three fundamental layers on which it operates (Fig. 6.2):

1. IoT device layer;
2. IoT gateway layer; and
3. IoT platform layer.

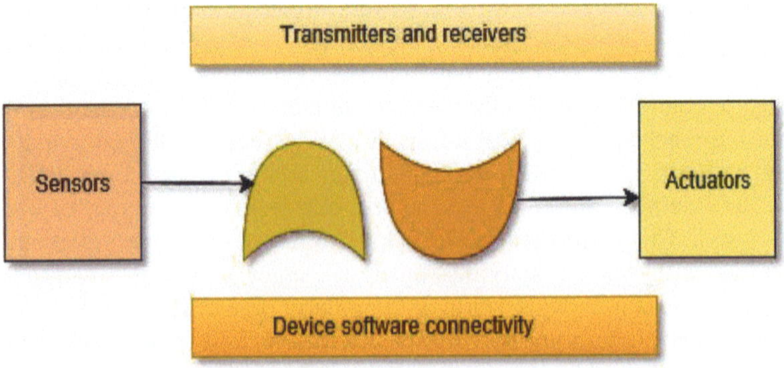

FIGURE 6.1 Schematic diagram of an IoT device.

FIGURE 6.2 Fundamental Layers of an IoT ecosystem.

The *IoT device layer* is the layer where an IoT ecosystem begins, and its accuracy has a huge significance in the performance of any smart product whose ecosystem depends on it. It interacts with the physical environment through the "things" in the IoT.

The device in this layer should contain at least the following IoT-enabled things:

1. transducers, sensors, or/and actuators and
2. device connectivity—transmitters, receivers, and device-operating software.

The *IoT gateway* is the pathway or bridge through which the transmitted data flow from the layer of "things" to the layer of operations. Edge computing/fog computing is the technology powering digital connectivity with abilities like high-grade processing and smart analytics. It is no ordinary bridge however, rather it is a smart and intelligent gateway housing technologies to aggregate connectivity for things to communicate while preprocessing their data, translating protocols, and ensuring network security.

The *IoT platform* could be called the control center of a smart system, though it obviously does much more than just that. This is where everything actually connects to create their intelligence! With cloud storage, smart analytics of big data, machine learning, high-speed messaging, and central processing this layer integrates a plethora of diverse technologies, clouds, software services, and APIs to ensure IoT interoperability device virtualization and real-time responses. It is the middleware between input and output in the IoT ecosystem.

6.1.1.3 Internet of Things ecosystem

With so many different technologies available, it is important for the IoT technology stack to combine the right kinds of technologies in an accurate, secure, and cost-effective way (Fig. 6.3).

An IoT ecosystem could power the entire world by providing enriched real-time information or smart solutions based. This layer aggregates

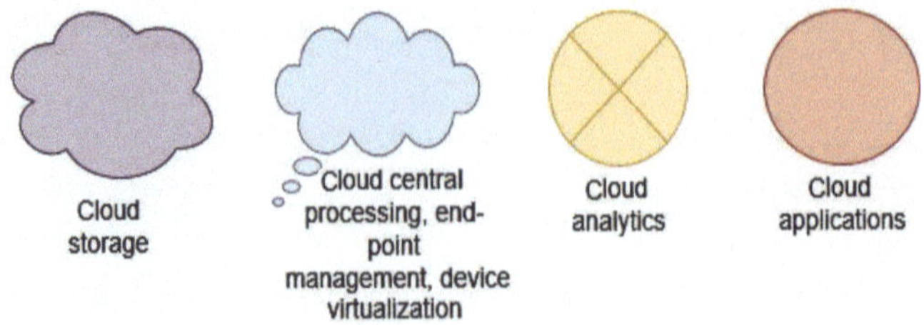

FIGURE 6.3 IoT and cloud computing.

inexpensive access for everyone to remotely control all parts of their digital life. In fact, even today the gateways, are capable of IoT interoperability and lucid access.

The best IoT ecosystem should ideally:

1. enable the platform to blend in with the applications used by devices and gateways;
2. provide flexible scalability crossways to any number of endpoints that a consumer may need;
3. safeguard end-to-end data stream encryption;
4. integrate high-security networks and private clouds for sensitive data; and
5. offer heavy-duty server clusters with maximized performance.

This is how the framework of the IoT technology is focused, though the layers change from time to time based on the applications. These three layers are an indispensable part of the IoT system, and their functioning is the focus of the next section.

6.1.2 Internet of Things architecture and reference models

"...[Architecture] standardization can be seen as a backbone for the IoT to create a competitive environment for companies to deliver quality products" [6].

As the IoT is gaining attention in the global market and is continuously expanding, there will soon be a need for a more flexible and layered architecture so that those billions of connections can be made easily with the help of the Internet. There are some projects that have been undertaken, such as IoT-A, which were completely featured on a common architecture, but no single model or architecture for IoT has been chosen universally. The primitive model, that is, the first one, contained only a three-layered architecture [6], including the perception layer, network layer, and application layer. The perception layer had incorporated sensors that sensed and gathered information about the environment, which sensed physical parameters (like temperature, pressure, etc.) or identified other things linked to them in the environment. Next was the network layer that managed the aggregation, routing of data, and filtering of data to the hubs and other smart devices. It was responsible for the connection of smart things, networking devices, and servers. Also, it was used for the transmission and processing of sensor data. Then was the application layer, which was completely responsible for delivering application-specific services to the specified user—basically driving data to the end-user [4]. Slowly, on the basis of application, more recent models have focused on greater abstraction to the architecture, which resulted in a five-layered model that was completely dependent on the demand of the IoT-based things and built the whole system (Fig. 6.4).

While focusing on and defining each layer, there is a need to consider each layer with its appropriate functions. The first layer from the bottom is the object layer, which is the foundation layer that includes all the physically incorporated sensors used in the IoT ecosystem that will collect all the data from the environment and process them simultaneously. This layer is called the object layer because it includes all the sensors, detectors, and actuators to calibrate weight, vibrations, temperatures, humidity, location, etc.

Next comes the object abstraction layer where the technologies like radio frequency identification (RFID), universal mobile telecommunications service (UMTS), Bluetooth, Wi-Fi, Zigbee, etc. help in transferring data from the objects to the next layer which is the service management layer.

The service management layer handles different functions concerning cloud computing and performs functions related to database management processes. It stores, processes, analyses, and retrieves huge amounts of data constantly. This layer is so named because there are services which are paired with the requests sent based on the addresses and names in the database. The application layer programmers work from different hardware platforms to manage the data from various sources and provide services to the requesters [6].

The application layer is the layer that provides services which are basically of high-quality for different IoT-based systems like smart homes, smart cars,

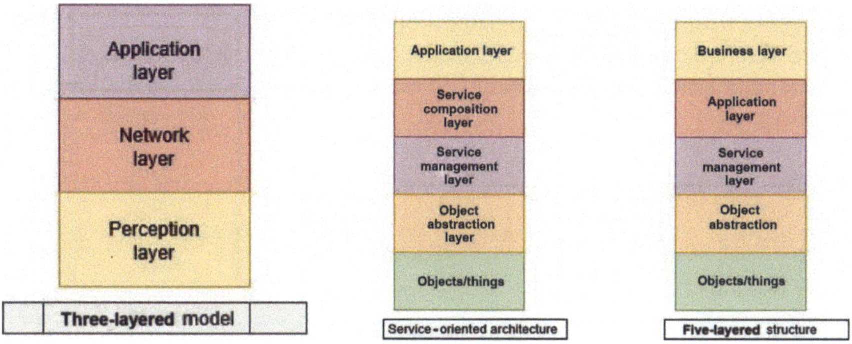

FIGURE 6.4 The architecture of an IoT based system.

home automation, smart buildings, and healthcare services. The business layer is the layer that manages the entire IoT system based on the application, maintains users privacy, and helps to build business models, flowcharts, tables, etc. from the data that are collected from the application layer. The complete package of developing a business model out of the data collected include designing, evaluating, monitoring, analyzing, and implementing the developed elements that are related to the IoT system. This layer, known as the business layer, also plays a significant role in decision-making procedures, which are mostly based on big-data analysis technology [6].

6.1.2.1 *Internet of Things elements*

"The heterogeneity of the IoT elements needs a thorough solution to make ubiquitous IoT services a reality" [6].

An explanation of how the IoT system delivers complete functionality is done by breaking it down into a multistage system where different steps are processed as follows: identification, sensing, communication of different devices, performing computations including services, and finally semantics [7]. This is the interconnected network for the management of information that keeps increasing the impositions on the IoT ecosystem, which includes data collection, data protection, and effective batch loading of data. At each and every phase, telecom administrations have the opportunity to thoroughly assess and revolutionize protocols to enable thorough competence [3] (Fig. 6.5).

1. Identification approaches deliver a clear unique identity for each object inside the network and counterpart services with their demand from clients. The distinction between an object's ID and address is critical as the identification approaches are not internationally valid. Several identification approaches, such as ubiquitous codes (ucode) and electronic product codes, are obtainable for the IoT [7].
2. Sensing and control IoT sensors include smart sensors like wearable detecting devices, or actuators, which control these devices. These sensors collect data from associated objects in a network and direct them back to a data cloud, database, or depository. Next, the collected data are examined, and precise instructions are carried out based on the services required.

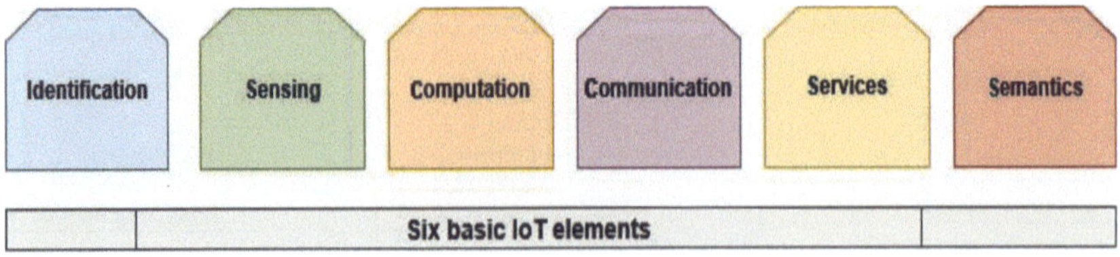

FIGURE 6.5 Basic elements of an IoT framework.

3. Communication technologies of the IoT provide detailed smart services by joining heterogeneous objects. These nodes commonly function by using low power in lossy and noisy communication associations [3].

6.1.3 Challenges and implication

The IoT is typically a universe of linked devices that provides keys for data by further processing the data sent to the cloud to distribute business information, and is able to do this while presenting a huge opportunity for many people involved in different industries and businesses. Acknowledging the demand for the IoT, a number of companies have started focusing on it and improving how they connect future products and services for better usage and data accessibility. However, before reaching the desired target of linking devices, there are three categories for the IoT to focus, as follows [4,5,8]:

1. technology,
2. business, and
3. society.

6.1.3.1 Technology

This is the sum of the technologies that are involved in the IoT, which are largely responsible for making the systems run smoothly. However, under this section alone there are a number of challenges encountered while designing it, including security, standards, connectivity, intelligent analysis, longevity, and compatibility [7] (Fig. 6.6).

6.1.3.2 Security

The daily data accessibility of an individual is widely known, and the threats posed by poor security are also common knowledge—this is a major concern that has been a focus for tech organizations and governments around the globe. There can be hacking of almost anything, from monitors to smart fridges, from thermostats devices to drug infusion pumps, cameras, and even the radio in a car. All of these are significant security issues that could be a result of IoT systems that will be made in future, as the increase in the number of IoT devices will lead us to set new nodes for each one of them, which then will be added to the network, where malicious attackers will look for opportunities to attack the system and create security holes.

FIGURE 6.6 Technologies involved in an IoT framework.

There is going to be a significant increase in security as the IoT is going to become a more deep-seated part of our lives. Apprehensions regarding the growth of IoT in the pharmaceutical industry is no longer restricted to mere speculations. Our lives and, most importantly, our health could be the target of IoT hacking attacks [4].

There is a number of causes behind the state of ambiguity in the IoT. This ambiguity is the result of the gold rush state in the industry, where all those involved wish to bring a pioneering linked gadget to the market before their competition. Under such conditions, the functionality of the gadget becomes the foremost factor, and safety takes a back seat [2].

6.1.3.3 Connectivity

With the increase in the numbers of devices, their connection with other devices is going to be a major task for the IoT sector, which will confront the assembly of current communication representations and fundamental technologies [5]. Currently, a centralized system is relied upon, where the server/client is required to endorse, approve, and link dissimilar nodes in a network.

The present prototypical idea is sufficient for existing IoT environments, where millions of devices are involved. However, when this network grows to join more devices, centralized systems will fail to function efficiently. The formation of better systems will entail massive funds and spending in preserving cloud servers that are able to handle such huge quantities of information interchange, and the entire structure could fail if the server turns to an unattainable system.

The forthcoming IoT will depend on decentralizing IoT networks. Decentralization of the IoT is conceivable by directing some responsibilities to the edge of the system, by means of fog computing representations where devices such as hubs take care of operation-critical maneuvers and cloud servers work on data congregation and logical accountabilities [5]. Additional options include the use of peer-to-peer communication, where procedures can be recognized and confirmed to each other in the right way and information can be interchanged without the contribution of a third-party agent. Every network that is build is basically a mesh creation and no single plugs so that the created network is not a failure. This prototypical idea has its own issues but these can be met with a combination of blockchain and the IoT [9].

6.1.3.4 Compatibility and longevity

The IoT is focusing its growth in varied areas and many different technologies are challenging to become the standard, which will lead to complexities and will require the placement of additional hardware and software while connecting the devices with each other.

Other problems include nonunified cloud services, the nonexistence of standardized machine to machine (M2M) protocols, and assortments in firmware and operating systems in IoT devices. Devices and technologies with time need advancements, once the system becomes outdated translation of amendments are needed so that the implementation of the system can result in success. This is particularly significant, as computing devices have a lifetime of a few years, while IoT applications (such as smart AC or TVs) tend to have a much longer service life.

6.1.3.5 Standards

Technology standards include:

1. network protocols;
2. communication protocols; and
3. data accumulation standards, which are basically activities that include handling, processing, and storing data that are collected through sensors [6].

The accumulation raises the value of data by growing, scale, scope, and frequency of data present for enquiry [2]. The challenges that are faced include:

1. The standard for handling unstructured data: structured data are kept in relational databases and the queries are put through SQL. Unstructured data are deposited in diverse types of NoSQL databases without a standard querying method [2].
2. Corporations that are extremely enthusiastic about leveraging big-data trappings face an absence of talent to plan, perform, and uphold systems [2].

6.1.3.6 Business

Whenever there is a business idea related to the IoT some factors related to the market and E-commerce should be kept in mind. The providers that are working in industries and providing services through cloud analytics will lead to greater monetization of a large portion of the IoT. Not only this, but some IoT-based ventures may also attract a modest revenue in the market.

While considering the challenges and implications in this sector, division of the IoT business sector can be done:

1. Commercial IoT: this embraces belongings similar to device chasers, associated medical devices, etc.
2. Consumer IoT: this category includes connected devices like smart cars, laptops, smartphones, etc.
3. Industrial IoT: this includes things like manufacturing robots, gauges, and other types of industrial devices.

6.1.3.7 Society

Societal aspects of the IoT differ for customers and regulators, who judge the services and help to create and widen the system, however it is not easy to meet all requirements because customers can change their demands regularly and may wish to have further reintegrated features and capabilities in their devices, for which a lot of time and resources are involved. The IoT is expanding and influencing society in different ways, creating a requirement to educate consumers about its proper use (Fig. 6.7).

6.2 What the Internet of Things means for pharmaceutical industries

Organizations striving for innovation do not have room to lag behind in software, artificial intelligence and machine learning, which are influencing and benefitting the pharmaceutical industry. The IoT is also providing the pharmaceutical industry with a range of innovative prospects that include improved regulation of drug development, preservation of equipment, and enhanced supply chain administration.

IoT sensors and trackers incorporated into an IoT system can easily enable improved circumstances for managing biomaterials and chemicals, ensuring high-quality equipment and also reducing drug scams, in what has become known as the Fourth Industrial Revolution. The earlier three were fundamentally the innovation of the combustion engine and automation, whereas the new era will be pushed through by digitalization. Progressively, our daily devices or equipment are being made "smart" with the incorporation of embedded technology.

The pharmaceutical industry recognizes that the shift cannot be an immediate one, thus they are developing progressively, for example, patients and costumers are being harmonized with smart devices and their feedback is utilised for future developments. A health tracker or board that is maintained

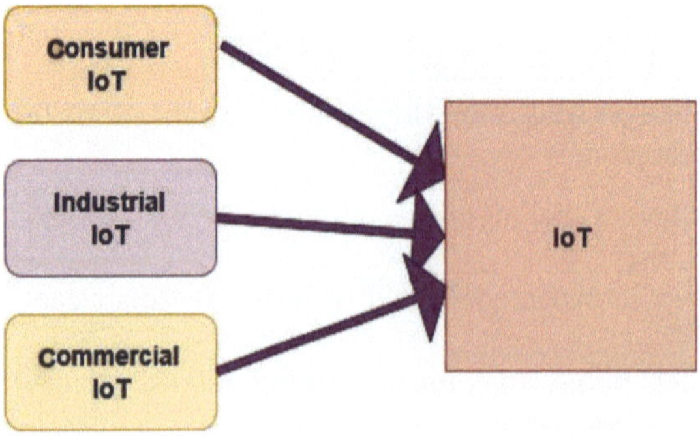

FIGURE 6.7 IoT application scenarios.

through these smart devices stores important health information of individuals, allows the healthcare workers and pharmaceutical corporations to identify how the manufactured products are performing, which makes the individual response more valuable and result in better efficiency of the system. This also helps in making better predictive decisions for patients and those within the supply chain.

Basically, it is the consumer that drives and shapes the supply chain of the system. Conventionally, in the pharma business, the approval of innovative industrial technology has been largely conventional for innumerable reasons, from supervisory to financial, which is why the future of the pharmaceutical industry will lead to the development of innovative and effective changes with the IoT. The IoT is a reliable technology for these systems, with great efficacy and good capacity to meet market demands.

6.2.1 Addressing challenges in the pharmaceutical industry

Digitalization is primarily driving health care toward modern changes. The pharmaceutical industry has always been an essential part of health care. Innovative technologies and modernization are enabling pharmaceutical corporations to improve medications and patient care [10]. Not only does the industry want to grow, but healthcare customers are also demanding better and efficient medications to deal with their health problems and to improve their quality of life. This development was necessary as the modern demands could not be met by traditional means [10].

We know that the pharma life science today faces unparalleled challenges, from goal selection to drug detection, clinical growth through engineering, supply chain, sales, and advertising. Pharma executives face many pressures, as there is less time and high productivity pressure with massive competition among the products manufactured; regulation and compliance are complex; emerging markets are uncertain; where there is more value-based pricing some drugs are still not covered by all areas and have strict action controls; drug discovery and development face many strict action controls; drug discovery and development can take many twists and turns resulting in a convoluted process; therefore whether the aim is to achieve efficacy or primary endpoint goals, clinical development and area risks can lead to significant financial pressures.

At the same time, new innovations and developments are forcing advancements in the industry and with this new challenges emerge, particularly in the global manufacturing and supply chain, where there should be a balance achieved between service and risk. Companies either produce their own products or use contracts for their manufacture, and packaging distribution models are always kept in mind. The healthcare sector is enormous but it is also changing rapidly at the same time. The entrance of new technologies is used to seize emerging opportunities. Customer groups are being created, drug selection decisions are shifting to centralized bodies, and technologies are being deployed. The pharma companies are utilizing multichain marketing and digital initiatives to gain access to markets in nontraditional ways.

The evolution of pharmaceutical engineering has reduced from a few years. Professionals advise that the market [11] has reached its saturation point. However, due to current technical developments, it is expected to be a part of further developments in the industry. Creations like wearable electronics, telemedicine, artificial intelligence, nanotechnology, and exactitude medicine are anticipated to alter the backdrop to pharmaceutical businesses. There remain some major challenges to the pharmaceutical industry that need to be addressed, as described in the following subsections.

6.2.1.1 Industry growth slowdown

Several reports have emphasized that the pharma industry will experience a slowdown in growth in the upcoming years because the major markets are reaching their saturation points. Traditionally, the pharmaceutical industry has experienced a periodic upsurge in development rates when generic medicines have focused on diseases like cancer and hepatitis [11]. Consequently, it can be predicted that the growth forecasts of the pharma industry are presently dependent on the development of new medicines.

6.2.1.2 Biosimilar showdown

Biosimilar medications are able to increase the efficiency of drugs and patient protection cost-effectively. Nevertheless, people involved in the pharma trade are reluctant to adopt biosimilars as it affects their inventive capabilities. [11].

6.2.1.3 Changes in payment modes

The healthcare industry is drifting from volume-based payments to value-based payments, which increases the value of the services provided and will improve the drug efficiency at the same time, making it more difficult for upcoming drugs to succeed in the market [11].

6.2.2 Adoption of the Internet of Things in the pharmaceutical industry

Pharmacy corporations are implementing IoT technologies in their engineering plants to attain optimization and for increased productivity. The techniques that were used earlier were basically used for management of data, which was a laborious task and was extremely vulnerable to human errors. Dissimilar data arrangements were used in engineering plants for diverse developments, which made data entry and its understanding a barrier to active communication. The technology used enables regulation with the ability to join equipment, webs, and organizations, so that the sharing of information can be easily done across a plant.

Pharmaceutical corporations use the IoT in their complex systems so that they can acquire access to immediate data and openness of processing, that is, from manufacture to delivery. One examples of this adoption is Johnson & Johnson, who adopted the IoT to acquire Food and Drug Administration (FDA) authorization for moving from Prezista, an HIV medicine, that is, moving from batch production to continuous manufacturing. The incorporation of sensor knowledge embraced by J&J [12] permitted them to eradicate the requirement for distinct challenging and sampling stages in the process. All of these separate processes were linked through the IoT and also ensured data reliability.

According to the survey, it was calculated that only 30% of the best 20 pharmaceutical firms have accepted the IoT in their work development to a certain level. Pfizer [12] is implementing the IoT with a fusion of supplementary technologies aimed at an integrated manufacturing course. IoT practices were used by Merck [12] to examine biologic progression data which led to an efficacious innovation of a problematic situation faced in a plant, which had very high rejection rates. An infrastructure in the IoT permits segmental mechanization of engineering plants, which can help in easy scaling up of pharmaceutical manufacture according to market demands. Correspondingly, the IoT is used to recognize blockages in the numerous operations that take place in a plant, in addition the IoT is also used for specifying whether any equipment is being overused or underused, resulting in increased operative efficacy.

Healthcare production is no exception, as revolutions in medicines, medical devices, patient nursing, and much more are quickly evolving. Efficacy and accuracy are increased through the use of smart technology in the healthcare industry.

6.2.2.1 *Patient monitoring*

Advancements in health care have been made through the IoT. It has provided the industry with the capability of developing a more accurate monitoring system for patients who are under surveillance and for those have already undergone a checkup. All of this can be easily done by *remote patient monitoring* which allows medical experts to observe and keep track of vital signs and evaluate bodily responses to cures given to the individual without being present at the patient's location. The devices that are used are completely dependent on the situation in which patient is, which can be anything from a cardiac device, an airflow screen, an associated glucose display, etc. The device gathers data and instantaneously these are sent to a databank for collation, and then sent to the clinician/surgeon for concurrent analysis, enabling the required action to be undertaken if anything is found to be disrupting the ideal level of the system. This is universally used for patients who have undergone surgery and can decrease hospital stopovers and the probability of readmission, as these complications can be examined and amended beforehand. Real-time data can also significantly assist medical specialists to further rapidly regulate and provide treatment opportunities to patients based on their individual requirements.

6.2.2.2 Medicine

An innovative invention in healthcare production in relation to smart technology has been the arrival of "smart pills." These smart pills [13] are consumed like systemic medicines, the only difference being that they are fortified with a method of monitoring that permits them to easily transmit data to a sensor placed on the body. The sensors placed on the body effectively monitor the patient's body in relation to their diagnostic condition.

The data collected from the sensors that are worn on the body are then transferred to a mobile app so that patients are also aware of their vital signs during and after treatment. Not only patients but also their specialist has access to daily data updates to measure the effect of the drug on the patient healing to indicate if any changes are needed to achieve the intended goal. This is also very helpful in recognizing whether or not the medication prescribed has any side effects.

In November 2017, a smart pill named "Abilify MyCite" was FDA approved and, its system is capable of recording the time of medicine consumption as well as other associated health records [13]. When the pill comes into contact with stomach acid and reacts, the reaction triggers a sensor which processes the information and then sends it to a mobile application. This will improve dosing prescriptions and add huge value to the knowledge of medical experts, helping them to understand patient responses individually. There is another pill known as the "robotic pill" which works after being inserted inside the body. Many companies are working on the development of pills that can navigate through the body, perform functions, and work in curing or fighting diseases or symptoms. Rani developed a pill, directed through the stomach which progresses to the small intestine where it transports an inoculation without revealing the confined material to peptic enzymes, and so protecting it. Once the inoculation is achieved, the remaining pill is broken down and assimilated. This alternative mechanism is beneficial for large medication molecules such as proteins, peptides, and antibodies [13].

6.2.2.3 Medical devices

Wearable equipment is interesting for patients of all ages as it can easily record and keep track of the vital signs of the body. The extreme popularity of Fitbit and Apple watches in the fitness market are widely acknowledged; these technologies are evolving not just in being able to record data but by performing a variety of functions based on commands or the apparent state. Smart bandages [13] are fortified with sensors that can evaluate the size of the wound beneath, so that it can keep track of how well it is healing, if there is an infection, and highlight necessary tasks if there is an infection.

Associated contact lenses have developed as a formula of wearable healthcare devices. Around 2014, large companies like Google and Novartis [13] initiated their plans to generate a *connected lens* which had the capability to examine blood glucose level through tears. The data collected by these lenses would then be transmitted to an insulin pump and notify the patient about their stage, if it was hazardous, and any desired modifications. This advanced observation system for diabetes patients could lead to life-changing actions for those unhappy to continue drawing blood numerous times a day. The novel application offers enhanced optimism to those suffering from diabetes [13].

6.2.2.4 *Medical facilities*

Although the main concern of the healthcare sector is to focus on how to benefit patients by providing the best treatment, at improving medical facilities is also an important factor. Some of the major goals for medical facilities are to develop more effective processes, preserve critical resources, etc. By mounting smart technology in hospitals and care amenities, medical specialists are able to observe the permanency and efficiency of technologies such as MRI machines, X-ray machines, CT and PT scanners before any conceivable faults or inappropriate use occur [13]. By using isolated sensors in equipment, amenities can minimize and probably eradicate the need for physical checks.

A recurrent problem in the medical profession is losing tools or possessions due to mutual practice. This can be a sever inconvenience in case of a crisis when tools cannot be situated. The use of Bluetooth low-energy position technology enables immediate locating of tools. This relatively minor invention could save many lives and is an inconsequential cost when compared with the related reimbursements.

6.2.3 Integrating Internet of Things solutions

Manufacturing mechanization and mechanism machinery are well-known in contemporary life sciences engineering. They empower the organizations in data creation and assessment by maintaining the quality of products, enabling industrial corporations to progress in overall equipment effectiveness (OEE) [14] and efficiency. Hence, the establishment of the industrial IoT has created a novel trend of input to the industry.

The benefit of the industrial IoT in a system like this is still unclear for corporations because they have not identified these in their applications and use cases so that it can be trialed as a proof of content [14] and then be scaled up to maximize the benefits. There are excellent prospects for all industrial corporations to engage in the industrial IoT journey. The distinctive initial idea centers on situation observing, prognostic preservation, and the most important aspect, analytics, however the value will be different for each company, generating prospects for innovative occupational representations and amenity returns.

Nonetheless, this will be possible until the connection of existing plants across varied platforms is made and, subsequently, the information is either sent to a cloud-based system like Azure or is used onsite. Once this first step is completed, analysis of the collected data is carried, and then the data are employed in such a way that various business benefits can be gained in a cost-effective manner.

The growing use of sensors related to manufacturing tools has enabled the industrialists to gain real-time data about on-floor processes and this has provided them with more dependable information and trustworthy data. Despite the positives, incorporation of the industrial IoT into engineering services does not come without drawbacks. For the IoT to flourish in pharmacy engineering, industrialized equipment from a diversity of producers must be able to attach to an extraordinary level of mechanization arrangements or procedures that can check connect to data banks or cloud facilities. Typically, this is achieved through associated control policies, such as programmable logic controllers (PLC). Regardless of the acceptance of mechanization over the past few years, many legacy systems that are in use are becoming obsolete.

There are certain prevailing structures that may interconnect to nearby systems without incorporation of the technologies such as networking or unsupportable communications conventions. An additional recurring problematic condition is that of big data. For the IoT-empowered services, the constantly accumulating data can lead to a dilemma. A system with increased accuracy and detailed production eliminates all doubt by proving how effective the system works. In situations like this, the evident solution would be to use cloud computing. All the data that are collected from industrial equipment [14] are sensitive and hence should be secured from illegal viewing. Corporations can focus more on keeping their critical information on their premises [14] and send limited data to a cloud, with the data only holding value when the two fragments are joined. This amalgamation solution meets the basic needs of creating less vulnerability to cyberattacks and maintaining connectivity.

6.3 Role of the Internet of Things

Product quality, compliance to regulations, and efficient operations are essential for pharmaceutical companies. Data and knowledge are required to plan investments in different areas responsible for production and supply, such as the shop-floor, production lines, inventory, and warehouse management systems. The IoT is a powerful grid of sensors and interconnected devices, that allows objects—from heavy pieces of machinery equipment to mobile and other pocket devices—to connect and share data with each other on a master system through the Internet. This allows outstanding levels of security and management by optimizing critical processes, including manufacturing and supply chain management. Establishment of the IoT has disrupted the decades-old model of pharma manufacturing. Modern consumers require an efficient and rapid response to their healthcare needs, thus it has become imperative for pharmaceutical companies to ensure secure regulations in manufacturing and transit of drugs and services. Along with maintaining speedy operations, well-managed organization at each level of the production is required. This includes, but is not limited to, detecting and analyzing faults during manufacture, to observing the supply chain to ensure minimum lag between different operations, and coherence in the manufacturing process.

Driving the shift from age-old practices, the IoT is being increasingly adopted within the infrastructure. Owing to this, pharmaceutical companies have been successful in collecting and utilizing data to improve their interdepartmental coordination. As a result, companies are able to oversee, manage, and control processes in real time and to their benefit, even remotely. This has greatly helped in the organization and assignments of workers also, causing a chain reaction in improving efficiency and management, and thus driving higher revenues for companies by minimizing losses and operational costs (Fig. 6.8).

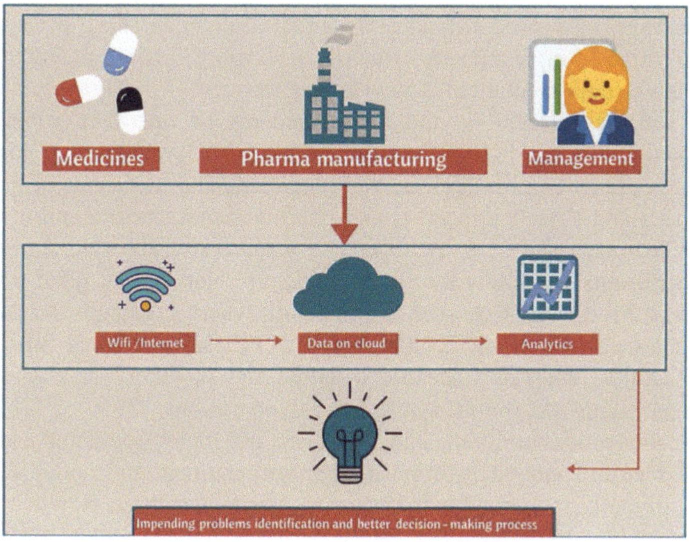

FIGURE 6.8 Application of IoT technologies in Pharma Manufacturing.

The connected technology has been implemented to different verticals, including planning, design, manufacturing, and delivery and transit. With the help of improved data collection capabilities and technologies, companies are able to minimize wastage and redundant operations. Along with production, the IoT is also changing the avenues of operations and logistics. Smooth and efficient logistics are challenging for any industry, and even more so in the pharma industry due to the sensitivity of the cargo and maintaining their inventory conditions, given the specific demands of some products. By retaining low-energy sensors, pharmaceutical corporations are able to monitor the conditions of goods in real time. IoT sensors can be used to monitor and control equipment. Vaccinations and medicines require specific conditions for their storage, and these conditions vary with the product.

6.3.1 Pharma manufacturing

Pharmaceutical research and development is viewed as one of the world's most sophisticated industries. Manufacturing and further processing of drugs, and then getting them onto the shelves involves a highly different area of technology, that is not as glamorous but is critical to avoid setbacks and failures. The goal of a pharma company is to produce quality drugs and make them accessible, all the while cutting costs and minimizing production time. These goals are met by introducing certain changes to the industry's operations. Thus, in the last several years, focus has been shifted toward increasing the quality of products and reducing production plant sizes and resource utilization [15].

A study by McKinsey suggested that leading pharmaceutical companies are on average 53% more efficient than small-scale companies. Much of this can be credited to higher levels of investment, more flexible manufacturing, and effective implementation of developing technologies. Thus, due to high demand for such flexibility and high standards of operations, there is a requirement for equally proficient data acquisition systems, and logistical and environmental control devices. With hundreds and thousands of independent devices and objects utilized in any pharma manufacturing unit or supply chain, it becomes crucial to establish a framework which will unify these independent units and allow for a communication network. A good IoT platform can be used to provide such a framework where it gathers and analyzes operational data to provide detailed insight. This also helps in introducing several features, such as traceability within the inventory and production floor, streamlining validation, and inspecting operations.

Given that production hubs and R&D labs are highly regulated and operated under strict guidelines from national and international authorities, the shift to a digitally integrated manufacturing process has been slower in adoption and yet the IoT and other Internet-enabled devices and technologies have transformed the industry. Drug companies are now exploring the possibilities of linking manufacturing units with their digitally enhanced labs. Manufacturing data are collected for research purposes and are used to validate performances of previously established parameters and protocols, allowing companies to study their operations in a real-time scenario and improvise. Furthermore, data collected from multiple points in an organization and levels of the production lines, testing units, and transit points, etc. can be used for the analysis of process automation. Eventually, the automation data can be modeled after segregation and used on cloud-based systems for real-time operations.

Communication and information-sharing networks between different units and dedicated labs have become an integral system to drive the digital revolution in all industries. Cloud computing and modeling engines, coupled with learning algorithms, are already being implemented in pharmaceutical labs to aid researchers and plant managers in bringing breakthroughs in R&D and manufacturing of drugs. In designing such information-sharing networks it is essential to design storage and analytical systems to handle huge amounts of data to effectively support plant automation systems.

6.3.2 The Internet of Things in supply chain management

A typical supply chain for a pharmaceutical industry may start right from the acquisition of raw material to the end operations, such as product dispensing and shipments [16]. It is necessary for the company to take into account multiple factors and risks while formulating their production strategy. Thus it becomes highly crucial to have an efficient supply chain system which encompasses all the key parts of production. Some of the most basic risks and factors include the life of perishable raw goods and uncertainty in the supply of raw materials. Thus, along with maintaining an effective flow of information and goods, it also becomes necessary to maintain and optimize the inventory of raw and processed materials.

In recent years, the introduction of IoT devices or sensors has affected the supply chain management facilities to a greater extent. RFIDs and sensor-enabled tags need not be scanned as barcodes allow for the transmission of data in real time. As the manufacturing processes are highly complex and data requirements are immense for the proper functioning of any plant, the benefits of real-time data transfer are essential to the supply chain. A zero error margin is expected of any responsible company while ensuring quality production, on-time transit, and smooth operations, all of which requires the aid of IoT devices to optimize manufacturing and operations to ensure productive asset allocation and utilization for these companies.

Thus IoT devices and applications for supply chain management have been a major investment for many industries and pharmaceuticals. The IoT connects equipment, helps in labor and material locating, intelligent packaging, and monitors cold chains. Internet-based solutions are helping pharma manufacturers to remotely observe cold chain ecosystems in real time by using sensors with tracking devices in warehouses, transit vehicles, and/or in shipments. Using Internet-enabled devices in drug inventories and management is already saving companies billions of dollars per year, and has allowed for better allocation of available manpower. This is particularly pertinent in businesses and regions where fake drugs commonly penetrate the value chain. Two-dimensional barcodes, RFID tags, and smart labels have made it possible to locate and track each transfer or movement in any supply chain, along with performing quality checks at each step. This has allowed companies to gather information about each step in the process, which results in a complete digital footprint and mapping.

IoT technologies have enabled placing temperature-sensing tags within the inventory and shipments, which has allowed for constant monitoring without disturbing the product. Tags can continually record temperatures and other environmental conditions which might be unsuitable for a human operator. These data are uploaded to the cloud and recordings are almost instantaneously available in the control room. For pharma companies, the IoT has extended visibility and control in virtually every area—from ideation and development to manufacturing, transit, management, and dispensing. The real-time information obtained when coupled with modern analytics tools can become the foundation for making rapid and more accurate decisions. Owing to its past successes and relative development in the field, an IoT-dominant ecosystem is sure to enhance business processes, notably in the supply chain. The IoT is the remedy for the risks that have been traditionally ingrained in the supply chain.

The IoT helps in improving the security of the supply chain and thus making it extra safe by promoting a bidirectional interaction among the seekers and the equipment, for example, in the tracking of inventory availability and the current situation of any container in transportation. The IoT also helps in tracking the movement within a drug inventory and thus saves stakeholders millions of dollars as it ensures genuine products are being utilized in the production, while identifying and deterring counterfeit products. It is important to ensure that the storage temperatures remain regulated as prescribed for the stored drugs both in inventory and in transit to guarantee that they remain stable and acceptable until they arrive at their destination. Another application of IoT-enabled packaging is that it constantly traces environmental conditions in cold chains during transport. This reduces wastage of medicines due to an increase in temperature in the vessels, assures acquiescence with regulatory requirements, and maintains the condition of products.

6.3.3 The Internet of Things in inventory and product management

A majority of manufacturers and industrialists rely on traditional bookmarking and recordkeeping methods, which are not only labor intensive but also are prone to human error. In practice, any inventory or warehouse relies on skilled managers and a workforce for replenishing, recording, storage, and distribution. However, as the inventory size increases and the items to store become more complex the management process can become highly disorganized and/or inefficient. In contrast, IoT-based inventory administration and asset trailing feature proposals for continuous perceptibility into the warehouse by collecting information procured by sensors and tags. Progress can be monitored at different sections of the storage [17].

Consequently, manufacturers can manage the on-hand inventory, improve allocation timings to machines and equipment, reduce lead time, and thus avoid unnecessary costs. In any inventory management operations that are grounded on the industrial IoT, an inventory item can be allocated an exclusive identification numbers that are comprised of encoded digital figures about the element. Tag readers and scanners can excerpt these data upon scanning and then spread it to the cloud for processing. In this process, along with ID numbers, the seeker also receives data about the time and location of the scanning, which adds to the security of the management system as every item and movement associated with it can be traced. Accordingly, based on these data, the system also locates the item position and shows real-time updates to the end solution users, allowing for constant surveillance at any point in time with any handheld device connected to the cloud.

Thus, the role of the industrial IoT in inventory management can be described as to collect, process, and further deploy the data extracted by tags and sensors into acumens roughly on inventory elements' condition, position, and engagements. For example, based on the availability of certain type of drugs in the inventory and referencing it with the current demand in the market, a system can forecast a cycle or period within which replenishment is required and can even notify the manufacturer to plan the purchase of raw materials. Thus the output provided by an IoT-based service can be of any form: it might be a tracking signal, data packet, or a message to the owner alerting the status of goods.

This provides IoT-based systems with an edge over traditional spreadsheets or tabulation-based software. However, like any other progressive technology, IoT-based solutions can be very efficiently utilized with existing frameworks to enhance the functionality of a system without requiring a completely new setup, thus cutting costs on expensive investments. This also allows for flexibility in merging an IoT-based system of one department with another without requiring any major refurbishments. As the an inventory worth can be a major part of an organization's possessions, inventory records are essential for the accounting areas to ensure that the corporation's finances are being managed properly.

The benefits of an IoT-based inventory management system are discussed below.

6.3.3.1 *Automation in processes of tracking and storage reports*

With RFID and other sensors, managers do not need to rely on manual trackers and invest in cross-checking strategies. The IoT extends beyond the limitations of the efficiency of human operators with regards to high numbers of stored items or large storage areas. Automated tracking procedures can save a lot of time and human labor.

6.3.3.2 *Visibility in operations*

Manufacturers can obtain the real-life location, the status of ongoing work operations, and association of the things, so that inventory supervisors can communicate strategies to improve resource, time, or labor allocations.

6.3.3.3 *Stock management*

The more data on availability, life, and demand of the current stock that are available the easier it is for managers to manage their inventory. With quality data and devices available to manage stock items, manufacturers can organize their end of productions and managers can optimize their stock distribution plans, while maintaining quality by monitoring the expirations and conditions of held items.

6.3.3.4 *Identification of bottlenecks and assessing possible solutions*

With data and information on each level of production and management, it becomes crucial to identify choke points and work toward improving these processes to guarantee effective operations. With real-time knowledge, any problems can be immediately identified and dealt with.

6.3.4 The Internet of Things in transforming the pharma industry

Digitization has led to improved connectivity of products, people, and processes pertinent to this industry. The industrial adoption of IoT services has led to major changes in the way this industry operates, plans, and delivers. Industrial IoT technologies have given manufacturers the ability to view the most intricate details within their management with little to no effort. This essentially means that the required data are made available at the tap of a screen. With such capabilities and many new emerging technologies, it has become possible to shape a local manufacturing hub into a large-scale production facility. This change has been easy as the pharmaceutical industry has always been among the forerunners in adopting new digital technologies to meet the constantly changing demands of the sector with reduced effort and the aim of increasing profits. Thus, within the industry, there has always been promptness in controlling manufacturing processes more effectively.

However, being one of the most regulated industries it demands a thorough understanding of manufacturing processes and strict compliance to rigorous standards that regulate these processes. Similar to any other industry, pharmaceutical manufacturing faces many challenges, including the need to improve product quality, process efficiencies, and reduce cycle times and wastage. Other motives include tracking of products and managing the supply chain, as discussed previously. With the benefits of industrial IoT and its rapid adoption, the dynamics of the industry are changing considerably. Constant connectivity ensures the inspection of manufacturing operations right from the factory floor. Digitization has made many such initiatives possible, but one of the many requirements for manufacturers will be to consistently upgrade their digital systems with advancements in Industry 4.0. Inside factories, machines are being connected and controlled by online systems such as Supervisory Control and Data Acquisition (SCADA). The industrial IoT allows management of such systems and to utilize the power of analysis to provide insights in overseeing plant workflows while maintaining the working of machines. One of the major advantages provided by the industrial IoT has been proper documentation and record maintaining services, simultaneously Industry 4.0 has driven a more paperless methodology.

Nowadays, the operations of an average production facility require terabytes of data to be assessed and maintained over the cloud, which can be very difficult to do with traditional methods. This becomes possible with production companies considering important decisions such as phasing out legacy systems and remodeling business paradigms. It can be characterized that the industrial IoT is at its nascent stages of growth, but its influence on pharmaceutical manufacturing is absolutely undeniable. Steadily its benefits are being recognized by many participants and researchers dedicated to the industry. It will not be a stretch to assert that IoT devices will become a standard for all vertices of the industry irrespective of their scale of operations.

6.4 Internet of Things applications in health care and medicine

Health care and medicine affect millions of lives daily. Managing IT systems and health devices is essential and sometimes even critical in saving peoples' lives. Medical institutions have hundreds of accessories and equipment that constantly monitor the activities within their premises, including the conditions of patients. Needless to say, it is critical to collect and store the data from these devices for patient records and further assessment. Thus reliable services become highly necessary. Another important objective of the monitoring process is to prevent any losses to the medical institutions' development and help doctors and practitioners to formalize progress for their patients. For example, the purpose of monitoring is to recognize whether the projected goals have been reached as defined in an action plan or strategy.

Accordingly, it is also necessary to monitor gaps and errors that might be present in the systems and correct them in a timely fashion, which is essential to improve services. It should be noted that manufacturing and industry represent only one end of the sector. In this age of technology and innovation, the demand for personalized and effective health care and medicine services has driven reform across the entire sphere of the industry.

The IoT's role in shaping Pharma Industry 4.0 has been immense, and in the same manner, its contributions to improving personalized health care are also remarkable. It can be safely stated that no other technology will create an influence as widespread as the IoT for the healthcare sector and pharmaceuticals as a whole. The global healthcare market is set to exceed $10 billion by 2024; this growing market also drives innovation and development—including precision medicines and monitoring devices. Advancements in private assistance technology, AI-based devices, and medical robotics are also driving lower healthcare costs. Techniques such as AI-based medical imaging, and prognosis/diagnosis are estimated to grow at more than 40%, exceeding US$2.5 billion by 2024 [18]. As a reference, the US healthcare and artificial intelligence market was valued at US$320 million at the end of 2016 and is forecast to grow by more than 38% compound annual growth rate (CAGR) in the near future. According to a few estimates, drug discovery applications make up to 35% of the healthcare market share. Key clinical health AI and machine learning applications could conceivably create $150 billion in annual savings for the US healthcare economy by 2026 and revenue in the AI health market is expected to reach $6.6 billion by 2021.

One of the world's largest markets for health care and pharmaceuticals, the United States, invests 18% of its GDP on healthcare annually. According to some reports, effective development of IoT-based healthcare transportation technologies could save over US$300 billion. Just as an example, before the IoT, doctor–patient interactions were limited to visits or telecommunication, within which constant monitoring and updates were limited. However, IoT-enabled equipment has made distant monitoring possible and easier for physicians to assess situations and make healthcare suggestions. It has also made patient satisfaction trends more positive and helped hospitals and staff to monitor and ideate upon their services. Furthermore, remote monitoring helps in reduced length of stay and lowers the readmission rate. Ultimately, it has created opportunities to lower the costs of operation and services provided by institutions.

The IoT has had an impact on healthcare applications that not only benefits patients and their families but also physicians, hospitals, and associated institutions.

Patients: With the influx of wearable technology and devices enabled with health monitors it has now become possible to fit a heart rate monitor, stress analyzer, calorie measurement, or fitness routine tracking facility to be incorporated inside a wristband or watch. Not only are these devices user-friendly, reliable, and affordable, they are also easily available due to market trends. Within the last decade, the rise in competition among major players has driven innovation, and wearable technology is only becoming more popular.

One of the major advantages of these devices is in their intuition and ability to work in the background without major energy consumption or bulky structures. This has also provided relief to elderly patients and people with mobility or communication-impairing diseases as the devices are also capable of sending emergency alerts and provide real-time data and updates to doctors.

Physicians: With the help of wearables and other on-body monitoring equipment embedded with the IoT, tracking vitals has become incredibly easy. Physicians can track their patients' adherence to the prescribed treatment strategy and when needed can immediately provide medical attention.

Hospitals: IoT devices, tags, and sensors are being increasingly used for tracking timely locations of drugs and medical equipment like tools, wheelchairs, oxygen pumps, and other necessary supplies. It is also possible to track the deployment of medical staff, a facility which can greatly improves assignments and coordination.

Infections and environmental monitoring are critical to any medical institution. IoT-enabled hygiene-monitoring devices help in timely sanitization and recording data about patients' conditions and their surroundings. IoT devices also help in logistics such as asset management, in pharmacies, or in managing inventories. The proliferation of IoT-enabled devices has unarguably redefined the healthcare and medicine industries by assuring better care, refined treatment outcomes, and lower costs for patients, along with better workflows, and improved techniques for accumulation and assessment of data.

An IoT system is generally a four-stage architecture wherein all levels are interconnected in a way that allows free flow of data from one stage to the other, with different stages handling different aspects of the process. A typical representation of this architecture is as follows.

Step 1: The first step is to collect data, for this several units of IoT-enabled devices and sensors are deployed.

Step 2: The data collected from these sensors are then converted into digital form and transferred to the next layers after processing.

Step 3: The data once aggregated and processed are moved to the cloud or another storage platform.

Step 4: Advanced analytics and solutions are applied to the data that are collected to drive the decision-making process.

The process as a whole allows for a comprehensive structure that is advantageous to any application with minimal setup or maintenance cost. One major advantage of digitization has been the regular updates that can be downloaded to the device without making any significant changes to the hardware for a very long time. This has allowed for lower replacement rates of physical devices, as now any update that occurs is carried through the device software or the Internet.

Other major advantages of the IoT in health care include:

1. Data allows physicians to make informed decisions and improve their strategies.
2. Constant monitoring and updates allow for preemptive measures as it becomes possible to identify and treat symptoms in the initial stages.
3. Management of tools, equipment, and drugs is a critical aspect of any institution.
4. The IoT makes the process easier as it helps in real-time monitoring and control of the environment and systems.
5. Overall, the IoT helps in error reduction and minimization of resource waste.

It should also be noted that even though the IoT in health care has revolutionized the sector, it is not without its concerns. The availability and free flow of data also open horizons for data theft and misuse. More specifically, many governments and organizations are skeptical of the processes involved with data management in healthcare institutions, as patient data and information are treated as a highly sensitive entity. Commercialization and the influx of private companies in this sector have further aggravated the situation. Many cases have been highlighted where insurance companies and private medical institutions have exploited the acquired information to the cost of their patients.

Health care is such a vast system, that both the challenges and potential of the IoT remain greatly unexplored and are seemingly endless. Currently, it is limited to remote monitoring and personal healthcare and medical devices. The IoT could be the key to not only keeping patients safe and healthy but also in reforming the way an institution or individual provides medical services [19].

6.5 Pharmaceutical data acquisition and management with the Internet of Things

Addressing the concerns which are presented up until now in this literature are only one step toward an IoT-led pharmaceutical and healthcare revolution. The success of any venture in this sector lies in realizing the demands and managing the resources, all the while maintaining a balance between life sciences and the operational aspects of the industry. Every day new technologies are changing the perception of health care, in both needs from the sector and its deliverables. Modern technology has made possible the development of bioimplants, wearable robotic equipment, digestible pills with microcameras and tools, etc. And the IoT has made the control and application of all such devices more plausible. Mobile apps have become common which can be used to monitor daily fitness trends of individuals. The availability of data has become crucial for the growth of the industry in both operational as well as academic aspects.

The success of the IoT has driven increased investments and research within this domain. Increasing numbers of companies and industrialists are pouring in resources to develop improved systems and technologies. Even small-scale ventures are exploiting the benefits of utilizing IoT frameworks. The IoT has become a catalyst in driving cutting-edge innovations that can impact the whole business and market irrespective of its scale. Its benefits interestingly incorporate already-existing devices, Internet frameworks and systems, and expand the horizon of machine-to-machine communications.

An IoT framework becomes a system of physical devices and different gadgets, programs, sensors, and many more things. The framework allows for effortless transfer of data among each of the elements. The example of Gartner can be considered to estimate the sheer scale of the influence of the IoT in data networks. Gartner is a leading research and advisory firm which in one of its studies anticipated that by 2020 more than 25 billion devices would be within the IoT framework. And with such a widespread network, the possibilities of devising newer methods for accumulation and use of data become nearly endless. Accordingly, with the proliferation of devices collecting health data, and the existing mechanisms of cloud computing a severe risk factor is introduced in the process as no framework can be guaranteed as 100% secure. Thus privacy and security of data have become one of the critical challenges to the IoT, especially in the healthcare industry and pharmaceuticals. The threat to security is not only from random actors but also from competitors of any institution, as well as from hackers.

Before the age of the IoT, any typical process required offline collection and management of data [20], which limited the access to data as these systems could be protected with local firewalls. However, since local systems were not part of a cloud or Internet-based service, the chances of any remote hacking were also minimized. With IoT, however, the number of devices with access to information has risen enormously. Any device connected to the grid is a gateway to potential hacking. The rise in availability of hackable medical instruments has pressured many national as well as international authorities to issue strict regulations regarding vulnerabilities to cyberattacks, remote communication, and hacking. It is very possible for any hacker to tamper with devices to steal information or modify its functions.

Thus "medijacking" or hijacking incidents related to medical institutions, information, and devices is an issue which should also be addressed. A report by TrapX stated that most of institutions are vulnerable to medijacking. The report also provided details about three major incidents that occurred in prominent hospitals in the United States. The severity of the situation can be understood from the fact that the hackers used malware to malfunction several breath analyzers in a hospital. Further, stolen sensitive information and passwords from the institute's database were tracked to Eastern Europe. At another hospital, hackers based in China infiltrated the radiology department systems to gather sensitive medical information and images.

An interesting solution to security and privacy concerns has been the creation of electronic health records. This is considered as the most far-reaching application of information and the IoT in medical and healthcare services. Every patient or person has their own record which incorporates their socioeconomics, medical history, information on their likes and dislikes, medical tests reports, etc. Often these records are required to be shared between different institutions or physicians, which can sometimes be a concerning procedure. As each record contains potentially sensitive information it becomes highly necessary to secure these documents from modification or leaks. Electronic health records can help in addressing these concerns by creating a system of tracking and notifications on any action that is performed with the digital documents. It should be noted that several governments including the US have steadily adopted electronic health records. The European Union also plans to introduce the EHR system in accordance with its 2020 mandate.

6.6 A case study by McKinsey & Company: how data are changing pharma operations

6.6.1 Introduction

Aircraft today can be designed and tested in only a digital environment. They can be designed on CAD software, and simulated and validated for real scenarios only based on the available data and knowledge. These models can be redesigned, individual machine parts replaced, and tested for multiple situations.

Science fiction? Yes and no. The technologies and software that are available today bring once perceived fiction to life, and this all becomes more possible with advanced techniques such as robotic process automation, artificial intelligence and the IoT—but their full potential is still unexplored in many manufacturing industries, including pharmaceuticals. Some companies are leading the industry with their earlier adoption of the IoT and advancements, but limited growth has resulted in limited results and a struggle to improve solutions beyond the pilot and ideation phases.

One question that arises is that, even with daily breakthroughs and technological progress, why is it so difficult to implement these new tools at a much larger scale? The answer lies in the current ideology of the industry. More than pragmatists it demands visionaries, and at the same time, organizations need to expand their horizon of expectations and risks while servicing expectations. Finally, they need to accept digitization and the IoT not just as tools or a means to an end but as the drivers of change in the industry and as a series of transformations over a gradual period of time.

Nevertheless, it should also be noted that, in recent years, life sciences and pharmaceuticals has been one of the earlier adopters of the IoT despite being restricted with stricter standards and regulations in comparison to other industries. As per a report by Mind Commerce Publishing, IoT in health care was forecast to become a $117 billion industry by 2020. Accordingly, it also mentions that medicine and health care make up the largest component of the entire IoT business, with a 15.1% market share.

As mentioned earlier in this chapter, the IoT is a network of thousands of interconnected devices that can share data with each other and allow for seamless control from a remote location. IoT application in the pharmaceuticals and healthcare sectors includes assessing the OEE, improving research and development, improving efficiencies, reducing lead time in manufacturing, and cutting down costs and wastage by reducing errors.

IoT facilities are updating the industry on an exponential scale. IoT devices allow the transfer of data and unrestricted communication between different sections of a production facility in manufacturing hubs, between doctors and patients in hospitals and managers with every aspect of their supply chain. The outcome of such a process is the generation of terabytes of data which can be further utilized by hospitals or research and development facilities for the purpose of studying insights about their operations.

6.6.1.1 The lessons from lean manufacturing

Lean manufacturing is a concept aimed at the minimization of waste. There is a lesson from lean manufacturing. This concept came into place after its inception in Toyota as a part of its Toyota Production System plan. The procedure helped Toyota to become the market leaders in the late 1990s with their improved manufacturing and production facilities and reduced generated wastage. This prompted other companies to imitate the idea of lean manufacturing also, which ultimately many of them failed at as their early efforts fell short. Their failure can be attributed to some small misunderstandings in regards to the concept, that is they tried to install individual systems and tools without really implementing any required organizational processes.

This plan was compelled to fail. In fact, the efforts of those organizations were reported as more disastrous than exerting no initiatives at all. *The Machine That Changed the World* describes the age as the "Lean-production systems on existing mass-production practices causes elevated distress and confusion." It took many years for some of these companies to realize that any new technology or innovation does not simply yield if installed just as it is. Rather, it requires informed planning to integrate it with existing infrastructure and investment in rewiring the entire system to adapt to the upgrades.

Similarly, the advantages of the IoT and AI will only be realized steadily with time and only if companies are ready to invest in them smartly and if they do not utilize them as individual fancy tools. Currently, a majority of the pharma companies are still testing their hypotheses with these tools and are limited to their pilot stages or experimentation. Eventually, if not progressed in a sustainable manner, this can also lead to more harm than good as results will be limited and investments become more threatening. It is essential for companies to take advantage of the current digital revolution and bring in more systematic changes.

6.6.1.2 *Why the future is different*

A comprehensive change is required in the industry as the proliferation of the IoT and other technologies will only create an upward shift in the sector, with a focus on the below-mentioned five aspects. Ultimately, these shifts will lead to a quantum change in pharma manufacturing and significant improvements in processes.

1. *Advanced product and process mastery*: As a start, companies can collect data and information about their machines, working trends, operations, and processes involved in manufacturing. They can weigh in various factors that affect the risk factors involved, quality, or outcomes of the process. Methodically they can map the inputs with their respective outputs, produce analytical reports, and with the utilization of historical data and information suggest the future course of action and better action plans.
2. *Real-time predictive analytics*: Production managers can use predictive analysis tools to better plan their routines. With options to assess real-time demand, production, and constraints they can better forecast working cycles, mitigate supply risks, and assign resources adequately.
3. *E2E digital twin simulation and optimization*: Companies and R&D centers can also develop and simulate their production processes, starting from individual machines to the entire factory floor, and test their units for different sets of simulations. Pharma executives can leverage data, study the effect of different external and internal indicators, test different team allocations, and optimize production parameters without any physical modeling or investing in heavy machinery beforehand. This way companies can save millions of dollars from investing in highly complex machines and performing physical tests over a larger period of time compared to comparatively inexpensive computer simulations within smaller samples of times.

   Operators can employ real-time AI-based tools to go beyond traditional rules like "frozen periods," rhythm wheels, or minimum order requirements. All this helps in saving time and cuts the need for inventory buffers and allow companies to run smoother operations.

4. *Automation of knowledge work and robotics*: One of the major advantages of the IoT is the emergence of tools and technologies like digital robots, automation tools, and self-learning algorithms. These tools are highly efficient in collecting and processing information, optimization, or corrective and preventive actions. In the case of a product deviation, language processing can be utilized to study and convert data into a Pareto diagram or other type of visualization. These solutions are found to cut run times by as much as 90% in some cases.
5. *Digital operations assistance*: One of the major hindrances to success in the pharma industry is human error. Statistically, human actions are proven to be up to only 92% accurate. Various unavoidable factors such as fatigue, poor health, or lack of concentration can affect the outcome of any job. Sometimes even a small mistake or error can lead to a massive loss to a company. This is especially true for processes which are noncreative or repetitive in nature. Automation tools thus provide a suitable alternative in this scenario as they cut down the need for human intervention by either taking over the tasks or by working as assisting devices.

6.6.1.3 The business case

Supply chains have been distinguished by lengthened lead times (up to 1 year for any median pharmaceutical) and expensive inventories (250 days). Although companies manage to keep customer service levels at upwards of 99%, they still struggle in balancing their supply and demands.

Therefore, what is the business case for digital? This suggests that any initial estimate made in terms of the growth technology is only an estimate. However, McKinsey research shows that companies that are aggressively boosting their tech departments can show earning before interests and taxes (EBIT) growth by 3.2% and revenue growth by 2.3 percentage points per annum [21]. This impact is generated from improved production times and operations within the production time. By leveraging better connectivity, reduced response times, and data from multiple sources, plant managers can better plan their production lines.

6.6.1.4 Three horizons to generate scale

Smart companies investing in IoT frameworks start with smaller setups within individual units and then scale up across the organization. Generally, the process is rolled out in three horizons, with a focus on the increase in productivity and profits as the scope of the project diversifies.

In the introductory phase, that is the experiencing horizon, the transformation is about driving initiatives that are highly impactful but restricted in their application, typically these initiatives are directed at a precise unit or stage of the process with the intention of generating momentum or laying a foundation. It is worth noting that with digitization, companies can report increased efficiency even at the initial stages. For instance, based on modern digital factory transformation case examples, a finding by McKinsey group observed the following statistics:

1. decreased variations of up to 80%;
2. increased lab productivity by up to 60%;
3. decreased enclosures due to variations of up to 80%;
4. increased OEE of more than 40% on packaging lines; and
5. decreased changeover times by more than 30%.

The second step is the exploring horizon, in which the company utilizes their previous experience to expand the framework. In this phase, companies work on "lighthouse" projects which are studied to display possibilities with the implemented technologies, its future, and working across the units. The scope for this covers the entire spectrum of the company and plays an important role in creating a unique approach for the company. Achievements in this horizon include prescriptive analytics, creating standards for data utilization across the units, and digital assistants to site operators. It is unwise for companies to push for value creation in this phase. Typically they can expect an additional 50% rise in terms of efficiency, quality, and flexibility metrics.

The final step is the envisioning horizon, in which the company introduces digitization across the whole value supply chain. This is possible with the help of predictive analytics models that can drive optimization processes. At this point, it is expected that the data are easily accessible and transparent, and secured with end-to-end encryption. Many tasks such as testing and verification are now simulated and advanced digital assistance tools are operational throughout the production floor (including both the supplier and customer ends). At the envisioning horizon, institutions can register as much as a 100% increase across their operational metrics.

6.6.1.5 *Key success factors*

The main reason any company fails in their digitization efforts is that they become overly dependent on their newly acquired tools and technologies without any improvisation or changes in their internal organizational structures to adapt with the changes. The following section presents a few areas where companies should focus their attention.

6.6.1.6 *Invest in capabilities and adjust the organization*

According to some surveys, up to 60% of tasks in a manufacturing unit can be automated. However, that does not indicate that 60% of the jobs in the unit will be replaced by machines, it simply means that there will be a shift in working requirements [21]. For example, some jobs performed by unskilled laborers can be replaced with devices, and resources can be redirected to training engineers, data scientists, and skilled personnel. This has caused fierce competition among companies to hire the best people in the field from various universities, introducing special training programs, certifications, and even capability-building programs. Internally, companies are driving cross-departmental teams by retracing organizational lines. Ultimately, job descriptions and requirements of the companies are rapidly changing or rather adapting to the growing digitization.

6.6.1.7 Apply strong change management

Even in today's age, launching a complete digital transformation at an industrial scale can be an ambitious task. Thus companies also need to take into consideration that half-planned or incomplete measures will only prove detrimental to growth. Any transformation strategy needs to encompass the entire value chain and management structure. Moreover, the alteration will disproportionately influence the intermediate layers and white-collar operators in expert roles. Thus, unless there is a vision to provide support and facilities to the different levels of workers within the company any such plan may stall.

This can be understood better with the following perspective: workers spent years acquiring and training for a specific skill set and then further years at a position within a company. Digital tools and technologies such as AI and the IoT provide a reliable replacement to one or even multiple positions. Keeping this in mind companies should focus on retraining or reallocating their workforce to complement the digitization efforts.

6.6.1.8 Rethink your approach to IT infrastructure development

Preferably than conceiving of IT projects as singular tasks, organizations should divert their support to realizing a connected framework that provides for secure and rapid passage of data (for example, through a data lake). The trick to connecting more data in less time? Start with business needs. Once a data lake has been established, it only creates the 2%–5% of connections that truly matter for the initial generation of use cases. Transformations typically require 20–25 use cases before they create ample momentum. This approach of a mindset shift is called the agile development of IT infrastructure and it is possible for it in some cases to not mesh with the existing IT framework that many companies already have in place. Unless clear hierarchy and goals are in place, this mismanagement can cause further delays to the system.

6.6.1.9 Break the budget cycle mindset

Over the past decade, pharma developments have concentrated on overcoming waste through measures like lean and six sigma. These efforts have led to exceptional reformations for companies and have introduced significant profits for these companies, but at the same time they have limited the companies to a brief-term schedule. Digitalization carries enormous potential, but it may not fulfill gains on an anticipated schedule that can be constituted into the budget cycle. Successes will require explicit ventures in IT, industrialization, and additional technological solutions. Upper management needs to assist their teams in differentiating within "budget tightening" plans and the resources demanded to get to a digitally facilitated prospective position. For this reason, many companies have started with separate teams to focus on the implementation of digitization efforts.

6.7 Conclusions

Digital manufacturing, IT, and the IoT are no longer a far-flung vision of a distant future but the reality of the pharma industry. Leading companies are strategizing in roadmaps and developing roadmaps to improve adaption of the IT infrastructure. What is more, they are remodeling their organizations to fully espouse innovation, and structure them to play in a future that will seem very unconventional from now [21].

6.8 Future directions

Everybody, from customers to businessmen, is implementing the variations that are carried by the revolution of the IoT. It has transformed the sphere in more methods than we could envision only a few years ago back. And the variations and progressions will remain in the future also—the IoT is our future shaper. Already, billions of sensors are associated with billions of devices and are redefining almost everything that comes under them. It is projected that approximately 75 billion maneuvers will be related by 2025, making the communication of items an easy task.

It has been estimated that US$6 trillion will be paid out on IoT solutions in the coming 5 years. The IoT has by revealed a huge perspective in very short time by proving benefits for all possible available advancements. Not a single industry has failed to be impacted by the IoT. From retail corporations to heavy businesses, everyone has invested heavily in IoT as they comprehend its importance in data-driven analytics and its subsequent effect on customer experience. Customers, on the other hand, are basically enjoying the services and experiences they have gained from this technology as it has influenced their life by not only making things easy but also by helping them keep track of almost everything possible.

The IoT is redesigning health care and pharmaceuticals. The use of wearable technology is in high demand as it can simplify the monitoring process. The sensors incorporated in the system are responsible for data extraction and can help doctors to analyze data efficiently. The IoT is going to drastically change this sector by making it easy for patients and doctors to interact. The technology will help to reduce the chances of death due to carelessness or misunderstanding of the medicinal effects on the body. Doctors will be able to prescribe the correct amount and the right type of medicine to the patient depending on their need. Not only patients but also pregnant women and newborns can be examined to correctly calculate the nutrients and supplements that will be required for them to maintain health. This revolution in the healthcare sector will change the overall situation, leading to a better future with healthier people, creating more prosperous societies.

References

[1] M. Lianos, M. Douglas, Dangerization and the end of deviance: the institutional environment, Br. J. Criminol. 40 (2000) 261–278. Available from: https://doi.org/10.1093/bjc/40.2.261.

[2] <https://www.pharma-iq.com/manufacturing/articles/the-role-of-iot-in-pharma-manufacturing-and-distribution>.

[3] M. Zahran, Smart grid technology, vision management and control, WSEAS Trans. Syst. 12 (1) (2013).

[4] <http://www.microwavejournal.com/articles/27690-addressing-the-challenges-facing-iot-adoption>.

[5] <https://blog.apnic.net/2015/10/20/5-challenges-of-the-internet-of-things/>.

[6] A. Al-Fuqaha, M. Guizani, M. Mohammadi, M. Aledhari, M. Ayyash, Internet of Things: a survey on enabling technologies, protocols, and applications, IEEE Commun. Surv. Tutor 17 (4) (2015) 2347–2376. fourth quarter.

[7] <https://www.linkedin.com/pulse/iot-implementation-challenges-ahmed-banafa?trk = mp-author-card>.

[8] <https://www.sitepoint.com/4-major-technical-challenges-facing-iot-developers/>.

[9] <http://iot.ieee.org/newsletter/january-2017/iot-and-blockchain-convergence-benefits-and-challenges.html>.

[10] <https://www.hgp-pharma-consulting.com/blog-en/internet-of-things-and-pharmaceutic>.

[11] <https://www.infinitiresearch.com/thoughts/pharmaceutical-industry-challenges>.

[12] <https://www.gep.com/mind/blog/adoption-internet-things-pharma-manufacturing>.

[13] <https://www.globalsign.com/en/blog/healthcare-of-the-future-rapid-adoption-of-iot-reshaping-healthcare-industry/>.

[14] <https://www.electronicspecifier.com/medical/industrial-iot-in-the-pharmaceutical-industry>.

[15] <http://www.pharmtech.com/internet-things-pharmaceutical-manufacturing?pageID = 2>.

[16] <http://pharmaceuticalcommerce.com/supply-chain-logistics/pharmaceutical-cold-chain-logistics-is-a12-6-billion-global-industry/>.

[17] International Society for Automation, ISA99 industrial automation and control systems security, Available from: <https://www.isa.org/isa99/> (accessed 12.03.16).

[18] <https://www.virtueinsight.com/pharma/2nd-Annual-Pharma-AI--IoT-2019/>.

[19] G. Moelgaard, The biggest challenges in a new pharma reality, NNE Pharmaplan, 2015, Available from: <http://www.nnepharmaplan.com/insights/on-the-move/the-biggest-challenges-in-a-new-pharma-reality/> (accessed 12.05.16).

[20] <https://www.spendedge.com/blogs/efficiency-pharma-manufacturing-iot>.

[21] <https://www.mckinsey.com/business-functions/operations/our-insights/how-data-is-changing-the-pharma-operations-world>.

Further reading

<https://www.linkedin.com/pulse/why-iot-needs-fog-computing-ahmed-banafa?trk = mp-author-card>.

<https://blogs.sas.com/content/hiddeninsights/2019/07/08/iot-use-cases-in-pharma-span-the-spectrum/>.

<https://www.42gears.com/white-papers/how-internet-of-things-is-impacting-life-sciences-health-care-industry/>.

<https://amiko.io/2019/05/06/amikos-digital-prescription-intervention-service-boosts-adherence-to-asthma-therapy-and-health-outcomes/>.

<https://www.naed.org/NAEDDocs/Research/Legal%20Issues/FreightWatch%202013%20Global%20Cargo%20Theft%20Threat%20Assesment%20Full_0.pdf>.

Industrial Internet Consortium, Deliverable, 2015, Available at <http://www.iiconsortium.org/IICDeliverables-May-2015.pdf> (accessed 12.03.16).

<https://www.peerbits.com/blog/internet-of-things-healthcare-applications-benefits-and-challenges.html>.

<https://www.alliedmarketresearch.com/iot-healthcare-market>.

Accenture Technology Review, Driving Unconventional Growth Through the Industrial Internet of Things, 2015, Available from: <https://www.accenture.com/us-en/acnmedia/Accenture/next-gen/reassembling-industry/pdf/Accenture-Driving-Unconventional-Growth-through-IIoT.pdf> (accessed 20.03.16).

<https://www.cognizant.com/whitepapers/the-internet-of-things-the-new-rx-for-pharmaceuticals-manufacturing-and-supply-chains-codex2437.pdf>.

<http://www.idc.com/downloads/idc_market_in_a_minute_iot_infographic.pdf>.

<http://www.bioethics.ac.uk/news/Intelligent-Pills-A-new-dawn-in-healthcare-.php>.

<https://tulip.co/resources/pharma40/>.

<http://www.mckinsey.com/~/media/mckinsey/dotcom/client_service/operations/pdfs/lean_and_mean-how_does_your_supply_chain_shape_up.ashx>.

Chapter 7

Internet of Things: from hype to reality

Arun Kumar Singh[1], Neda Firoz[2], Ashish Tripathi[1], K.K. Singh[3], Pushpa Choudhary[1] and Prem Chand Vashist[1]

[1]Department of Information Technology, G. L. Bajaj Institute of Technology and Management, Greater Noida, India, [2]Departments of Computational Biology and Bioinformatics, Sam Higginbottom University of Agriculture, Technology and Sciences, Allahabad, India, [3]Department of ECE, KIET Group of Institutions, Ghaziabad, India

Most of us will have heard the famous quote from Shakespeare:

> *Our bad neighbor makes us early stirrers, which is both healthful and good husbandry*
>
> William Shakespeare (1564–1616), King Henry, in Henry V, act 4, sc. 1, l. 6–7.

> *Breach and fraud are always possible;*
>
> Securing them is our prime concern in IoT.

The era of the Internet of Things (IoT) is sweeping over and replacing the Internet creating a world where smart things exist connected to each other intelligently. This was predicted by Eric Emerson Schmidt, the former C.E.O. of Google over 20 years ago. The physical world is now connecting to the digital world so quickly with the emergence of the IoT that it seems the Internet will become invisible soon, meaning the physical world will be connecting to the digital world seamlessly. The world will enjoy smart

An Industrial IoT Approach for Pharmaceutical Industry Growth.
DOI: https://doi.org/10.1016/B978-0-12-821326-1.00007-3

connectivity in the same way that the city of Barcelona has emerged to be the smartest city in the world. We are moving toward system-to-system connection, with smart networking reaching its peak. The idea of software-defined autonomous machines is about to become hugely important, which will become ubiquitous. With the advent of the IoT, we explore how it is becoming a reality and whether it has any limits. Maciej Kranz in his book on the IoT explains the very essential detailed and inclusive idea of the IoT, with IoT expanding to businesses, and covering and impacting on a variety of technology areas. Artificial intelligence (AI) and machine learning have a huge scope because of the enormous data generated by sensors and devices connected through the IoT. We will explore in this chapter the hype around the IoT and the reality. We will also discover improved metrics in the IoT that is allowing it to be a leader in the technological world. We are witnessing the fourth revolution in the digitization world and discuss the reasons behind its exponential growth. The protocols that differentiate it and a new set of protocols have evolved. This also creates security concerns and data are described as the new oil, raising further challenges of data privacy.

7.1 Introduction to IoT

The IoT has gained a huge share in the IT industry and academia. The main cause of this rising interest is the hidden potential that can be explored and exploited in the future. It has the capability to create a comfortable brand new world where everything is linked over the Internet conversing and operating smartly. The ambition is to be able to see this efficient connectivity and smart interaction among objects at a reasonable cost, creating opportunities for a healthy environment, in which ordinary objects will have enhanced performance, such as:

- Allowing and influencing automatic sensing;
- Speedy analysis of innumerable service- or product-related constraints;
- Triggering action automatically before any service or its operations are hindered;
- The IoT embraces gathering and evaluating of gigantic amounts of structured and unstructured data from many in-house and outdoor sources, including social media;
- There is also a commitment to acquiring advantages by offering improved services and enhanced corporate opportunities.

These features and commitments sound bold, nevertheless one can contemplate the tremendous influence of the Internet in various fields including education, corporate business, climate control, science, government, communication, and human civilization as a whole. Most experts believe that the IoT has the power to create a leading opportunity for technology. The term "Internet of Things" was first used by Kevin Ashton in a demonstration at

Procter and Gamble in 1999. He linked the novel impression of radio frequency identification (RFID) in Procter and Gamble's supply chain to the then red hot topic of the Internet, in a way that gained executive attention. He went on to mention that "The Internet of Things is inclusive of valuable potential to change the world, just as the Internet did, maybe even more so." The MIT Auto-ID Center later presented their insights about IoT in 2001. Shortly after that, the IoT was officially announced and published about in the International Telecommunication Union (ITU) Internet Report in 2005 [1].

The use of the IoT is accelerating in the modern world of communication, with Internet advances in the form of smart devices that are capable of interconnecting with each other and that can cooperate with far off structures. However, the main ability of the IoT is its wide impact on our day-to-day lives. Health care is also improving, including health monitoring, pharmaceuticals development, drug designing, and drug suitability. The IoT is now described in detail.

7.1.1 Internet of Things

It is very important before discussing the noteworthy components of the IoT to give a definition. The IoT can be defined [2–4] as a part of a complex network of physical elements enabled by the following:

1. Object sensors: to gather information and data;
2. Identifiers: to classify the cradle of data (e.g., sensors and devices);
3. Software: to process and analyze the data, and
4. Internet connectivity: to communicate, alert, and advise.

The IoT can be defined as the network of smart things, having concise and strong component recognition and identification, rooted with the intelligence of computer software, sensors, and ubiquitous computing capability, and linking to the Internet.

The era of the IoT [1] has brought about innumerable possibilities for improved quality of life. The data are cheap, and streaming and access are feasible and cost-effective. The IoT era, as we may call it, offers wonderful opportunities. The subsequent efficiency, precision, and economical advantages have revolutionized its technical impact on the world. It enables Human's efforts to encompass inanimate objects in the communication world uniquely (Fig. 7.1).

The main opportunity is the connection of physically visible objects which can be anything/everything (e.g., devices, sensors, people, machines, animals, trees, tables, chairs, etc.) across the largest network of networks, the Internet. The goal is to monitor, control, and rule the inanimate world. The target is to utilize the artificial intellect of inanimate objects for personalized goals and advantages, by taking control, granting access, and enriching the services and intelligence of the physical objects around us.

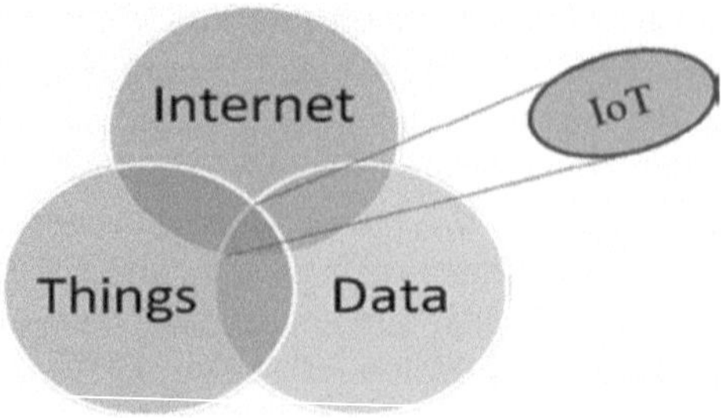

FIGURE 7.1 IoT Structure

In this chapter, we further explore the IoT framework, cloud computing, and its merger with virtualization and the IoT. We give a comparison and understanding of the IoT and the Internet. We describe a better means of communication in a network, which uses packet-switching technology. The seven basic layers of the open systems interconnection (OSI) model and how they merge with IoT layers for a new IoT model development are described. An understanding of IoT protocols and the layer stack are also illustrated. The application protocols and a survey of their feasibility with the IoT layer model stack are presented. There is also a discussion on the application interoperability of the IoT model.

The sensors and devices that are used in collecting the data are described. The importance of the IoT from the hype to a virtual reality mirror, tracing its importance and growing use in drug identification and monitoring in the medical world are discussed. The final point of the discussion in this chapter is on security concerns [5] in the IoT model when providing services to the customer.

7.2 Internet of Things framework

The IoT framework divides the IoT solutions into four main categories, which in this book are discussed in detail. The four levels [6] are:

1. IoT objects (smart devices);
2. IoT infrastructure and network for transport of data;
3. IoT platform providing services (e.g., software establishing the connection among the devices with applications and carrying out all the management); and
4. Applications of the IoT (particularly business-oriented applications, e.g., CRM).

Control is granted from one level to the lower levels, starting from the application level and proceeding to the device level, then continues to support the levels in the hierarchy. Sensors and actuators that make up the IoT objects are a part of the IoT device level. The application services channel of

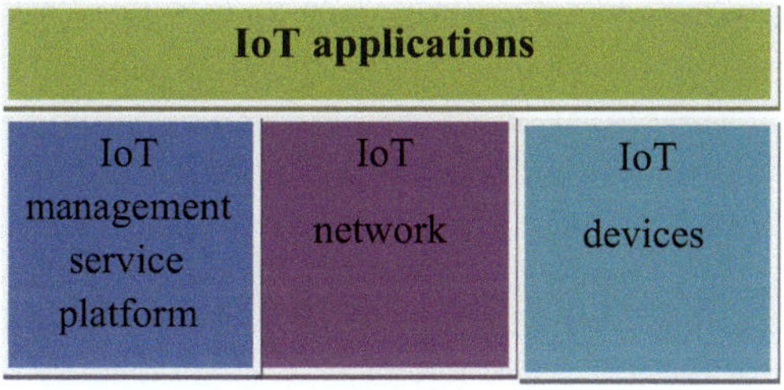

FIGURE 7.2 IoT levels of connections in the IoT.

the IoT consists of crucial management and smooth interaction of software functions to enable communication between the IoT devices and the network channels. This enables the main tasks of establishing the connection of the things and different levels of the network through the application layer. The IoT application level has all functions within the IoT network. Fig. 7.2 illustrates the layout of the different IoT levels, showing the way in which the information is moved from one IoT module to the next level.

The advantages of the IoT-based model are:

1. Complexity is abridged: The level breaks down the elements of the IoT and the overall communication into reduced and naive components, indirectly helping IoT components to develop prematurely and design, along with the troubleshooting process.
2. Consistency of the IoT components and modules: The model makes the specific components consistent and unravels each level and their modules between the various levels. This, in turn, allows various vendors to merge and join many solutions using the common models of support.
3. Unit engineering: The level breakdown allows a variety of IoT hardware and software components to link across the nodes.
4. Operability among nodes and levels: This increases interoperability between manufacturers by safeguarding the expertise of technical blocks that are capable of working together and operating in links.
5. Increases novelty: This allows designers to focus on finding solutions to the main problem. It helps to remove the worry of the basic roles of implementing ideas across various commercial verticals.
6. Ease of access in education: The complex IoT solution is broken down into smaller and more achievable components, which, in turn, eases the education and learning process.

7.2.1 Merging cloud computing: virtualization of the giants in the Internet of Things

Until the 21st century, businesses used to deploy and manage their own physical computing infrastructure. With the advent of cloud computing, businesses have expanded and the deployment of computing infrastructure has become cheaper, and is now accessible from anywhere in the world, and more safe and manageable. With its introduction in 2008, outsourcing fully or partially of infrastructure and reliability to businesses like Amazon and Google has increased. The service providers like AWS of Amazon, Microsoft's Azure, and Compute Engine of Google, have the largest share of the virtual infrastructure extension and their maintenance, and this share is expected to increase rapidly. They have reduced the cost by up to 99% when compared to the data from 2008 and 2018, by providing virtual platforms of computing and making them affordable even to individuals [6]. This decline illustrates our reliance on virtual platforms and virtualization of infrastructure and cloud computing, and hence forces us to adopt the smarter ways of computing without the overheads of physical infrastructures. The model has proved itself over the decade and suggests an increase in reliance on it. The cloud providers take the cloud services to the comfort of our homes, offices, businesses, etc., cheaply, and simplifying expansion like never before. Enterprises have to pay according to their requirements, including bandwidth consumption, storage capacities, CPU cycles, etc. Firms retain the choice of deploying personal cloud solutions at their personal data centers and can also carry on offering personalized computing services to any subusers or organizations. These models are flexible and accessible, while keeping the idea of personalized control, management, and monitoring credited to their personal IT sections.

Cloud computing is also flexible in offering hybrid cloud replicas, which contain public and private clouds that allow organizations creativity in scalable solutions though they are using the public cloud while keeping full control over critical data. Cloud computing is attractive to many enterprises in the context of drug identification; companies can self-provision services owned by them for any type of workload in drug identification in real time. Eventually, they grow and with minimum expertise they paying only for things that they use, that is, similar to a subscription charge.

Cloud computing is structured in a way that we can study is in three different categories:

1. Infrastructure as a Service (IaaS);
2. Platform as a Service (PaaS); and
3. Software as Service (SaaS) (Fig. 7.3).

The key points for cloud computing include:

1. Pay as you go;
2. Access from anywhere, anytime, any device;
3. Economy of scale;
4. Flexibility;
5. Expertly managed.

FIGURE 7.3 The three components of cloud computing.

The key idea is to lower energy consumption and improve resource performance, efficiency, and competence. Cloud computing has gained coverage through the initiation of virtualization technologies. Instead of different dedicated IT infrastructures like servers, storage capacities, hardware networking, they are placed into a single firm entity. Virtualization has opened the scope for cloud providers by breaking down one machine into multiple virtual machines, thus giving way to a remote server which may be correctly called a virtual memory storage machine and virtual network store, for each company. They all run over the same physical and distributed infrastructure, which can be shared without interfering and giving a seamless performance.

Virtual machines are actively participating in encouraging the virtualization process of services, with the IoT creating trends that help in running the OS on similar hardware and server devices. The containers are other means of virtualization similar to virtual machines. Here the virtualization layer executes on top of common kernel of the OS. The OS and its kernel run on the hardware with many separated guest groups placed on top. These remote guest processes are known as containers. Although they share the same kernel as the OS, they are separated from each other at the application level. They are designed to run in isolation for various applications.

These giant technologies are being merged to utilize their capacities for obtaining the best possible solutions for the IoT. The cloud computing and virtualization process has gained tremendous interest and scope over the years. It has enhanced users' experiences and improved services in terms of interfaces. The interaction between humans and machines or users' experiences can be thought of as the most crucial element of the IoT and its success. The core principle is to meet the desired results that are set for the product designed with minimal annoyance or stress. The basic IoT user's experience determinant is that it meets the desired requirements for its usage and installation without experiencing any problems. Too much effort in engineering products and overdesigning them may prove to be a waste of time.

The frustrating and slow retrieval of information at user interfaces can lead to customer abandonment. For example, a toaster should be able to make good toast. However, if we overload the design and input extra engineering with information overload, confusing the ordinary customer, then we risking the product's reliability for users. The customer may become frustrated and not want to use or recommend the product to others. These days we are reinforced by tremendous technologies for doing the same task which allows engineers to adapt policies toward users' experiences. Thus it is required to keep in mind the KISS principle (Keep It Short and Simple). The KISS principle states that if systems are designed and kept simple then they are expected to perform best [7]. The main user experience principles include the following:

1. *Naïve and ease of access principle*: The best user experience system is one which is simple including the interface for users. Simplicity is always

the first priority in any design. Removal of unnecessary complexity adds to this principle. One should make sure that cognitive workload whenever possible is reduced. Also, we need to make sure the user's interface is reliably stable and intuitive and establishes a clear hierarchy for visual reception.

2. *Principle of context*: This principle ensures that users are aware of the context and know their place or position within the system.
3. *Principle for humans*: It has to be ensured that the user interface provides human interactions and prioritizes them over machine-like interactions.
4. *Principle of participation*: This makes sure that users are fully engaged at the user interface, it delivers value by providing a strong sense of information.
5. *Principle of beauty and delight*: This makes the user experience enjoyable with elegance in providing services.

Thus we conclude that following the benefits of the virtualization process we need to use the power of the cloud and get the maximum output by merging these technologies on the same platform. The concepts of virtualization, cloud IoT, and their merger are emerging trends on which efforts should be focused toward the development of a cloud IoT virtual platform. This will boost the capacity of our IoT tools and sensors, with improved work patterns and sensed data, enabling analyzing data from anywhere in the world and many more innumerable benefits.

7.3 The Internet of Things and the Internet

7.3.1 Chronicles of the Internet

The main communication technology and networks were primarily based on circuit-switching methods long before the dawn of the Internet, using telephone lines where all nodes or stations were allocated end-to-end dedicated electronic connections for the fluent transfer of data [8]. Fig. 7.4 shows the main differences between these two giant technologies.

As technology advanced it was realized that circuit-switching methods were not very appropriate for the establishment of connections, called networks, among computers, as the connections were growing at a huge rate. This began with the use of electronic computers in the 1950s, where a new concept was introduced of packet switching rather than using the same circuit-switching concepts. Several versions of packet switching came about in the 1960s. The TCP/IP was developing and entered the market during the 1980s. With the advent of the Internet, the World Wide Web evolved after a decade (during the 1990s). The concept of the web of things evolved during these times and is a part of the IoT [9]. To illustrate the significance of these technologies, an example is given, where we have two computers, A and B. A is in New Delhi, and B is in New York connected by a circuit-switched

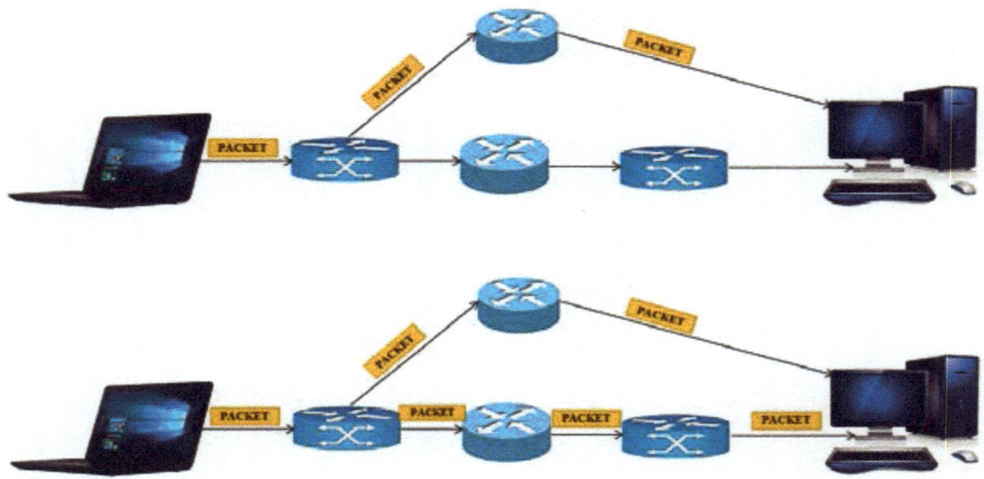

FIGURE 7.4 Circuit-switched versus packet-switched methods of communication.

network. The simplest method is to choose a naïve path connecting both computers that are located miles apart. The chosen path in a circuit-switched model, referring to a dedicated path connection between A and B only for the message exchange time. The problem with this method is that the line of connection stays tied up, regardless of the amount of message exchanged, which means no other device can use that path. This creates overheads and delays for other devices that may want to connect using the same path even if the bandwidth is free. The circuit-switching model creates an exchange of messages that is very bulky and not smooth or consistent [10]. The two computers in this type of network might exchange a file and then lose the connection after an exchange of files for some time, as others might be using the same line. The packet-switching technology was announced as a substitute to circuit-switching methods to make communication between computers easier. It was later reported in some papers that packet switching evolved during the Cold War. The main motivation for its design development was to design a network that was protected from nuclear attacks. Later it was found that such theories were false and it was factually denied by the Advanced Research Projects Agency Network [11], which was the first network that implemented the Internet protocol suite TCP/IP and a very early user of the packet-switching network technology. In the 1990s, the work on the Internet emphasized its robustness in enduring overloads and overcoming losses of any kind of Internet networking.

In order to understand the fundamental principles of packet switching let us consider an example. Let us assume we want to send a goods container from place A to place B (in this case New Delhi to Allahabad). Instead of sending the whole container using one route, we divide them into different subpackets and send them using different routes. Packets are then accumulated and sent in a specific fashion so that they can be:

- Numbered and reunited and assembled back in the right sequence when reaching the endpoint;
- The packets should contain both the addresses of the sender and receiver at each point;
- They are transferred over a network of routes which make way for it to travel easily;
- Then these packets are sent forward across the decided routes, however they may change the routes as required according to traffic updates. Alternate routes can always be chosen so as to deliver the packet as quickly as possible.

Packet switching is a simple terminology used in computer networks and communication. It is not a specified protocol. The Internet protocol is TCP/IP, which was founded by Robert Kahn and Vint Cerf. They founded a really well organized protocol that we still use today. The IP protocol is used for logical addressing and TCP is used for data delivery at the destination.

Therefore we can say that the Internet protocol obtains the physical address and TCP helps to deliver the data to its destination.

We can appreciate that both these technologies helped to build the foundation stone of the modern-day Internet. The earliest days in computer networks developed ideas that allowed general conversation amid computer users; built by J. C. R. Licklider of Bolt, Beranek, and Newman (BBN), in April 1963 [12], while examining the notion of the "intergalactic computer network," in a memoranda. These wise concepts built the modern-day Internet into a form that has enabled discussion of the IoT today. Licklider was appointed the head of the Behavioral Sciences and Command and Control programs at the Defense Department's Advanced Research Projects Agency (ARPA) in October 1963. He persuaded Ivan Sutherland and Bob Taylor about the network concept and its importance and the development possibilities, but later Licklider left the organization [13,14] without any new assignments for the development of this project. The devices that use the internet should be able to utilize the IP stack. The packets that use the internet protocol specified lists are termed as datagrams. These datagrams are composed of header and data information. To understand the concept of header and data separately, take an analogy of a letter that goes into the post office in an envelope. Information such as the entire size of the packet, source and destination address, IP address [15], the time to stay, and error-correcting information reside in the header.

1. The packets are autonomous and do not depend on the type of hardware. The way in which they can be ported from one hardware type to another is through encapsulation into different types of frames. Now the trick is that these frames are understood by the hardware individually because of the frame format which delivers the packet to the destination.
2. TCP is the acronym for transmission control protocol. The TCP protocol helps in collecting the packets in the exact order, while also keeping checks for any missing ones. If packets are found to be missing, new ones are requested by the TCP endpoints which asks for them to be resent. It also keeps a check on duplicate packets. The TCP endpoint allows the session to connect the computers over a network. It is fascinating to see the TCP and IP protocols perform their respective jobs accurately together.
3. A very crucial facet of packet switching is enabled by the forward and return addresses labeled in them. The address for a computer is very different from the physical addresses that we use. Computers only deal with binary information. Therefore we have the IP version 4 based on a 32-bit IP address and IP version 6 based on a 128-bit IP address.

The most multifaceted responsibilities in large-scale networks in the Internet are IoT-OSI, TCP/IP, IPv4, and IPv6, including the routing table which need to be reliable and efficient for seamless communication.

Almost all data networks that are used these days rely upon the OSI model of communication [16]. This OSI model came into being via the International Organization for Standardization (ISO), in 1984, in order to resolve networking and communication establishment problems. ISO is one of the global federations for organizing national standards organizations and represents around 100 countries. This model is intended to elaborate, explain, and standardize critical communication functions that may be used by any telecommunication or computing system. It takes critical care of the underlying internal design and structure, the technology in use so as to implement it on every platform of digital communication which is available. The main aim is interoperability [17] of many communication systems using one standard or set of protocols. This OSI model is conceptual and defines the various components so that they can communicate in data-based networks. This utilizes the idea of "divide and conquer," so that they can virtually break down the present network communication commitments into small functions that are called layers. Thus they are much easier to learn and develop [7]. The well-defined standard interfaces among layers help the OSI model to back up the concept of modular engineering methods and multiple device interoperability.

7.3.2 The open systems interconnection model

The OSI model is important in IoT model operation. There are seven layers to the OSI model and these are discussed briefly here [18]. It consists of seven layers, namely:

1. Physical layer
2. Data link layer
3. Network layer
4. Transport layer
5. Session layer
6. Presentation layer
7. Application layer

Each stratum or layer is accountable for providing some kind of distinct and structured services to the associated end-to-end layer either upwards or downwards in the stack. While the division differentiates each layer with its separate function, they create the flow of data without disconnecting from each other by working independently and simultaneously (Fig. 7.5).

We now discuss each of the layers and their individual capacities:

1. *OSI layer 7*: This layer is known as the "application layer," and starts at the top. It provides abstraction to the user level of the intrinsic details of all the layers below it. It also specifies protocols for sharing and the interface methods in use by the hosts within a network. This is a place

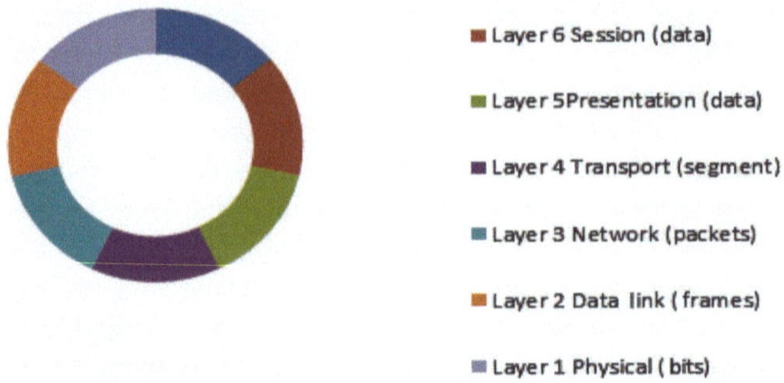

FIGURE 7.5 OSI layers description with data format.

where a user interacts with the network utilizing advanced level protocols [19]. The DNS, HTTP, Telnet, SSH, FTP, SMTP, RDP (remote desktop protocol), etc. are used at this layer.

2. *OSI layer 6*: This layer is known as the presentation layer, and it lies beneath the application layer. In this layer are OS services like Linux, Unix, Mac, Windows, etc. The presentation layer is committed to delivering and formatting of information in two steps toward the application layer, which may or may not perform additional processing. It takes care of any issues that may arise as a consequence of sending data from one node to another. The presentation layer removes the concern of the application layer in dealing with syntactical differences to represent data residing in end-user systems [20]. Examples include the translation of an EBCDIC (extended binary coded decimal interchange code)-coded file in the form of text to an ASCII (American Standard Code for Information Interchange)-coded file, while also adding some encryption for security, such as secure sockets layer (SSL) protocol.
3. *OSI layer 5*: Session layer is beneath the presentation layer in the OSI model. It has the task of dealing with the communications in order to establish a session among the two network nodes [21]. Examples include the establishment of a session between a computer and the server.
4. *OSI layer 4*: This layer is the transport layer. It helps to establish an end-ways communication through the endpoints. This utilizes the notion of windowing, in which one is responsible for the decision of the amount of information flow or the data to be sent at one point in time connecting the two nodes in a network [22].
5. *OSI layer 3*: The layer three is the network layer. The routers operate at this level. This layer helps to put the data into packets which we may call IP datagrams. They have IP address information of the source and destination address which is transmitted to the hosts and over the

network. They also help in routing of IP datagrams that are marked with IP addresses. There is a specific routing protocol which enables routers to communicate with each other. The transfer of information among two nodes is enabled by this, and the routing algorithms help to specify the routes. Every router contains the prior information of the networks to which they are attached. Then there is a routing protocol, which helps to share the coming information to the immediate neighbors and then to the other ends within the network. The routers learn this way around the existing topology of the network they are working in. Though layers 3 and 4, that is, the network and transport layers, are different, they are closely knit together in practice. The name of the Internet protocol TCP/IP comes from the transport layer protocol (TCP) and network layer protocol (IP) [2–4]. Here the host enables the message transfer in the network by placing over it the data to be sent, marked with a destination address in the hope that it will be sent to the specified location. The packets containing messages could arrive in an altered order, and not as specified during the transfer. The next task of the higher layers that exist at the destination point is to rearrange the order of the packets and then supply them to the applications stratum that features applications functioning.

6. *OSI layer 2*: This stratum is known as the data link layer. It helps switches to work on the data link layer. This layer decides when to deal with the delivery of frames amid devices that are placed on the same LAN and uses media access control addresses. The frames are not allowed or they do not have the function to cross the borders of a local network. The internetwork routing is controlled by layer 3. This helps to permit the data link protocols to emphasize their attention on local carriage, giving the address scheme and media intercession. Therefore the data link layer is comparable to a local neighborhood traffic cop, which does the arbitration amid parties struggling for entry through the medium, without the input of the final destination. For instance, the data link protocols include the Ethernet used in multinode form in LAN and the point-to-point protocol [23].
7. *OSI layer 1*: This is the basic stratum of the OSI model and is called the physical layer. This physical layer describes the physical world interface with the virtual world using an electrical or mechanical interface for establishing the physical medium. This contains the idea of basic networking hardware transmission technologies. The wiring and cabling are part of this layer. This layer is committed to outlining the patterns of diffusing raw bits over a physical link that binds together network nodes, copper wires, fiberoptic cables, radio links, etc. The physical layer governs stream of bits and their placement from the data link layer on to the pins for a USB printer interface, an optical fiber transmitter, or a radio carrier, etc. [24].

7.3.3 Internet of Things networking protocols and services

There is a strict need to devise and implement separate IoT networking protocols [17]. The success of the Internet has been attributed partially to the Internet protocol. These offer basically two kinds of services:

1. A layer is known as the normalization layer or the IP layer, which assures the application entities work independently and are simultaneously interoperable. This feature allows it to adapt to the assembly of data link layer expertise. Moreover, with a surfeit of application protocols, they are the most successful. The IP layer maintains its standard of surviving and competing in the market.
2. Layered abstractions: The layered application programmatic interfaces (APIs) are abstracted by the abstraction protocols form the complexities of each other. Such abstractions are now very well defined and helpful in enabling seamless operation, and they define contracts or "slip surfaces" which in turn allows novelties to develop in any layer autonomously.

With the growing research, development, and new technological advances, scientists have increasingly engaged in the IoT world. IoT has proved itself to be simple in engaging with it and straightforward such that it can defend itself during crisis situations and carry on the layered approach benefits to next levels of achievements.

Nevertheless, there has been a great deal of debate over the new IP stack, to ensure the suitability of the stack matched the requirements. In the late 20th century and early 21st century, scientists who were in the domain of wireless sensor networks critically denounced IP networking as inappropriate for the application [7]. Progress in the IP stack was made in the last two decades, which has addressed the quoted prerequisites for sensor networks. The limitations of IP technologies of older times were also addressed.

7.3.4 A pyramid view of Internet of Things protocol stack

We can visualize the IoT protocol stack as the expansion of the TCP/IP protocol, which is also a layered model. It contains the layers as illustrated in Fig. 7.6.

The noteworthy application layer which is a part of the TCP/IP protocol stack is elaborated in two variants of the IoT protocol pyramid. We present the protocol functions and their details in Fig. 7.7 [25].

The technology gaps remains, although the IP protocol stack for the IoT is continuously striving to meet the API standards, and providing the common application services platform for the IoT [26]. In the context of the searching and detection powers, the support of the IoT application services stratum aims to deliver back up support services for the following:

1. The means through which the things in the IoT or the devices and their associated applications can discover each other automatically and trace the middleware nodes of service providers [27].

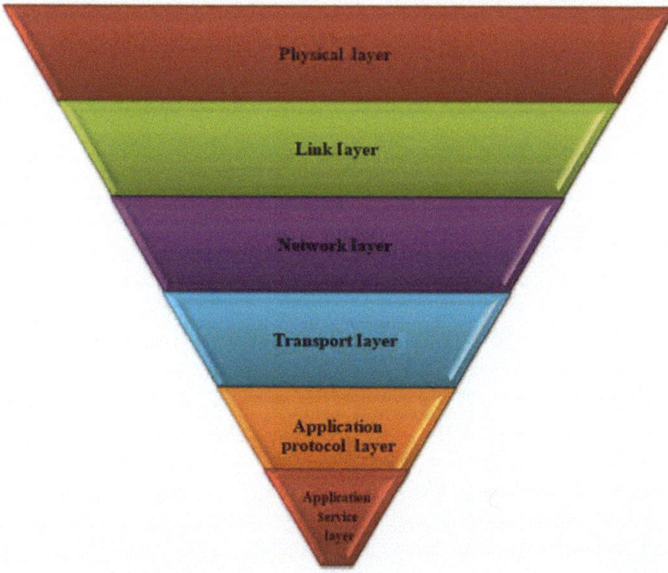

FIGURE 7.6 Stratum of the IoT pyramid.

2. Processes that help the applications in searching the devices tagged with precise attributes or contexts such as sensors of a specific type or inside a certain distance or stated range.
3. Features for the seamless interaction of IoT applications that are capable of executing a search algorithm for any type of data which is grounded on its special attributes. Examples include semantic annotations and spatial or temporal contexts.

7.3.5 Internet of Things application interoperability

IoT application interoperability is now discussed because when we develop applications with so much effort and connect the things together in the IoT we want them to be able to operate over all devices, OS, networks, and computing environments seamlessly. Application interoperability is not a new concept, it has existed for more than two decades in the computer world. Thus the vision is extended to the IoT application interoperability. Though they are yet to realize their true potential, their development, deployment, and management are key challenges [17]. This encompasses a challenge for industry experts, computer network experts who are constantly trying to build the services profitably for industry, and also consumers simultaneously, by specifically tailoring the requirements, developing concurrent and

Protocol	Function	Primary usage	Transport	Format	SDO
CoAP	REST resource manipulation via CRUD Resource tagging with attributes Resource discovery through RD	LLNs	UDP	Binary	IETF
XMPP	Manage presence Session establishment Data transfer/(text or binary)	Instant messaging	TCP HTTP	XML	IETF XSF
MQTT	Light weight Pub/sub messaging Message queueing for future subscribers	Enterprise telemetry	TCP	Binary	OASIS
AMQP	Message orientation, queuing and pub/sub Data transfer woth delivery guarantee/(at least once, at most once, exactly once)	Financial services	TCP	Binary	OASIS
SIP	Manage presence Session establishment Data transfer(voice, video, text)	IP telephony	TCP, UDP, SCTP	XML	IETF
IEEE 1888	Read/ write data into URI Handling time series data	Energy & facility management	SOAP/ HTTP	XML	IEEE
DDS / (RTPS)	Pub/ sub messaging wih well defined data types Data discovery	Real time distributed systems	UDP	Binary	OMG

FIGURE 7.7 Survey of IoT application protocols.

different verticals for consumers and service providers, providing space for speed and cost reductions, and simplicity of robust applications. This intricacy is driving the overall cost of constructing the IoT with feasible and quick solutions. The subsequently developed solutions are characterized universally and ubiquitously and have strong coupling amidst the application and its entities. An application entity represents the endpoint in the IoT which is accountable for creating and/or consuming data and it extends to the entire range of data over the cloud application [28] from the sensors or actuators built into them. The bolted class of existing IoT solutions makes them expensive to execute at the initial level, but they become more expensive, as a result, it is difficult to sustain them and grow them with time. The reason behind this is the fact that the application code frequently requires changes, updating variables, or a changed state, wherein a device is swapped with another device which is equally competent and functionally equivalent although they are manufactured by another company, or a scenario in which a new device is installed and integrated into the solution. This leads to a new type of challenge, which is called IoT application interoperability.

This requirement can be shown as two separate verticals:

1. Prerequisites for constructs and standard APIs;
2. Prerequisites for semantic interoperability.

7.3.5.1 Requirements for abstractions and standard application programming interfaces

To realize the dream of IoT we need to understand the difficulty that lies in its implementation until the APIs control the various verticals of functionalities. They help the devices, things, or smart objects to follow and understand common terms which enables interoperability. To achieve this, the industry needs to be able to reach a full mode of interoperability using APIs. Companies should cover mechanisms and build solutions to recognize the data which are to be shared by application entities and the procedures used for allocation. The smart APIs show data that have the potential to enable distinct devices to come together on common platforms, to create innovative approaches toward novel, innovative, and interesting internet workflows. The advancing accessibility of these regular APIs will make it feasible for the abstractions at certain levels to be deployed in IoT functions, associated with the following:

1. Management of multiple devices such as activation, message trigger, authentication pass generation, granting authorization, updating related software/firmware, etc.
2. Management of huge data generated, such as delivering, writing, editing, subscribing, notifying, reading/sending alerts to delete, operations, etc.

3. Management of the associated application and its interfaces, such as start, stop, debug, upgrade, etc.

The abstraction constructs are an important addition to the APIs and interfaces for a seamless performance, to deliver the logical representations of the functions, and still are able to hide the nuances and other possible variations involved during implementation efforts. The abstractions elaborate the service contracts and these contracts are helped by the fact that they understand well the intricacies like syntax and semantics of the APIs, are able to stipulate methods used in interaction with modules, and supplement them with any necessary functions [29] (Fig. 7.8).

The use of standard APIs is beneficial and brings about a change by introducing "slip surfaces." These slip surfaces eradicate coupling in modules that are functionally discretely provided in the IoT solution. This enables the modules to boom to its full engaging the available traders or vendors of the IoT to work seamlessly interwork and assimilate into a unified system. We discuss here a module used for illustration purposes which could be substituted by any competent vendor in the market that promises to fulfill the regular API administering the related slip surfaces of the unified edifice slabs of the system [2–4,30].

7.3.5.2 Requirement for semantic interoperability

Semantic interoperability provides assurances of accessing and interpreting data unambiguously for the entities in the application related to the IoT. The ability to give a stream of unambiguous data descriptions helps to guarantee that it can be processed by the computing capability of the machine. It can

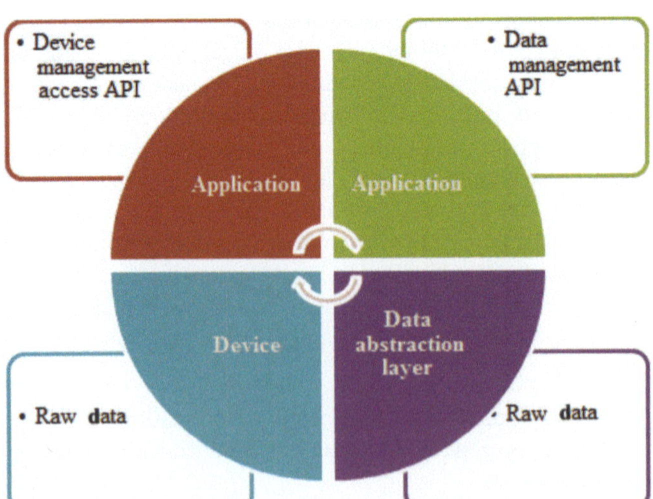

FIGURE 7.8 Abstractions and APIs.

be further decoded by the application entities and this gives a sign of enabling the automated information communications and exchange of critical information in the IoT. In order to share IoT data in a beneficial way, semantic interoperability among communicating systems is now a must. The process guarantees a common standard and language between machines or devices in the IoT. This gives accuracy, efficiency, reliance, and interoperability to the application systems. Hence, a fluent and seamless interaction of machines is expected. There are variants that help to boost the abilities of IoT devices for interoperability. An analogy can be drawn with a pyramid to capture the vision of application interoperability where the base serves as the syntactic or structural area. The format of data exchange is properly defined to avoid confusion.

It is now an obligation to successfully execute structural interoperability between various applications on the platform. The peak of success is achieved by semantic interoperability and it can also be well reserved for futuristic approaches. It engages with the message content and interpretation of meanings and is not limited to data formats.

Semantic interoperability can be achieved in innumerable ways, one of them being the strong, universal, persuasive, interactive, and informative models. These should have the ability to capture knowledge and learn by themselves, backed up by the power of AI and machine learning.

The other method could be through using semantic intermediaries, or decipherers, who may convert the information into a suitable form such that the entities in the application are able to decode or understand them (Fig. 7.9).

7.4 Internet of Things sensors for drug checking

Connecting devices, which are used for communication in IoT, include some devices such as sensors, actuators, and RFIDs, which are described in more detail here.

7.4.1 Sensors

A device (electronic) capable of sensing events, deviations in its corporal atmosphere such as heat, sound, vision, temperature, pressure, parameters pertaining to the chemical and biochemical environment, and giving an equivalent output is called a sensor. Sensors take inputs in analogue format and give output in digital format. The analogue-to-digital converter is essential for understanding the data from attached sensors. Sensors are analogous to the five human senses. Sensors form the first set of devices that interact in the IoT device connection. Sensors are crucial in the IoT concept establishment. They make a channel of connections between physical objects with the Internet. Sensors have the only function of sensing the data and transmitting

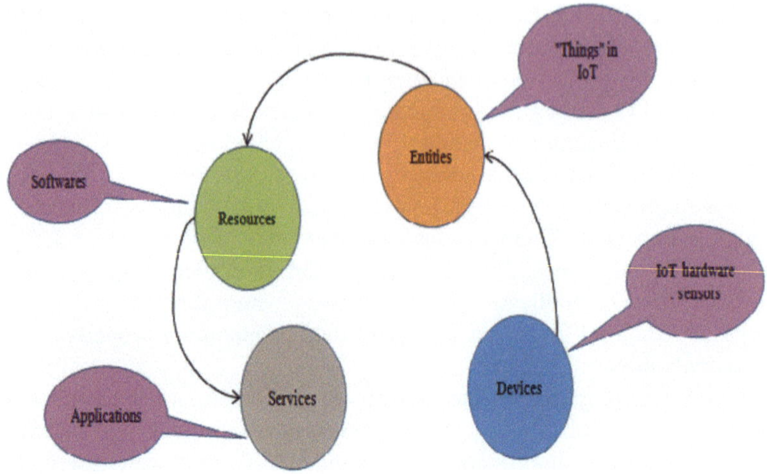

FIGURE 7.9 Concept of IoT ontology.

them with accuracy over the Internet. The core function is limited here in sensors, yet they may also notify the gateways for duplicities. The sensors in this specific case would be the drug checker that may take a snapshot of a drug, and store it in a database for future retrieval whenever the history of the patient is required. These specific sensors may produce the desired results when trained to sense the correct data. This may require some sort of programming logic to be present within the sensors for maintaining the accuracy of the sensed data [1]. In this specific scenario, an IoT sensor needs three components at least—sensor(s), microcontrollers, and connection establishment—in order to send the user data on the attached gateways or other system blocks (Fig. 7.10).

Smart sensors are flexible in collecting large blocks of data at any time or site, and send these data in real-time over an IoT network via an IoT gateway. The data of the patient are sent for analysis and to establish a correlation with other medical-oriented intelligence databases. This gives detailed case insights and brings the patient under proper treatment control. This helps give the opportunity for doctors to treat patients more efficiently and in a patient-oriented manner.

Now we have a little idea of what a sensor looks like and the possible help that one may get from using sensors in the IoT. The main purpose of the sensor is taking data from its surrounding environment and delivering the output to its contiguous tools or gadgets (e.g., gateways and actuators) or diligent applications on the product. Radars are sensors that basically do the sensing job smartly, and use their capacities for sensing data to form physical interfaces (inputs), then converting the incoming signals into electrical signals (outputs). This sensing is understood by both the communication as well

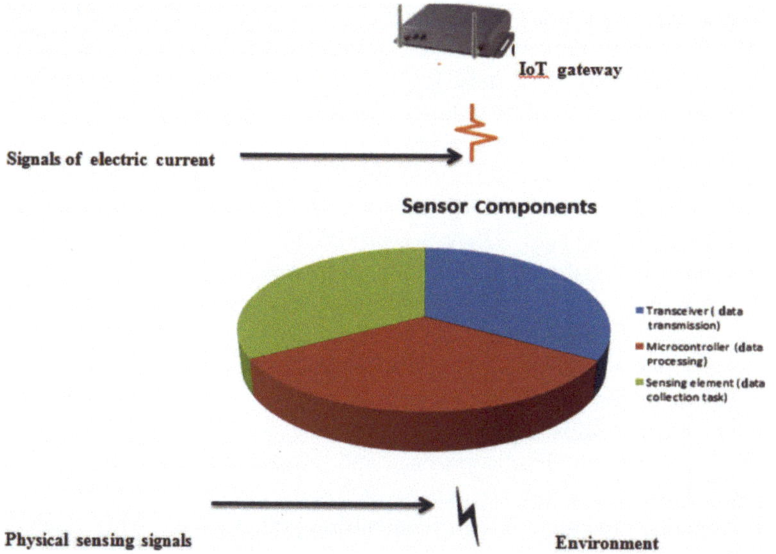

FIGURE 7.10 Components of a smart sensor.

as the computing device. The output signals via getaways are further processed and realized by the applications.

7.4.1.1 Types of sensors used in the Internet of Things

Having talked about sensors, we now focus on their types, of which there are two main types: proprietary and nonproprietary sensors. The existing trend in the IoT world follows a few very good sensors. These are now ready to embrace IP-based sensors. This helps to improve the connectivity over wired and wireless sensors on the Internet. IP-based sensor networks desire a space where each device can be uniquely identified and is assigned an IP address such that its identity is clear, especially on a larger network. Starting to build an infrastructure for IP-based sensors from scratch is a tough task. There are many types of sensors, some of which are described here [31].

1. *Temperature sensors*: To measure the external temperature we have temperature sensors. This is estimated by environmental factors such as physical, chemical, mechanical, electronic, and biological systems showing variations with a rise or fall in temperature. There are basically four kinds of temperature sensors:

a. Thermocouple sensors: This is a type of sensor that contains two distinct and unlike conductors in contact. The power is created with the help of the thermoelectric effect. The sensor is comprised by assembling two disparate metals on one end.

b. Resistance temperature detector (RTD) sensors: These are radar equipment that senses temperatures, and even minute changes can be recorded and resisted. They were used for a long time for recording temperatures in laboratories and other industrial processes. They have proved themselves for accuracy and efficiency.

c. Thermostats: These are similar to the RTDs, they have no resistance to changes in temperatures. They are made from semiconductor materials.

d. Semiconductor sensors: These are of various types, like the voltage output, the output for current, digitally supported output, resistance output, and silicon-diode temperature sensors. The new types of sensors are accurate and highly efficient.

2. *Pressure sensors*: Such sensors aid in measuring the pressures of solids, liquids, or gases and they also include water levels, and the flow, speed, and height of fluids. Examples of such sensors are those used for compressors, pumps, hydraulic systems, and refrigerators, etc.

3. *Imaging sensors*: These are complicated sensors that are deployed in digital cameras, computer vision, medical imaging such as X-ray scans, and night vision tools. They give a precise record of information by capturing the image that supports great lens capacities and then convert the variable attenuation of waves into signals.

4. *Noise sensors*: These help to edit the high noise which may damage human ears (e.g., permanent or temporary deafness, cardiovascular diseases, stress, hypertension, etc.) and even animals (for instance, deafness). The noise may be caused by appliances at home, machines, construction sites, airplanes, trains, and loudspeakers, etc. and are basic causes of noise pollution. Many firms have started to install sensors to keep a track of noise pollution. Ambient sensors for noise monitoring are also used in health care for patient comfort in hospitals. Noise level changes are explicitly observed and the signals of change are transferred over the device for appropriate action which may be automatic or manual.

5. *Air pollution sensors*: The rising pollution concerns are giving way to the installation of such sensors that monitor and even regulate air quality in urban areas. Air pollution sensors detect and monitor the presence of air pollution in the surrounding environment. They are basically concerned with five components: ozone, particulate matter, carbon monoxide, sulfur dioxide, and nitrous oxide.

6. *Proximity and displacement sensors*: These sensors aid in detecting the existence or absence of an object by utilizing electromagnetic fields, waves of light, or sound. There are various types that suit specific applications and surroundings.

a. Inductive sensors: To monitor very close-range detection of ferrous elements;

b. Capacitive sensors: These are meant for nonferrous materials detection;

c. Photoelectric sensors: These are to detect long-range targets;
d. Ultrasonic sensors: These serve the purpose of detecting the long-range distance of objects with very rough surfaces.

Fig. 7.11 illustrates the details of specialized sensors.

7. *Infrared sensors*: These sensors help to track the movement of an object. Heat helps to detect infrared rays and this eases the process of detection.
8. *Moisture and humidity sensors*: These are also known as hygrometer sensors. They enable us to report and calculate the degree of relative humidity of the atmosphere. They tend to rely on electrical capacitance to discover the degree of capacitive measurements.
9. *Speed sensors*: These sensors are generally used to measure the speed of means of transport, such as those in air, water, or land. Wheel speed sensors, speedometers, Doppler radar, and laser surface velocity meter are examples of these sensors. There are other sensors available in the market including sensors that measure acceleration, biosensors, sensors for gas and chemical reactions, mass sensors, tilt sensors, force sensors, etc.

7.4.1.2 Sensor physiognomies

There are multiple characteristics of sensors that we use for the functionality of IoT. These are modern sensors with very advanced capacities. They help to collect enormous amounts of data and quickly transfer them over the network for processing and analyzing. They help to consume less power, with a good battery life, and a rapid response time. Sensors are expected to be smarter than humans these days [32]. They are more versatile, dynamic, and are the most important aspect of IoT drug checkers which are capable of collecting real-time data. These sensors, in general, are either autonomous or manually operated. The gateway in IoT sensors seems to filter unwanted data and provide security at the very first level.

Sensor Technology Name	Range in mm	Usage
Inductive	4–40	Iron materials, close-range detection
Capacitive	3–60	Non iron elements close-range detection
Photoelectric	1–60	Martial long-range target detection
Ultrasonic	3–30	Martial long-range target detection with tough surfaces having multiple colors

FIGURE 7.11 Types of proximity sensors.

IoT sensors are expected to behave in a certain manner. They should be able to comply with the following characteristics so that they may be described as modern IoT sensors.

1. *Data filters*: The main task of these types is to filter unwanted data. This data collection and transfer should be very accurate and secure. They need not do any computing or analytical processing.
2. *Reducing power utilization*: There is a crucial need to reduce power use by inbuilt IoT sensor equipment. The various sensors for IoT verticals (e.g., railways, highways, smart grid) need to be installed in sites that are complex in nature and their updating, replacing power sources, and correction from time to time can be difficult.
3. *Compactness*: There is very limited space for the IoT verticals that are installed in various sectors. Thus there is a need for sensors to be accommodated in limited dimensions and to fit the space given.
4. *Smart detecting operation*: The sensing needs to be smart and remote operation should be as smooth as possible. This comprises collecting information without touching or making contact with it. The operation can be performed broadly in three ways:
   a. Via presence/proximity: The sensor should be able to determine the existence or deficiency of an object.
   b. Measurement of speed: The exact position of the object despite its speed, for monitoring traffic, etc.
   c. Detecting the range: Range detection is very important, especially to avoid collision.
5. *Sensitive sensors*: The measurement of the ratio of even small changes from electrical to physical signal is called sensitivity. For instance, the voltage changes during a change in temperature.
6. *The measure of linearity*: This is used to measure the level at which an output is proportionate linearly to the input.
7. *Dynamic range*: This is a range of signals taken as input that could be transformed into electrical signals by the sensors dynamically.
8. *Accuracy*: These sensors are required to be accurate between measured or detected data and the ideal data.
9. *Hysteresis*: The width of the predictable or error expected is known as the hysteresis in terms of the calculated numbers. A similar output is not returned when the input is hastened and fed into the system.
10. *Limited noise*: Sensors should not make noise or create noise pollution. Smart sensors are those that filter noise and noisy data. They are spread across a spectrum of varied frequencies.
11. *The high dimension of resolution*: The resolution of a sensor needs to be such that it captures the slightest fluctuations.

12. *Minimum interruption, higher reliance, and ease of use*: Sensors should be able to function normally without any errors. They should be trustworthy and reliable. Access to the sensors and easy operation are plus points and add to sensor functionality.

7.4.2 Radio frequency identification

The most common method after sensors to capture data or information from "things" is by using RFID [33]. RFID can be understood as a mechanism that captures the data already embedded inside a tag of things or objects via radio waves. It contains two basic parts, a microchip and an antenna. The microchip stores the information and processes it. The antenna is used to receive and transmit the signal over the network from the things or objects. Thus, for example, in the case of a drug checker, the RFID would check the pharmaceutical brand, date of expiry, composition labeled on the medicine, its reactions, dosage, etc. and could transmit this information to the patient's care regime.

The tag consists of serial numbers for unique specific objects. The information is read by the reader on a given tag via a two-way radio transmitter–receiver. The reader then transfers read and delivered outcomes to the computer program of an RFID. The RFID sensors supersede older sensors based on the reader-tag. These include:

1. The RFID tag is not needed to reside in the reader's direct line of sight and is readable at a distance of 12 m. The range of battery-powered tags is around 100 m.
2. The data on the RFID tag can be altered depending upon the business. However, the barcode data are difficult to alter once installed.
3. The RFID tags are tough. The printed barcodes are clearly visible.
4. The RFID data are coded via an encryption algorithm on the tag so that unauthorized users are unable to change or manipulate them.
5. The tags in RFID can read many thousands of tags concurrently. This is one of the important steps in retail stores because time is saved and critical task performance is enhanced (Fig. 7.12).

There are some disadvantages to RFID technology which can be ignored due its greater advantages. The main uses and applications of RFID technology are:

1. *Controlling access and management*: Many agencies implement RFID tags for identification purposes. The authorized people gain access and can control to manage the business. Examples like electronic toll collection will give access to the vehicles of only registered owners.
2. *Passport*: The RFID technology helps to recognize frauds on airports. They can scan the passports designed with electronic tags holding information in digital format.

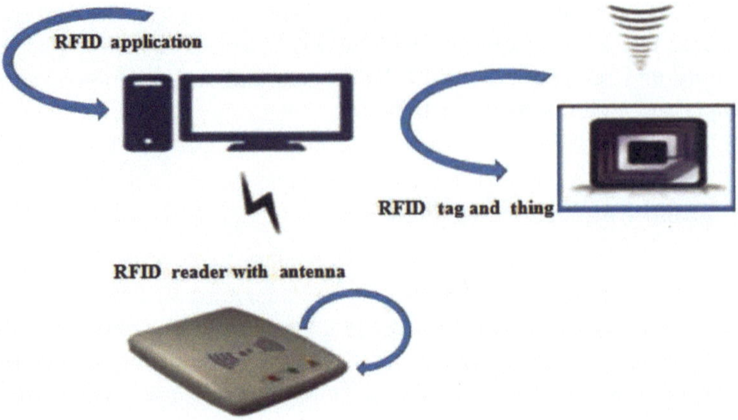

FIGURE 7.12 RFID main components.

3. *Healthcare improvement*: In the healthcare sector, RFID technology is very useful. RFID-based solutions in health care are trending and are proving their efficiency in private and public hospitals all around the world. With mobile devices becoming cheap and with easy real-time data tracking, healthcare facilities can be improved.
4. *Logistics and supply chain tracking*: Logistics supply using RFID has become easier. RFID tags at shipments help to improve the existing supply chain management. They help and permit retailers in using automation.
5. *Athletic support*: The ease of tracking the precise timing of athletes in sports events like races and swimming marathons is critical. Athletic timing is a major implementation use for RFID.
6. *Animal tracking*: RFIDs have became crucial in the control of animal diseases such as "mad cow disease," by aiding in animal identification.
7. *Other functions*: RFID is important in tracking logistics, such as airport baggage, health tracking, drug tracking, marketing, tracking the attendance of IT assets, library management systems, and real-time location tracking.

7.4.3 Internet of Things actuators

The actuator can be defined as a device operated with an inbuilt motor which controls the actions taken by a system. The source of data or energy is taken by the device in various forms (e.g., hydraulic fluid pressure) and the data or energy is translated into motion to control the system.

7.4.3.1 Need for actuators

Sensors are accountable for sensing changes in their environment, collecting crucial data, and helping in monitoring systems by providing them with these data. The collection and display of the data onto a monitoring system are not useful until it is transformed into intelligence, which can be later utilized for controlling and governing an environment before an entity is affected by the impact [34]. The actuators exploit the collected and analyzed data by the sensor and other forms of translated intelligence for the control of IoT systems. For instance, the shutting off of gas flow on its own by realizing that the pressure reading is below the appropriate threshold.

7.4.3.2 *Types of actuator*

1. *Electrical actuators*: Electrical actuators work on the principle that they convert electrical energy to mechanical energy. The torque produced helps to regulate some devices and equipment. They are also cast-off in engines for controlling valves.
2. *Mechanical linear actuators*: These actuators are mechanical in nature and convert rotary motion into linear motion. Screws and chains are used in such types of renovation. Examples of such actuators are described as "screw," with screw jack, ball screw, and roller screw actuators all operating on the same principle.
3. *Hydraulic actuators*: These actuators are very simplistic tools with machine parts expanded on valves which are linear/quarter-turn based. They work on the principle of Pascal's law, which states that "when there is an increase in pressure at any point in a confined incompressible fluid, then there is an equal increase at every point in the container." These hydraulic actuators are made up of cylinders or fluid motors using hydraulic power to permit a mechanical process. This mechanical motion provides the required output in terms of linear, rotary, or oscillatory motion. They can be operated manually, for instance, a hydraulic car jack, a hydraulic pump, etc. [35].
4. *Pneumatic actuators*: These actuators work on an identical notion relating to those of hydraulic actuators while they differ in the fact that they use compressed gas in place of liquid.
5. *Manual actuators*: These actuators have levers, gears, or wheels to help movement, while there is a power source that is external in nature that helps to activate the valve mechanically. In remote areas, valves that are operated by power actuators are needed.

7.4.4 Internet of Things device control and regulation

There are multiple ideas for monitoring and regulating the IoT devices and we will discuss here two out of them. The local and global control of IoT devices, helps in controlling the system above all other philosophies. The first method of local control needs a local intelligent controller while the other approach is to transfer the regulating power on to the cloud for global operation. They embed and install easily available sensors which are also cost-effective. The extension of this would be to pull the controller boards out of the furnace and air conditioner and connect their inputs and outputs to the Internet, so that a cloud application can directly read their states and control their subsystems.

Thus, very clearly, the approach with more devices connected on the IoT platform that are capable of providing the likelihood of strategic control is almost impossible in isolation. One would use ambient forecasts for weather conditions, and the specified addresses of the residents as inputs, for example, to detect or devise the optimum strategy for providing comfort to the surroundings and elevating the standard of living while also reducing power consumption. We conclude that this requires the combination of local and global approaches contingent on the verticals and the IoT environment specifically.

7.4.4.1 Identification of things in the Internet of Things

We discussed above that the most expedient way to recognize every IoT device is by assigning a unique IP address to each of the sensors and actuators in the network. IPv4 addresses are expensive and limited, which is a disadvantage. IPv6 addresses are not feasible or deployed yet in the network. Also, there are still many sensors and actuators in the network that are not enabled by the IP. The IP address of IP gateways is exceptional with a unique IP address. Thus, non-IP-enabled sensors and actuators could be easily recognized via the linked gateways.

7.5 Internet of Things services in drug identification

7.5.1 IoT services

IoT is no longer a dream. We are realizing the dream of building lively interactive smart objects and things around us. Health care is a very important sector where we are attempting to achieve the best possible health services possible. Human health is of prime concern when we talk about sensors and machines. At no point should a machine fail to identify the correct mechanism to deal with human health. The smallest of errors can lead to devastating and life-threatening effects on small to large scales depending upon the type of error. We are proposing the kind of health sector service for identifying the smartest IoT tools to assist doctors in their efforts. The current rate of deaths caused by wrong medications and erroneous diagnoses by doctors is alarming. Therefore we are trying to develop learning and interactive models for drug checking as a kind of assistant for doctors to make their job easier and provide them with the most possible details via these smart objects. Concurrent research is taking place in this field for drug identification and drug checking. Drug compliance has now become an international matter of concern for healthcare providers, as we try to build solutions and secure drugs from any possible threats. IoT is now deployed at even the smallest of places by placing sensors and RFID devices to sense and collect healthcare data. These sensors examine the data of drugs that are sensed or fed as input to continue the type of treatment using the previous history of the patient [13,14]. It also helps to detect the side effects of the drug on its recipient, the possibilities of reactions, allergies, etc. The World Health Organization is taking up the new IoT solution to combat deadly diseases and supporting research in this regard. The technologies are designed to integrate and be installed even on handheld devices, mobile devices, Personal Computers, laptops, Personal Digital Assistants, etc.

The IoT is the prime topic of this chapter which describes IoT drug checkers and identification, or it may be described as drug tracking for a patient's medical history. The sensors are responsible for collecting data and provide strong support for accidental handling of data, medicine care, timely reminders, etc. The other fields where the IoT could be extended to in medical and pharmaceutical industries include:

1. Medical equipment and drug monitoring and management;
2. Medical device and pharmaceuticals anticounterfeiting;
3. Complete real-time monitoring;
4. Medical waste information management;
5. Patient information management;
6. Medical emergency management;
7. Drug storage;
8. Patient information management;
9. Fighting pharmaceutical errors;
10. Medical equipment and drug tracking;
11. Connected information sharing;
12. Newborn antikidnapping systems;
13. Alarm system.

7.5.2 Technical problems

There are some technical problems that the medical IoT faces and these are discussed in the following points:

1. *Management of dynamic networking and node mobility in huge network compositions*: Whenever there is an expansion of IoT tools the size of the networks can be intimidating or overwhelming because of the incurred complexities of infrastructure involved in their cost, deployment, etc. Thus the responsibility increases to manage the huge network topology.
2. *Data completeness and data compression*: The nodes may work without any breaks for 24 h and collect huge data so this huge amount of collected data needs to be regulated and stored safely for easy retrieval, by using or developing complex algorithms. The data compression algorithm

is quite costly for sensors because of the spatial complexities. Compression algorithms also cannot lose the original data, otherwise the system could misdiagnose a patient's condition.

3. *Data security*: Data security is the major concern in IoT data-handling models. We discuss this vertical in detail in the coming paragraphs. We now know that the threats to wireless sensor network nodes are real and their vulnerability is always high. The computing power of a sensor node is limited, thus there is an incumbent and immediate need to cater to the data security with high priority. There is a need to develop a separate encryption algorithm suiting the requirements of sensors.

7.6 Internet of Things extended to artificial intelligence and machine learning

The IoT has tremendous potential when enhanced by AI, predominantly with the rise of new standings such as "machine learning," "deep learning," "genetic algorithms" and other new terms that are being coined regularly. The main discussion here is restricted to AI and machine learning for obvious reasons. The main point that we want readers to understand is that:

> *Artificial intelligence will be functionally necessary to wield the vast number of connected "things" online, and will be even more important in making sense of an almost endless sea of data streamed in from these devices.*

Thus we come to the acceptance of the notion that the real world is now romanticizing the power of artificial intelligence combined with the IoT. This can be discussed in many ways, one of them being the potential future of the IoT combined with the power of AI [6,36].

Cognitive technologies like machine learning and AI certainly have timely proven themselves to be an indivisible part of the IoT. They are potentially able to help make sensors AI-enabled, make them smarter and, this implies increase their value. These emerging technologies provide support human workers with extended care and improve their decision-making abilities and thus the overall business performance.

It leverages the support of cognitive technology to enable the learning process to be much simpler and more user-friendly. The platform integrated with AI is bringing value to its customers, and it is constantly committed to serving the user with seamless facilities. This is where cognitive technology and AI are growing in value among users. Thus one can say that these are truly exciting times for customers and developers alike.

7.6.1 Security and privacy of the Internet of Things over the Internet

Securing the Internet normally means dealing with computer security, explicitly related to the Internet, sometimes relating to browser security and also to network security on a more general platform as it applies to other applications or operating systems altogether. The objective is to set up rules and methods to prevent hostile attacks on the Internet [16,36]. The Internet has vulnerable channels for the exchange of information with high risks of intrusion or fraud, such as phishing. Various methods have been deployed to guard the transfer of data, including encryption and methods of ground-up engineering.

Internet security is the method of fortifying communication on the Internet platform and includes explicit protocols pertaining to securing such as:

- Internet security protocol (IPSec);
- Secure socket layer (SSL).

7.6.1.1 Internet security protocol

This entails a set of protocols devised by the Internet Engineering Task Force. It helps at the network level to provide security and also aids in creating for IP stratum authenticated and confidential packets.

7.6.1.2 Secure socket layer

The SSL was designed by Netscape Communications Corporation. It is one of the security protocols in wide use and provides safe access to the transport strand. It looks after the following security issues:

- Privacy
- Integrity
- Authentication

7.6.2 Potential threats

There are multifaceted threats to Internet security that have the capability to create enormous threats from various sources, and that can even disrupt other Internet-connected devices. Cyber crime is on the rise and criminals have increasingly evolved to take the opportunities to exploit this area. They use several techniques to endanger the integrity of businesses, bank accounts, and organizations, including:

1. Mobile worms
2. Malware
3. PC and mobile ransomware
4. Large-scale attacks like Stuxnet that attempt to destroy infrastructure
5. Hacking
6. Spam
7. Phishing

The IoT works in combination with the Internet and thus it potentially needs to extend its policies toward securing its activities and data over the network from fraudulent attacks and approaches. The medical data of any individual are a very private matter and they need to be protected from infringements and breaches. The needs of patients should be met with the utmost security and privacy. The IoT and its services are required to extend their concerns toward the security while dealing with patient information and healthcare history. This the security and confidentiality of its users need to be guaranteed.

There are offline systems working with pure confidentiality in the medical world. Ideally, applications should also follow the same principle of guaranteeing privacy. The IoT has many thousands of users connecting on the Internet every day and thus it opens up the threat of attacks to its users. Such targets have the multidimensional aspect of substantial damage and at times the threat can be life threatening. Efforts to intimidate the IoT are imposing a potentially larger threat, which needs to be dealt with the utmost responsibility. The IoT demands analysis and risk management and its assessment to measure and keep track of such threats in this ever-changing world.

A major area in providing security functions to the IoT is to authorize and authenticate the identity of its users. Many IoT sensors are limited in their capabilities to rebuff such profound threats. They lack the required processing, storage, memory, and power needs. This supports the state-of-the-art proprieties for authentication. The state-of-the-art protocols of cryptography and authentication are required. such as AES encryption methods and RSA digital signature support. These protocols have proven themselves repeatedly when deployed appropriately. They have only two shortcomings, one is that they need a high computing environment and that they are hungry for resources. They need to be appropriately reengineered to adjust to constrained sensor devices. The other inadequacy is that the protocols for authentication and authorization require a lot of installation and configuration. In basic formats for IoT disposition, access to the device is limited and so there is a requirement for new protocols of authentication and authorization. This leverages the experience of strong encryption/authentication algorithms and, to our amazement, they are still capable of running on sensor devices.

Encryption and decryption protocols sort out the problem to an extent. However, with time the effectiveness of these algorithms decreases for many reasons, one being Moore's law. This creates an interesting dilemma for the use of the algorithm of encryption in IoT sensors. For example, a smart sensor in a hospital can operate for 50 years while the encryption protocol decreases its lifespan to about half of this time.

The additional facets that need to be thought over in IoT sensors include:

- Building strong authentication mechanisms and strong identities;
- Privacy of data levels and limiting the accessibility at specific points;
- Strong foundation for networks and its supporting infrastructure, for example, DNS, DHCP, etc.;
- Delay the protocols implementation in order to reduce the transient connectivity.

IoT attempts to preserve privacy extended from personal to national levels for user data, especially in the case where a patient's personal health care is collected for a variety of purposes. The sensors generate sensitive personal and it is the moral responsibility of those holding the data to protect them from any potential threats. It is very important that the owner's identity remains hidden, yet that ownership authentication and validation are supported. The mechanism that supports this kind of implementation is known as shadowing. It enables the user's object to play the part instead of the user himself, storing the data in the database on their behalf. Thus, the information creates a hidden layer or shadow on the owner's part implementing the authentication methods as well without the need for revealing the sensitive data. For instance, the data stored via biometric sensors jointed with a physical object could be used to grant access by unlocking a door. The prominence of issues like confidentiality, integrity, and authenticity in the IoT cannot be overstated.

7.6.3 Security and privacy in the Internet of Things

The IoT makes a promise of making our lives easier and more comfortable by turning physical objects around us into sensor devices, and helping us thereby deal with multiple situations. Though this provides us with conveniences that one could have never imagined previously, they pose serious challenges pertaining to data security and privacy. Disregarding the issues of security and privacy can lead to serious repercussions on many areas of our lives as individuals, including our homes, the vehicles we use, and in the specific context of the outcomes that could eventually exploit and harm our health as they are dealing with our medical data [37].

If the homes that we live in or where we spend most our time, do not have a smart meter currently, it is quite possible that it will have many such meters soon and these will control and monitor our power consumption for the benefit of ourselves and the power-providing firm. From the heating systems for winter seasons to the lighting of your house, everything is turning into a subscription basis. Thus there is no doubt that smart gadgets will be found all over our homes, for example, the smart camera notifying of any suspicious or other kinds of movement over smartphones, remotely opening smart doors, smart refrigerators notifying about a shortage of vegetables, fruits, or cakes. Therefore, in a nutshell, the level of devices going smart is accelerating at its highest rate ever. Now let us imagine how much of the data are being sensed every day from our dealing within our private areas that affect the way we do our businesses. Thus it is not difficult to realize the control which could be gained by an attacker by simply hacking tools and systems, if we overlook the potential threats. In fact, the damage that they may cause is unimaginable. The cyber attacks in the present IoT era would have the greatest impacts in human history, gradually leading to the devastation of our lives and properties. Just one attack on the IoT drug checkers, would have a severe impact. Our bodies would be victims of cyber attacks, which sounds weird but that is the threat to human civilization if we do not implement adequate security measures in time. Researchers have debated and reported that an attacker may control health devices, pacemakers, dosage and medicine plans, etc. for any individual by hacking the data transfer link and stealing the data to gaining control or even imitate the patient. This would grants the transfer to attackers the greatest imaginable powers, like what and when to inject insulin, for instance, and its dose, leading to grave health hazards that may lead to killing of patients using a smart health device. These concerns have forced doctors to immobilize the wireless proficiency of the heart pacemaker of the ex-US vice president, Dick Cheney, so as to shield him from such harmful attacks [6,38]. There are extremely high vulnerabilities when we talk of security alone, and this is personal medical security which will have extremely serious repercussions on human lives. If by chance an attacker of was to gain access to smart objects used in medical and health care, then it could give partial or full access to the sensing capabilities of those smart devices. This influential access could then be used to cause serious harm.

Information is viewed as very important prized asset to advance knowledge and to make a prosperous and flourishing society. In order for characteristics to achieve this perfect goal, we pay specific attention to a secure and accurate process for the delivery of information. Other than knowledge, we are responsible for building harmony in society by following the principles of knowledge that lead us to the truth.

When we look at information security, we realize the importance of the information itself. The more we value assets, the greater is the need for security. The most common need is protection from a wide range of threats and reflecting information as assets to protect business continuity. CIA (i.e., confidentiality, integrity, availability) describes the meaning of information security. Information security is thus defined as "the protection of confidentiality, integrity, and availability of an organization." The right time to address threats to information security is now and always. A wise quote that we are reminded of goes as "When communication is free from any impediments, the information will be safely delivered and comprehended and thus directing us to the light of truth." As is wisely said that: "prevention is better than cure."

The ethics of the security practices need to be well coded and no conflicts must occur amongst the following:

1. *Multiple technologies*: The platform at the IoT combines various technologies which is a great benefit, for instance, RFID, wireless technology for networks, cloud computing, virtualization, simulation, etc. Each of these comes with a separate package of vulnerabilities.
2. *Manifold verticals*: The paradigm of the IoT has many applications (also known as verticals) extended to various fields like health care, education, industry, businesses, etc. Every vertical has its own requirements for security.
3. *Scalability*: Scalability is a very important issue, as it is about the building of defensive mechanisms for securing data in the IoT framework. The scaling should be done cost-effectively.
4. *Big data and availability*: The enormous amount of data generated needs proper and secure handling. Availability means that the sensors are available for operation, with never-failing systems being required.
5. *Resource limitations and remote locations*: Sensors need to be smart in their capabilities, yet resources are limited due to their physical limitations, and because they can cause serious attack vulnerabilities and service disruption. Sensors should possess the ability to remotely retrieve and locate data, fit in small spaces, etc., giving good and smooth operability from remote locations.

7.6.4 Internet of Things security requirements

We conclude and review in this segment the security prerequisites for the IoT [37,39]. These requirements include the following:

1. *Confidentiality*: This vertical safeguards trade of messages by and to the envisioned parties.
2. *Integrity*: This vertical assures that the traded messages are neither altered nor meddled with.
3. *Authentication*: This helps to assure that the entities that perform operations are for real and not fake or masquerading.
4. *Availability*: This ensures that the services continue without any interruption, no DoS attacks, and safety from external spamming threats.
5. *Authorization*: The assurance of proper control permission so that the operations are performed in time. This grants proper permission to do a task.
6. *Freshness*: This assures that data are fresh. Replay attacks can use this property to set a target by using a message and replaying it to return some entity into a previous state.

7. *Nonrepudiation*: This assures that the action will not be denied once it has been performed.
8. *Forward secrecy*: This factor safeguards the interest of maintaining privacy once the entity exits the system of networks.
9. *Backward secrecy*: This enables the assurance that the new entity will be unaware of the communications that took place before it entered the room.

Therefore, on a concluding note, we list the following points which would describe an ideal situation.

The user of a computer should be aware that:

1. Never use computers to harm other people physically, emotionally, and/or by any other means;
2. Never intrude on other people's work on a computer or online;
3. Never spy around in other people's files;
4. Never steal to serve the harmful intentions of others;
5. Never misrepresent the information shared by others online or any other platform via computers;
6. Never lose view of the social consequences that are harmful or negative of the tasks accomplished with the computers.

7.7 Conclusions

The IoT application and its increasing use are witnessing drastic changes. It is increasingly used in all aspects of the healthcare industry. In this chapter we have explored how the IoT trends affect health care by adapting to current needs and are emerging as the personal assistants of the doctors of 21st century. We have also explored the IoT from its birth to implementation stages that covered the birth of the IoT to its reality in different areas of human lives, and have intensified our focus only on the healthcare sector.

We are by now aware of the influence that the IoT already has in the world of connectivity. There are many design issues in the implementation of the IoT at basic levels, with the installation of infrastructure being one of these major challenges, and the costs that will be incurred in developing from the start is a daunting task. The drug checkers will gradually revolutionize the medical world once they are operated in their full capacity, and they will save lives. Though the challenges are many, we are looking forward to the greater aspects of IoT implementation and contesting in the market has already begun, which is a positive start to the concept.

References

[1] K. Krishnakumar, J. Gubbi, R. Buyya. A Framework for IoT Sensor Data Analytics and Visualisation in Cloud Computing Environments. cit. on, 2011, p. 12.

[2] A. Rayes, S. Salam, Internet of Things—From Hype to Reality, Springer, 2017.

[3] A. Rayes, S. Salam, The Internet in IoT—OSI, TCP/IP, IPv4, IPv6 and internet routing, Internet of Things From Hype to Reality, Springer, Cham, 2017, pp. 35–56.

[4] A. Rayes, S. Salam, The things in IoT: sensors and actuators, Internet of Things From Hype to Reality, Springer, Cham, 2017, pp. 57–77.

[5] Green Book, The Green Book, Four Types of Actuators, 2011. Source: <http://www.the-greenbook.com/four-types-of-actuators.htm>.

[6] D. Evans, The internet of things: how the next evolution of the internet is changing everything, CISCO White Pap. 1 (2011) (2011) 1–11.

[7] P. Barnaghi, W. Wang, C. Henson, K. Taylor, Semantics for the Internet of Things: early progress and back to the future, Int. J. Semantic Web Inf. Syst. (IJSWIS) 8 (1) (2012) 1–21.

[8] Q. Zhu, R. Wang, Q. Chen, Y. Liu, W. Qin, Iot gateway: bridgingwireless sensor networks into internet of things, 2010 IEEE/IFIP International Conference on Embedded and Ubiquitous Computing, IEEE, 2010, pp. 347–352.

[9] S. Andreev, O. Galinina, A. Pyattaev, M. Gerasimenko, T. Tirronen, J. Torsner, et al., Understanding the IoT connectivity landscape: a contemporary M2M radio technology roadmap, IEEE Commun. Mag. 53 (9) (2015) 32–40.

[10] S. Madakam, R. Ramaswamy, S. Tripathi, Internet of Things (IoT): a literature review, J. Comput. Commun. 3 (05) (2015) 164.

[11] A. Kulkarni, S. Sathe, Healthcare applications of the Internet of Things: a review, Int. J. Comput. Sci. Inf. Technol. 5 (5) (2014) 6229–6232.

[12] J.A. Swets, The ABC's of BBN: from acoustics to behavioral sciences to computers, IEEE Ann. History Comput. 27 (2) (2005) 15–29.

[13] A. Rayes, S. Salam, IoT vertical markets and connected ecosystems, Internet of Things From Hype to Reality, Springer, Cham, 2019, pp. 239–268.

[14] A. Rayes, S. Salam, The Internet in IoT, Internet of Things From Hype to Reality, Springer, Cham, 2019, pp. 37–65.

[15] J. Yick, B. Mukherjee, D. Ghosal, Wireless sensor network survey, Comput. Netw. 52 (12) (2008) 2292–2330.

[16] P. Sethi, S.R. Sarangi, Internet of things: architectures, protocols, and applications, J. Electr. Comput. Eng. (2017). 2017.

[17] D. Estrin, R. Govindan, J. Heidemann, S. Kumar, Next century challenges: scalable coordination in sensor networks, Proceedings of the 5th Annual ACM/IEEE International Conference on Mobile Computing and Networking, ACM, 1999, pp. 263–270.

[18] F. Yang, D. Hughes, N. Matthys, K.L. Man, The PnP Web Tag: a plug-and-play programming model for connecting IoT devices to the web of things, 2016 IEEE Asia Pacific Conference on Circuits and Systems (APCCAS), IEEE, 2016, pp. 452–455.

[19] A.H. Alhamedi, V. Snasel, H.M. Aldosari, A. Abraham, Internet of things communication reference model, 2014 6th International Conference on Computational Aspects of Social Networks, IEEE, 2014, pp. 61–66.

[20] J. Deogirikar, A. Vidhate, Security attacks in IoT: a survey, 2017 International Conference on I-SMAC (IoT in Social, Mobile, Analytics and Cloud) (I-SMAC), IEEE, 2017, pp. 32–37.

[21] H.M. Aldosari, A proposed security layer for the Internet of things communication reference model, Procedia Comput. Sci. 65 (2015) 95–98.

[22] V. Gazis, M. Görtz, M. Huber, A. Leonardi, K. Mathioudakis, A. Wiesmaier, et al., A survey of technologies for the internet of things, 2015 International Wireless Communications and Mobile Computing Conference (IWCMC), IEEE, 2015, pp. 1090–1095.

[23] S. Tayeb, S. Latifi, Y. Kim, A survey on IoT communication and computation frameworks: an industrial perspective, 2017 IEEE 7th Annual Computing and Communication Workshop and Conference (CCWC), IEEE, 2017, pp. 1–6.
[24] T. Lee, The hardware enablers for the internet of things-part i, IEEE Internet of Things Newsletter, 2015.
[25] R. Krasteva, A. Boneva, V. Georchev, I. Stoianov, Application of wireless protocols bluetooth and ZigBee in telemetry system development, Probl. Eng. Cybernet. Robot. 55 (2005) 30–38.
[26] C. Chen, A. Knoll, H.E. Wichmann, A. Horsch, A review of three-layer wireless body sensor network systems in healthcare for continuous monitoring, J. Mod. Internet Things 2 (3) (2013) 24–34.
[27] Z. Shelby, S. Krco, C. Bormann, CoRE Resource Directory, draft-ietf-core-resource-directory-02, 2014.
[28] M.L. Sichitiu, Wireless mesh networks: opportunities and challenges, in: Proceedings of World Wireless Congress, Vol. 2, 2005, May, p. 21.
[29] N. Finn, P. Thubert, Deterministic networking problem statement, draft-finn-detnet-problem-statement-05, 2016.
[30] W. Steiner, N. Finn, Deterministic ethernet: standardization in progress and beyond, in: RATE Workshop, December, 2013.
[31] J. Gubbi, R. Buyya, S. Marusic, M. Palaniswami, Internet of Things (IoT): a vision, architectural elements, and future directions, Future Gener. Comput. Syst. 29 (7) (2013) 1645–1660.
[32] L. Mainetti, L. Patrono, A. Vilei, Evolution of wireless sensor networks towards the internet of things: a survey, SoftCOM 2011, 19th International Conference on Software, Telecommunications and Computer Networks, IEEE, 2011, pp. 1–6.
[33] J. Thrasher, RFID vs. barcodes: what are the advantages? RFID Insider, April 2013. Online: <http://blog.atlasrfidstore.com/rfid-vs-barcodes>.
[34] H.A. Khattak, M.A. Shah, S. Khan, I. Ali, M. Imran, Perception layer security in Internet of Things, Future Gener. Comput. Syst. 100 (2019) 144–164.
[35] K.S. Mohamed, The era of Internet of Things, 2019.
[36] A. Rayes, Enabling IP-based smart services, Network-Embedded Management and Applications, Springer, New York, NY, 2013, pp. 179–194.
[37] M. Dabbagh, B. Hamdaoui, M. Guizani, A. Rayes, Toward energy-efficient cloud computing: prediction, consolidation, and overcommitment, IEEE Netw. 29 (2) (2015) 56–61.
[38] K.P. Dyer, S.E. Coull, T. Ristenpart, T. Shrimpton, Peek-a-boo, i still see you: why efficient traffic analysis countermeasures fail, 2012 IEEE Symposium on Security and Privacy, IEEE, 2012, pp. 332–346.
[39] M. Dabbagh, A. Rayes, Internet of things security and privacy, Internet of Things From Hype to Reality, Springer, Cham, 2019, pp. 211–238.

Further reading

IoT Services for Medical Imaging Equipment Market, MRI, Xray, CT scanners, and tomography: ABI research. Online <https://www.abiresearch.com/market-research/product/1021023-iot-services-for-medical-imaging-equipment/>.

Chapter 8

The Internet of Things: looking beyond the hype

Devender Bhushan and Rashmi Agrawal
Manav Rachna Institute of Research and Studies, Faridabad, India

8.1 Introduction

There has been a lot of attention paid to the domain of the Internet of Things (IoT) by policy makers and very large businesses. It is apparent that private sector organizations and the government sector are keen to take full advantage of the existing opportunities and are keen to protect them and also to further develop new returns as there is great potential for the IoT in almost every economic activity. The IoT can be considered as a paradigm shift that creates an environment in which objects are able to communicate with each other through the wired and wireless connections. This communication is done to create new services or applications which have the common goal of converting the world into a smart world. However, this representation is not simple. There is a lot of debate about the actual IoT versus the predictions made about it. There can be enormous incongruities between the actual numbers and the forecasts about the growth of IoT devices. The most incredible fact about the IoT is that it has no universal definition. Conventional wisdom claims that the "Internet of Things will change everything," be that garbage trucks, children's school bags, or home appliances. However many questions arise in this respect, such as: Are we getting the real information? How certain is it? At what speed is this information generated; and Is it in real time? All these questions are valid and interesting and provide a need to further investigate the world of the IoT to discover its reality. The aim of this chapter is to discuss the various developments achieved so far in the impacts of the IoT in various industries.

8.2 Internet of Things: common definitions

The term IoT is widely used by manufacturers, whether in the home or industrial sectors. Today, from coffee cups to sleeping issues, claims are made

An Industrial IoT Approach for Pharmaceutical Industry Growth.
DOI: https://doi.org/10.1016/B978-0-12-821326-1.00008-5

about use of inbuilt IoT. The reason for this being the promised IoT capabilities. At individual level everyone can predict the future world where all things are connected and communicating with each other over internet and seamlessly communicating over a common protocol to work intelligently. The main goal for the same to enable objects around us operate intelligently and effectively and more over economically to create human surrounding more favorable and operable without explicit instructions from the humans.

IoT business claims are very ambitious. IoT can carry out automatic sensing and analysis of thousands of services or product-related parameters and automatically take action before the relevant service or operation is impacted. The IoT also collects huge amounts of structured and unstructured data from various internal and external sources, such as social media, to gaining a competitive advantage that enables better services and improved business processes. The Internet's impact in the fields of education, communication, business, science, climate control, government, and humanity are widely felt. According to some commercial reports, it is widely believed that the IoT will become the largest technology opportunity ever.

The term "Internet of Things" was first used by Kevin Ashtom in a presentation he made in 1999 while working at Procter and Gamble. His idea was to link RFIDs with Procter and Gamble's supply chain, which became a hot topic on the Internet possibilities. He added that "The Internet of Things has potential to change the world just as the Internet did. Maybe even more so." In 2001, MIT Auto-ID center presented their IoT vision. International Telecommunication Union (ITU) formally introduced the IoT with their Internet Report in 2005. IoT adoption is increasing with the evolution of modern wireless telecommunications, resulting in the increasing presence around us of smart objects like smartphones, smart glass, smart campuses, etc., which can communicate and collaborate effectively and smartly with each other.

The business community is already employing IoT to develop new business models and improve existing processes to reduce costs and improve business results. Personal lives are improving with enhance learning, monitoring, and security, etc.

8.3 Internet of Things: vision

The IoT began soon after the birth of the Internet in 1989, probably as a vision of communication between machines. There was first direct communication with a machine when the telegraph was developed in 1930. With the evolution of defense advanced research projects agency (DARPA) in 1962 and advanced research projects agency network (ARPANET) in 1969, the Internet became a significant component. By using the ARPANET, the global positioning system (GPS) became a reality in 1993 [1]. The Trojan coffee pot was possibly the first IoT application, and the first Internet device was created by John Romkey in 1990 [2]—a toaster connected to the Internet with a TCP/internet protocol (IP) protocol.

It is astounding to think that the IoT was only a concept prior to 1999 and that the term "Internet of Things" was introduced by Kevin Ashton, a visionary technologist, in 1999. In order to gain attention from his senior managers, Ashton made a presentation and used the term IoT during his work with supply chain management, but real recognition was not achieved until 10 years later. LG electronics created a refrigerator with Internet Digital data intensive operating system (DIOS) that was able to keep track of food items stored inside it based on their RFID tags, however it was not successful commercially due to its high cost. The concept of IoT gained traction around 2010. The famous "hype-cycle for emerging technologies" was invented by a market research company and Gartner included the emerging phenomenon "the Internet of Things." In January 2014, Google bought Nest for $3.2 billion, which created an awareness of the term IoT by the mass market.

The IoT can be considered as a revolution, which came into existence through the Internet, in which objects identify themselves and where it is also possible to acquire information through communication between these objects. The IoT provides an integration of energy, health care, tourism, security, transportation, environment, and many other personal and industrial sectors.

A recent initiative by the European Commission is the Alliance for Internet of Things Innovation (AIOTI). This supports interactions among the various IoT players. It is hoped that this initiative will provide prodigious support in making future standardization policies. For creation of a dynamic ecosystem of the IoT, all members of AIOTI will work jointly to build a research cluster for the IoT (IERC) to transform IoT ideas into IoT solutions.

8.4 Internet of Things applications

1. *Wearables*: The IoT and wearables are an exciting current topic. Wearables are smart devices which can be easily worn on the body to carry out specific tasks. These wearable devices are able to perform computing tasks in the same way as computers/laptops/mobile devices. Although wearable devices refers to devices which can easily be put on or taken off from the body, there are now devices which can even be planted on the body, like the smart tattoo.

   In the history of wearable technology, the first device was used in 1961 by MIT professors Edward O. Thorp and Claude Shannon, inside shoes. They designed a timing device to predict the accuracy of a roulette wheel at a gambler table. The first electronic wristwatch calculator was "Pulsar" in 1975. In 1994, the first wearable wireless webcam was designed which opened the path to wearables for researchers.

2. *Homes and buildings*: IoT with smart homes applications is very useful and effective. With the help of sensors and controllers smart homes monitor and automatically trigger services/events to save time in the case of emergencies including fire, intrusion, or gas leak. Smart building solutions also enable video monitoring, light control, air condition control, and power supply control.

   The following are examples for smart building enterprise solutions:
   - *Safety monitoring and altering*: In this example we consider noise-level monitoring in urban zones such as hospitals and schools, and sounding in real time. Industries can monitor and control the levels of harmful gases like CO_2 emissions.
   - *Smart lighting*: To minimize energy consumption the IoT can be used to adopt weather-adaptive lighting.
3. *Health and wellness*: The IoT enables a wide range of applications in the health and wellness sector. It has vigorous connection opportunities including patient-to-doctor, patient-to-machine, device-to-machine, doctor-to-machine, mobile-to-human, tag-to-reader, object-to-object, and sensor-to-mobile. This connects humans, machines, objects, sensors, and devices in an intelligent way to provide an effective healthcare system.

Personalized healthcare, a term in IoT health care, is integrated practice of the well-being of an individual based on an individual's unique characteristics (biological, physical, and psychological). The IoT is able to provide data that are more extensive than typical analytics with the help of instruments which are skillful in accomplishing prospective research. Therefore, the IoT supports health care by providing additional real-world and consistent data, which produce enhanced solutions with the detection of previously unknown information. Some of the popular IoT applications in health care include:

1. Medical research: Real-world information is critical in nature and the IoT unlocks a pool of important data and information through real-time data capturing and analysis.
2. Medical hearables: Hearables are hearing aids which have helped people who have hearing loss.
3. IoT for patients: Devices like wearables (fitness bands and other devices which are connected through the Internet) give access to personalized data for the patients. Through these devices one can measure blood pressure, pulse rate, heartbeat, calorie count, exercise, etc.
4. IoT for physicians: By using monitoring equipment which is embedded within the IoT, physicians can keep track of their patients' health in an efficient manner. It enables physicians to be vigilant and active for their patients to provide the best treatment in a timely manner.
5. IoT for health insurance companies: Through intelligent connected IoT devices, health insurance companies can trace claims and detect

fraudulent claims. For involvement in health data generated by IoT devices, insurance companies can also offer incentives to their customers.

6. IoT for hospitals: In addition to the monitoring of health data of customers, the IoT is useful for hospitals in other ways also. The most important uses include real-time tracking of IoT-enabled medical equipment like nebulizers, oxygen pumps, wheelchairs, and defibrillators.

   Although the IoT in health care is not limited to its uses as listed, researchers are putting constant efforts into providing the maximum societal benefits. Some of the advantages of the IoT in healthcare, as shown in Fig. 8.1, are:

   - Proactive treatment
   - Drugs and equipment management
   - Reductions in cost
   - Improvisations in patients' treatments
   - Reduced time to disease diagnosis

Healthcare applications interconnect a number of devices using the Internet in which security is a major challenge as these data must be secure. Yehia et al. [3] discussed the hybrid security techniques for IoT applications. They discussed the success of healthcare applications as being solely dependent on its security as the patient data must be secured for both legal and ethical reasons. Some wireless medical sensor network-based secure healthcare applications are used where temperature, vital signs, blood pressure, and heart rate are collected and visualized by a body sensor network [4]. For a diabetic patient, the body area network [4] can be helpful for autoinjecting whenever the insulin level is high.

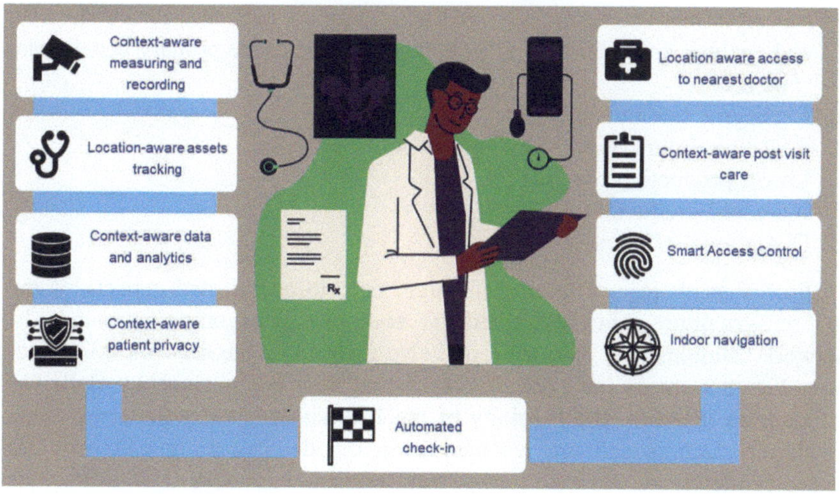

FIGURE 8.1 Applications of the IoT in health care.

For smart healthcare systems an IoT-aware architecture was created by Catarinucci et al. [5], who provided a smart hospital infrastructure which was composed of:

1. User interfaces which were used for data management and visualization;
2. Smart gateway for the IoT;
3. Hybrid sensing network (RFID based).

   Although a lot of work has been done in the area of health and wellness, researchers are still struggling to achieve unresolved issues and the security and privacy issues of these applications have also become a challenge.
4. *Mobility and transport*: Due to industry regulations, logistics industries are forced to implement cost-effective and scalable solutions. Most transportation companies have already discovered the benefits of IoT solutions in offering services, and improving efficiency and security to control real-time visibility of operations. The major uses include:

a. Intelligent parking;
b. Smart roads and traffic congestion control;
c. Connected rail.

5. *Agriculture*: It has recently been projected that the global population will reach 9.6 billion by 2050. Therefore to fulfill the food requirements of this colossal population, a revolution in agriculture is required. The agriculture industry has identified this need and has embraced the IoT to help meet this challenge. In agriculture, the various challenges faced include extreme weather conditions, impacts of the environment on crops, sudden changes in climate, etc. The IoT is now helping the agriculture industry to meet the vast demand for food by addressing these challenges. Smart agriculture covers agrotech, agriculture, and farming. The revolutionization of this sector has been done with the use of wireless networks, data collection, and analysis and use of cloud technology. Some of the primary benefits are (Fig. 8.2):

a. Intelligent cost management;
b. Real-time information;
c. Lower production costs;
d. Increased production;
e. Operational efficiencies.

Stočes et al. [6] published an article describing selected aspects of the IoT in agriculture. They described IoT equipment categorization, standards, network solutions, and platforms, and also focused on the network infrastructure. Fig. 8.3 shows the LoRaWAN schema which was given by LoRaTM Alliance to represent the topology of the IoT network using the LoRaWAN technique. Data which are received from the devices using gateways are stored in the cloud, and then further transferred to end users. This is a secure and expandable network.

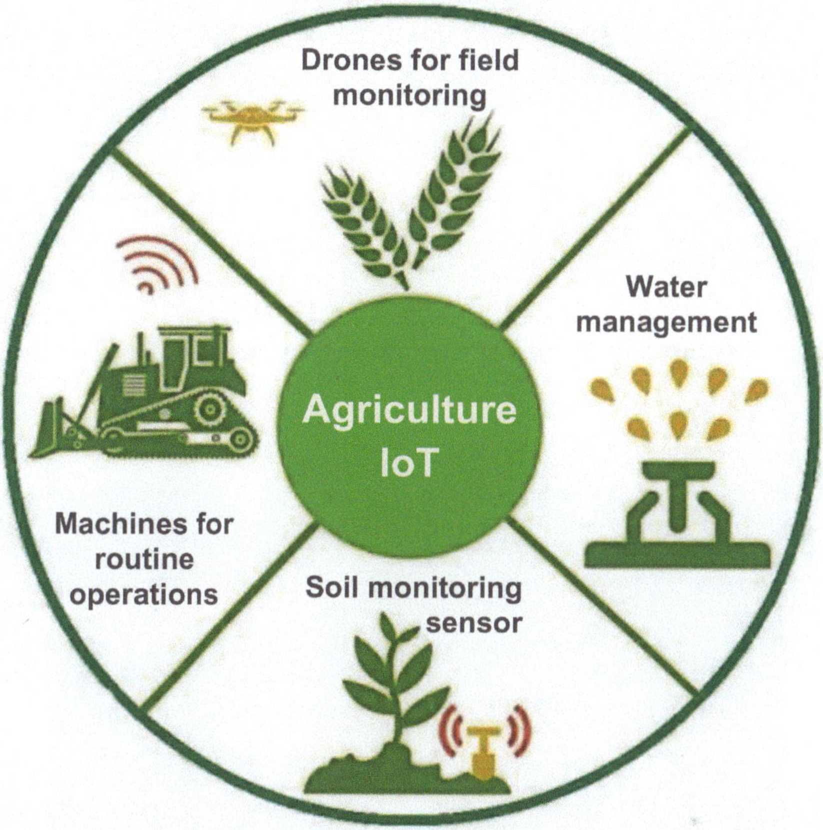

FIGURE 8.2 The IoT in agriculture.

For the IoT in agriculture, one of the leading examples of utilizing open software and hardware is FarmBot [7], which is illustrated in Fig. 8.4. This is an open source automated farming machine. It is suitable for mobile or fixed raised beds in schools, research labs, and in small backyards. FarmBot uses the free FarmBot web application and can only be programmed and controlled using this web application.

Internet connection is a necessity for all IoT device operations, and in almost all cases a wireless connection can be deployed using various standards and utilizing several parameters. Some of the most important parameters are:

a. Uplink and downlink data rate
b. Frequency band
c. Channel width
d. Energy consumption
e. Topology

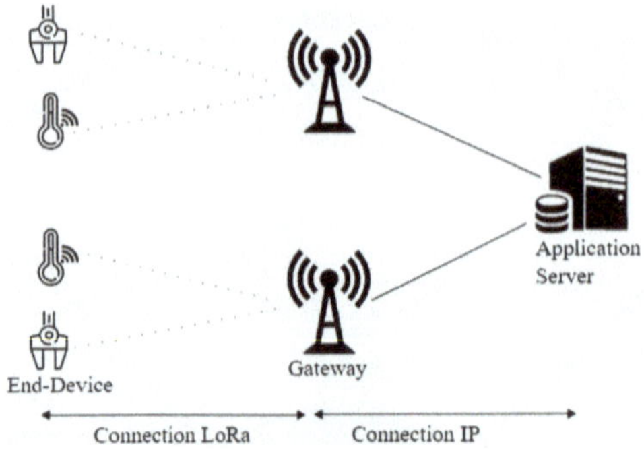

FIGURE 8.3 LoRaWAN schema. *From https://commons.wikimedia.org/wiki/File:Architecture_lorawan.svg.*

FIGURE 8.4 FarmBot. *From https://commons.wikimedia.org/wiki/File:FarmBot_Genesis_with_solar_power_and_compost.jpg.*

In a general sense, if any device is connected to the Internet it is known as an IoT device. However, the following criteria are used to classify distinct devices:

a. Type of device (server, sensor)
b. Device dimensions
c. Purpose
d. Energy consumption

Fig. 8.5 represents the complete ecosystem for the IoT in agriculture which illustrates how cloud is used to connect devices in agriculture. The use of the IoT in agriculture empowers farmers by providing them with smart decision-making tools.

8.5 Internet of Things reference framework

The IoT is a collection of generic components that enables physical objects to interact with other elements in the system. These components are:

1. Sensors
2. Identifiers
3. Software
4. Internet connectivity

Collecting all these components together, the IoT creates a network of things with element identifications, embed software with sensors, and connectivity to the Internet. The IoT enables things to exchange and share information with connected devices or operators in the network (Fig. 8.6).

There are many frameworks in which we can design IoT solutions, for reference IoT is divided into four levels.

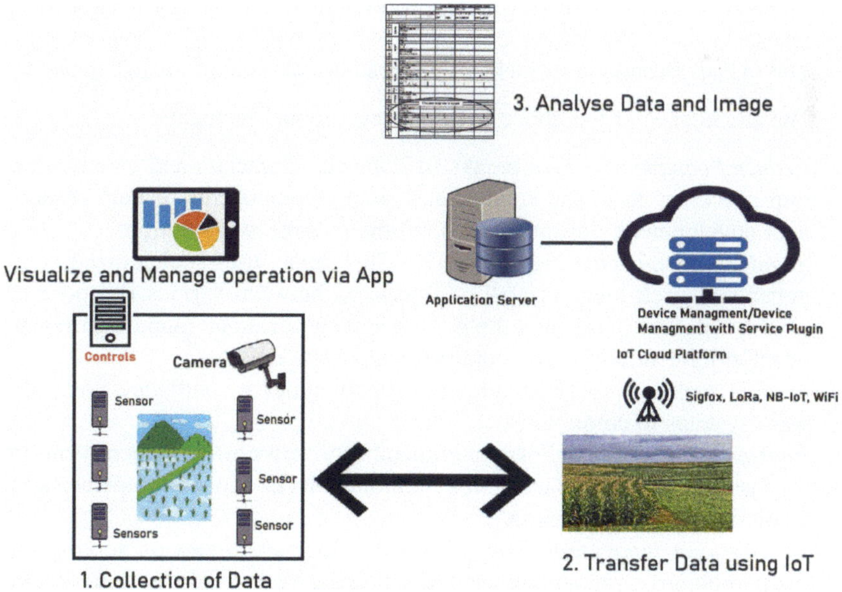

FIGURE 8.5 IoT ecosystem for agriculture.

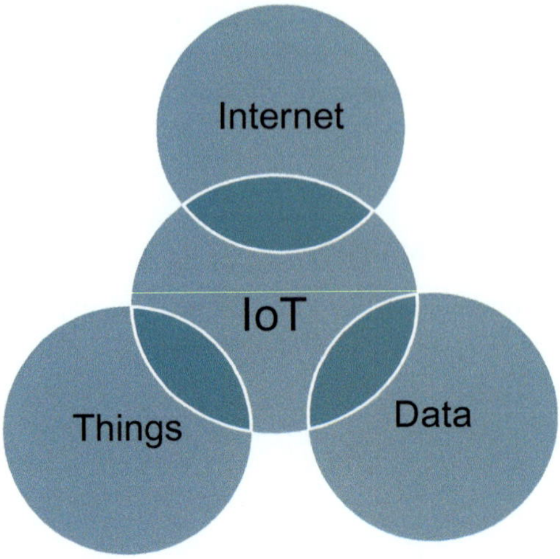

FIGURE 8.6 IoT definition.

Information control is passed down from one level to the next, starting at the application level and proceeding to the IoT device and back step up.

1. The IoT device level includes all IoT sensors and actuators.
2. The IoT network level includes network components like the gateway, routers, and switches.
3. The IoT application service platform level includes the key management areas.
4. The IoT application level includes applications working in the IoT network.

The advantages of the above IoT four-level model include:

1. *Reduced complexity*: This breaks IoT objects (elements) and communication processes in to the smaller and simple components to help component development, design, and troubleshoot developed solutions.
2. *Standard components and interfaces*: This model uses standardized components at each level as well as interfaces between various components of the solution. This will help in allowing different manufacturers to develop joint solutions and common support models.
3. *Module engineering*: This allows different types of hardware and software systems to communicate.
4. *Interoperability*: This allows interoperability between component vendors by ensuring the various technology-building blocks can work together with commonly defined protocols.
5. *Accelerating innovation*: This allows developers to focus on solving the main problem without worrying about basic functionalities that can be implemented across the various businesses.

6. *Simple education*: The solution splits overall complex IoT solutions into small more manageable components that make learning easier and faster.

8.6 Twelve driving factors for the Internet of Things

When discussing business transformation, IoT has become a significant force and its impact has already been felt across all verticals and society. This also leads to the creation of new businesses and social requirements (Fig. 8.7).

8.6.1 Convergence of operation technology and information technology

Operation technology (OT) is the term use to define the world of industrial plants, controls, and automation equipment that includes machines and systems to run a business, and the controller's actuators and sensors. Information technology (IT) is the term used to define the world of end-to-end information systems focusing on computing, data storage, and network support to business operations such as enterprise resource planning (ERP) systems, and designing systems like Auto-CAD (computer aided design) and human resource (HR) systems.

Historically, IT and OT were always managed by two separate organizations with different cultures, philosophies, and sets of tool or technologies. IT departments were developed by companies to create efficient networks, internal communication, and web hosting, and to develop and manage business apps like customer

FIGURE 8.7 Twelve driving factors for the IoT.

relation management (CRM) and ERP. On the other hand, OT relies on real-time data that drives security, safety, and process control, and depends on tested and trusted processes. Many processes need to run 24 × 7 with zero down time, and the production environment does not tolerate shutdowns for software updates or upgrades. IT is more lenient with software updates for introducing new updates in technologies, securities, or processes (Fig. 8.8).

The IoT plays a major role in OT and the traditional IT operational model. With the growth of business-specific technologies, traditional IT operations are not able to scale up and keep pace with the evolving demands or provide the required expertise. Due to the pressure of IoT technology adoption, OT is forced to accept integration with other functions, hence IT and OT functions are expected to work together to minimize risks.

8.6.2 Creative internet-based businesses

8.6.2.1 Uber

Most will be familiar with the story of Uber, where the cofounders attended a conference in Paris in 2008. Garrett Camp and Travis Kalanick were discussing the problem of finding a cab, especially when it was raining. When they tried to solve this problem they came up with three main requirements: the solution should be Internet based (request and track service from mobile), it had to provide a service fast, and rides had to pick up from any location.

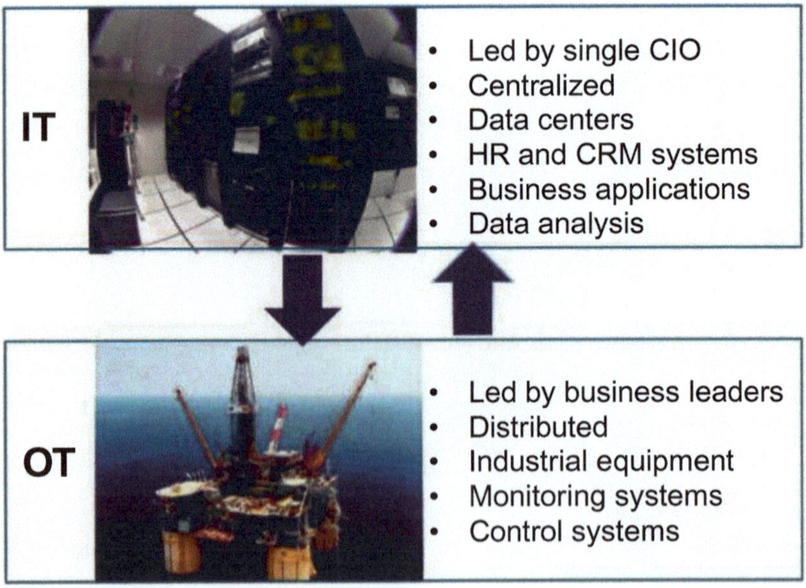

FIGURE 8.8 Operations (IT and OT).

The main component in this solution is the Internet-based platform connecting customers (passengers) with car drivers. In this scenario consumers are not Uber's employees and there is a huge number of drivers who can join the platform. Uber has the ability to scale at a very fast rate and at zero marginal cost.

To overcome this Uber uses sensors in the driver's smartphone to track their behavior. From tracking speeds, braking too hard, or taking excessively long routes, all information is tracked by the Uber team. They use GPS data to track route details, with gyrometers in smartphones used to measure small movements, while GPS combined with accelerometers shows how often a vehicle starts and stops.

These are some more examples of successful internet businesses:

1. Zomato
2. PayTM
3. Amazon
4. Facebook

8.6.3 Mobile device expansions

There is a huge number of new things connected to the Internet each day, not only mobile devices or sensors, but also things which have never previously been connected to the Internet but are now part of the network. This is due to the vast adaptability of mobile devices. The growth of mobile data traffic is expected to be 18-fold in the next 5 years. According to Cisco Visual Network Index, smartphone traffic will grow from 1.74 exabytes per month in 2014 to more than 18 exabytes per month in 2019 (Fig. 8.9).

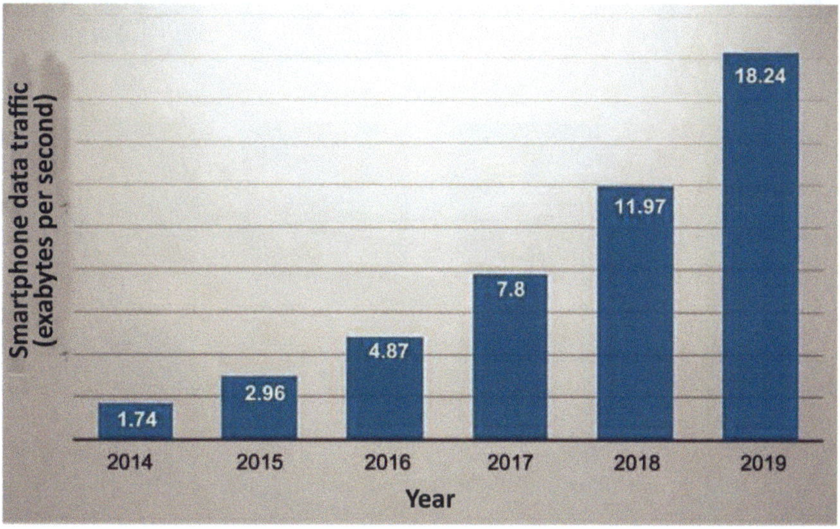

FIGURE 8.9 Smartphone traffic over time.

8.6.4 Social network expansion

Social networks, such as Instagram, Twitter, and Facebook, and the cloud-based services like Amazon's AWS and Google GSUITE are examples of the large scale migration to the cloud across all industry verticals (Fig. 8.10).

8.6.5 Analytics

To understand the different versions of analytics the following terms must first be defined: big data, structured data, and unstructured data. Big data refers to the extremely large data set generated by IT systems (software) as a result of an operation. This large amount of data can be used to analyze and identify patterns and give an insight into the operation of the system.

Structured data refers to the organized data that can be fit in the columns and rows of an relational database management system (RDBMS) system.

FIGURE 8.10 Major social channels.

Unstructured data, on the other hand, are difficult to organize, such as videos and social feeds.

Analytics 1.0 refers to a process to collecting structured data from various sources and send them to a centralized location to be correlated and analyzed using predefined queries and views. Analytics 2.0 basically collects structured and unstructured data from various sources and send them to a centralized location to be correlated and analyzed using complex queries with forward-looking and predictive views. Analytics 3.0 allows to collect, parse, analyze, and correlate structured and unstructured data at or close to the edge (source of the data) (Fig. 8.11).

8.6.6 Cloud computing and virtualization

Earlier enterprises used their own computing infrastructure to deploy and manage their IT system. Cloud computing was first introduced in 2008, and allows enterprises to outsource computing infrastructure to cloud providers like AWS, Microsoft Azure, and Oracle Cloud, either fully or partially.

A public cloud provider delivers on-demand cloud services over the Internet, and enterprises need to pay only for the computing, storage, and

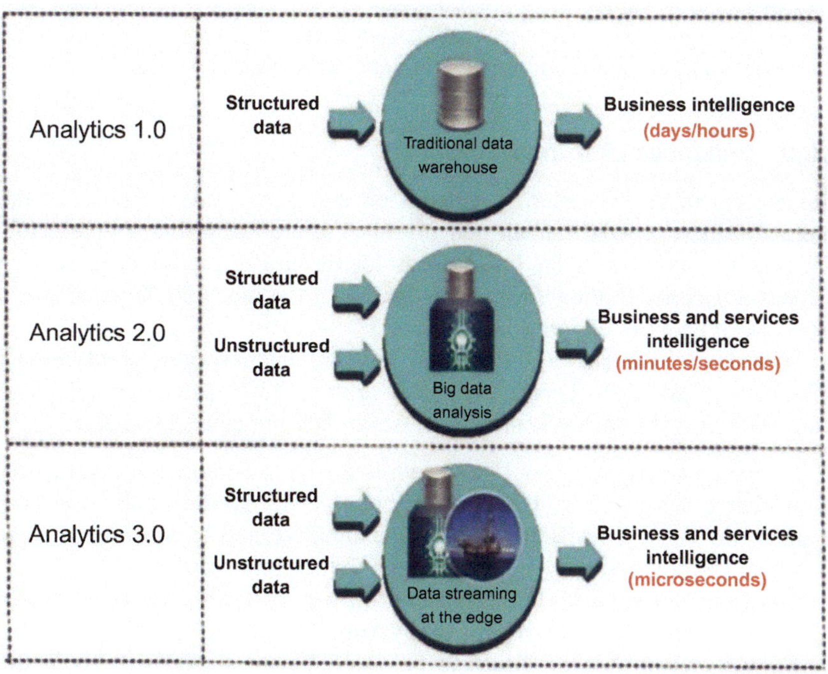

FIGURE 8.11 Analytics 1.0, 2.0, and 3.0.

bandwidth consumed. On the other hand, private cloud solutions use their own data center to deliver computing services to internal business users. Another model is the hybrid cloud which is a combination of these providers.

8.6.7 Technology expansion

IoT hardware like sensors, affordable computers like Raspberry Pi, open source microcontrollers such as Arduino, and software technologies are not only being developed faster than ever before but also at lower prices. These developments are already transforming user's behaviors and developing new opportunities for businesses. Business leaders now understand that unless their organizations adapt to such changes quickly, they will become inefficient in a competitive marketplace.

8.6.8 Digital transformation

Digital transformation initially started with a very limited scope, that is, it was a move to paperless operations to reduce the destruction of trees. However, the digital transformation was soon adopted by industries with extended goals of moving to digital operations. Data captured from various sources including devices and the processes that are enable by digitization can be used to improve the processes and analytical results. Many healthcare companies use digitization to improve the patient experience and reduce cost.

8.6.9 Enhanced user interfaces

The success of an IoT solution depends on the user experience (UX) or the interaction between the human and machine where applicable. A basic IoT UX principle is to meet the basic needs for the use of a product or service without difficulty. Overengineering or including too much intelligence into a product can backfire and discourage users.

With the help of markup languages, software engineers can adopt the UX principles called KISS (keep it short and simple). KISS states that most systems work best if they are kept simple. Major UX principles include:

- *Simple and easy*: A better UX system is a system without a use interface. A simple design without unnecessary complexity should be the best way to work. Being consistently stable and intuitive and establishing a clear visual hierarchy are other major factors to take into account.
- *Contextual principle*: The design must ensure that users are contextually aware of where they are within the system.
- *Human principle*: User Interface provides human interactions about the machine like interactions, and should be sleek.

- *Engagement principle*: UI should fully engage the user to deliver the best value and quality information.
- *Beauty and delight principle*: The UX should be enjoyable and engage the user with the system or service.

8.6.10 Internet of Things adoption rate

The IoT adoption rate is approximately five times greater than that for electricity and telephony. Mobile devices and tablets are replaced at a faster rate than ever before. The number of connected devices now exceeds the planet's human population. These are not only laptops, mobiles, and tablets, but also sensors and other objects which were not connected previously. Adding all these physical objects to IP networks requires new network models. Information technology and communication systems need to deal with these requirements (Fig. 8.12).

8.6.11 Security requirements

Data protection within systems has been an issue since the inception of data networks. With Internet commercialization security concerns have expanded to cover privacy, financial transactions, and cyber threats. We are buying and deploying smart gadgets all over homes, therefore to safeguard and secure data has become more critical in the IoT era as the impact of damage is greater with connected devices as they can affect directly physical objects connected to the IoT.

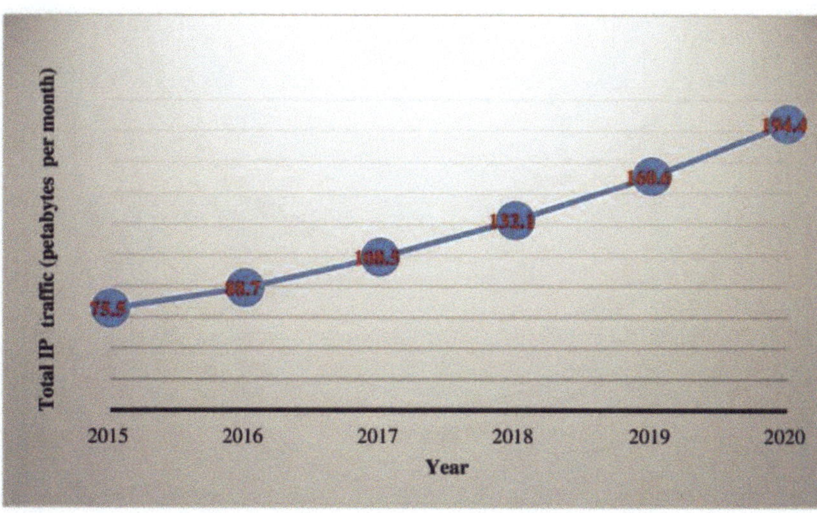

FIGURE 8.12 Global IP traffic growth.

8.6.12 Moore's law

There are three major observations for Moore's law:

1. Over time, computer power has approximately doubled every 18 months. According to this fact the number of transistors in a circuit has grown twofold every 18 months.
2. The size of storage technology transistors has continued to shrink over the years and has now reached the atomic level. We also are able to put more storage onto a device of the same size. A very good example to understand this is that in 1995 the total number of transistors in all PCs was about 800 trillion, but today 800 trillion transistors are included in a weekend's sales of Apple's iPhone.
3. The price of transistors has been reduced by more than 50% per year.

There is no accurate figure for the estimated IoT revenue for the next 10 years, but forecasts by industry experts claim the opportunity is huge (Fig. 8.13).

8.7 Functions and requirements for Internet of Things service platforms

Billions of sensors, devices, and applications are expected to join the IoT over the Internet. The service platform is one of the most critical and important prerequisites for successful, scalable, and adoptable IoT solutions. The service platform provides the abstraction across the diverse devices, data sources, and their management with complete control of the range and system process.

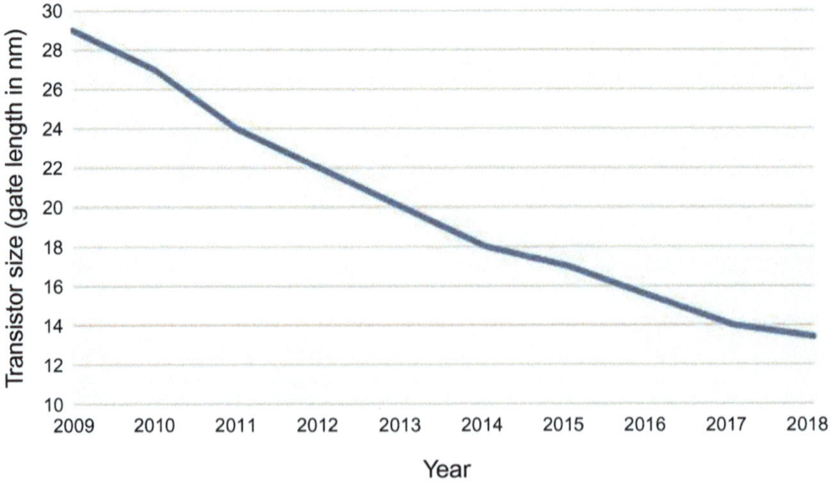

FIGURE 8.13 Transistor sizes over time.

The service platform must adopt solutions by integrating all required technologies and components into a common, open, and multiapplication environment. It must include the capability to configure, deploy, troubleshoot, secure, manage, and monitor all IoT devices connecting to the system. This also includes the ability to manage application (software/firmware) installation, upgrading, patch uploading, starting/stopping, debugging, and monitoring. Fig. 8.14 shows the IoT service platform management functions.

The term FCAPS—fault, configuration, accounting, performance, and security—is often used in network management studies as a useful way to break down the multipart network management functions. While FCAPS still applies, the management functions of IoT solutions are more multifaceted than traditional networks due to following factors:

- The IoT system need to include new devices like sensors, white-labeled gateways, and switches. Some of these devices are not very expensive and do not include the type or level of instrumentation required for traditional management functions.
- The IoT system mostly uses recent technologies like GPS to track the exact location of the device, which was not considered by traditional management solutions.
- More than two dozen access protocols are supported by IoT solutions, and network management for each of these may vary.
- Multiple verticals are supported by IoT solutions, each of which has a different set of management, quality of service, and grade of service requirements.
- IoT solutions support fog computing with new and challenging network, computing, and storage requirements.
- Many companies and service providers outsource operations and this requires multisourcing key parts of the network and management functions.

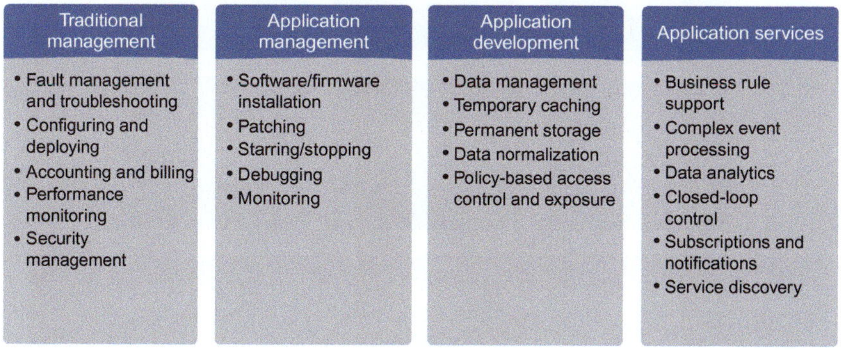

FIGURE 8.14 IoT service platform management functions.

8.7.1 Common Internet of Things service platform functions

The common functions of the IoT service platform can be categorized into 11 areas:

1. IoT platform manager
2. IoT discovery and registration manager
3. Firmware manger
4. IoT data manager
5. Communication
6. Topology
7. Group manager
8. Billing and accounting manager
9. Subscription and notification
10. Application program interface (API)
11. Element manager

Fig. 8.15 shows the common functions for an IoT service platform. It does not contain the multicity of the entities nor the relationships among them.

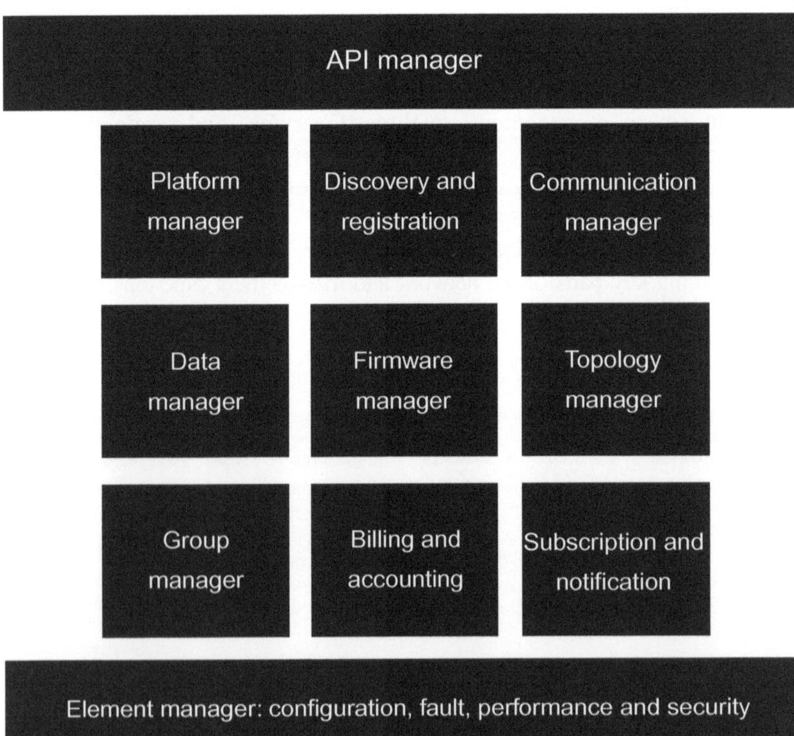

FIGURE 8.15 IoT service platform functions.

8.7.2 Internet of Things security and privacy

The IoT promises to make our surroundings smarter and make our lives more convenient with the help of smart objects which sense the environment and communicate to perform automatic actions based on automatic responses from other smart objects. This makes our lives more easy and convenient but this convenience and connection to IoT devices brings a new set of security risks and privacy issues that must be attended to properly. Ignoring these issues can affect different aspects of our lives including our homes, cars, or even our bodies.

IoT security problems are extremely serious when it comes to businesses. If an attacker hacks any of the smart IoT objects used in an enterprise network, it could be critical, such as an attack on a gas company IoT network that controls air pressure could lead to a loss of human life.

8.7.3 Security challenges

Cyber security threats include the following:

1. *Multiple technologies*: Multiple technologies like RFID, sensor network, cloud/fog computing, and virtualization can be combined with the IoT solution. Each component has its own vulnerabilities. The main problem with an IoT paradigm design is to secure all possible accesses to all of the objects in the chain.
2. *Scalability*: 26.3 billion devices were forecast to be connected to the Internet by 2020, as predicted Cisco. This large number makes it important to design solutions to help and scale such capacity with efficient defensive mechanisms.
3. *Multiple business verticals*: The IoT paradigm has many applications including smart buildings, health, smart cities, etc. The security requirements of each vertical are different.
4. *Availability*: Availability refers to a system that is continuously operational for a long period of time. It is measured relative to 100% operational availability.
5. *Big data*: IoT paradigms have huge numbers of objects which generate a huge amount of data over time. This makes it important to come up with efficient mechanisms that can secure these huge amounts of data.
6. A widely held but difficult standard of availability to achieve is known as the five 9s (99.999%).
7. *Resource limitations*: IoT devices generally have limited resource capabilities like CPU, memory, storage, battery, and range to transmit data. This makes devices vulnerable to denial-of-service (DoS) attacks, where an attacker can use the limited resource capability to cause service disruption.

8. *Remote location*: IoT devices in many verticals are installed in locations where they are difficult to reach. Devices can easily be hampered from working correctly without being seen or noticed.
9. *Mobility*: There is an expectation of smart objects to change their locations in the IoT paradigm. This add extra difficulties in efficient defensive mechanisms.
10. *Delay-sensitive service*: Most IoT applications are expected to be delay sensitive, and the implementation of a protection mechanism degrades their service time or may cause disruption.

8.7.4 Security requirements

The security requirements for IoT include:

- *Confidentiality*: This ensures that exchange data can be understood only by the intended entities.
- *Integrity*: This ensures that exchanged data are not tampered with or changed by a third party.
- *Authentication*: This ensures that only the entities involved in the operation are able to access the system.
- *Availability*: This ensures that service is continuous and not interrupted. DoS attacks target this requirement.
- *Authorization*: This ensure that the entities have the required control of permission to perform the operation they intend to do.
- *Update*: This ensures that data are updated or new. Replay attacks target this requirement by sending old data to control an old state.
- *Nonrepudiation*: This ensures that an entity cannot refuse an action that it has performed earlier.
- *Forward secrecy*: This ensures that when an object leaves from the network, it will not understand and be responsive to the communications that are broadcast or exchanged after its removal.
- *Backward secrecy*: This ensures that any new object that joins the network will not be able to understand the communications that were exchanged or shared prior to it joining the network.

8.7.5 Three-domain architecture

The three-domain architecture needs to be considered in security analysis.

1. *Sensing domain*: This domain is made of all smart objects that have the capability to sense the environment and start sending sensed data to another device in the fog domain. The smart object in the sensing domain is expected to change the location over time (Figs. 8.16 and 8.17).
2. *Fog domain*: This domain consists of fog devices that are located in areas that are highly dense with many smart objects. Each fog device is

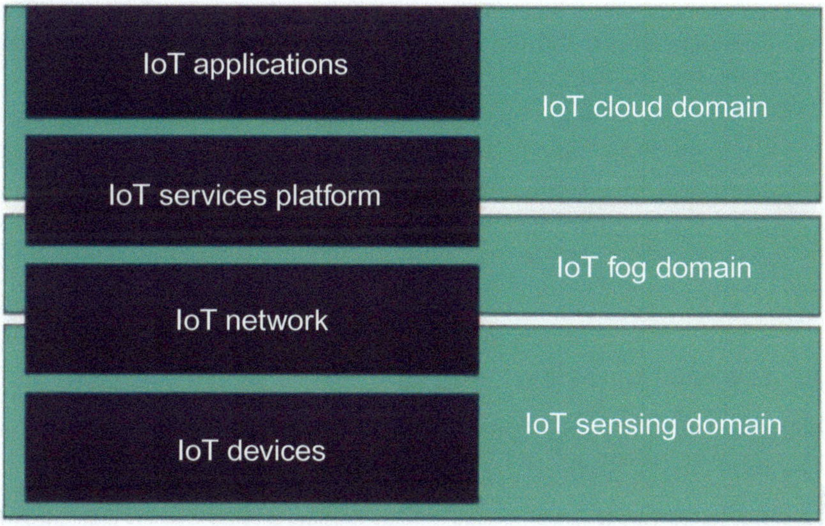

FIGURE 8.16 IoT domains mapping.

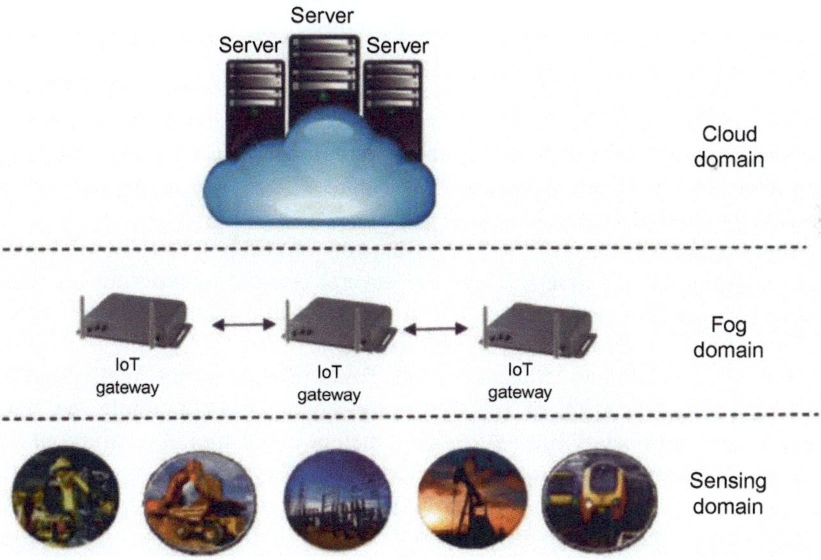

FIGURE 8.17 The IoT domains.

allocated a set of smart objects where the allocated objects report the sensed data. The fog device performs operations on the data including aggregation, preprocessing, and storage.

3. *Cloud domain*: This domain consists of large numbers of servers that host the application and are responsible for processing on data reported from the fog devices.

For each one of these domains, we identify the common security attacks, then describe how these attacks are launched, what vulnerabilities are exposed, and what technique can be used to prevent a reoccurrence.

With the above study we understand the purpose, applications, implementations, and security challenges. We now discuss the reality for the future of the IoT.

8.8 The road to reality

There are many other challenges in addition to the core hardware, software, and security in current IoT solutions. The volume of data is one example of the challenges facing the global fiber network and that it needs to manage. Cisco forecasts in its white paper "The Zettabyte Era" 2017 that global M2M (machine-to-machine) IP traffic could increase at a compound annual rate of 49% over the next 5 years, from around 2 exabytes in 2016 to more than 14 exabytes in 2021. This amounts to only 5% of the projected total IP traffic from all sources, and is twice the share represented by M2M traffic in 2016. This shows the IoT's potential for consuming the bandwidth in the coming years.

Important issues have also been raised due to the intersection of IoT technology with humans. Driverless cars must process continuously huge amounts of complex data to ensure a smooth and safe journey, which requires emerging deep-learning, neural-net approaches to compute. One more major concern is security, as covered above.

The success of the IoT will also depend on finding ways to interact between different IoT assets. There are several companies working together to develop standard protocols and software to enable interoperability, however so far no universal standards have emerged.

According to Blaauw "Success on this front, though, might finally lead to the long promised vision of IoT connected homes." For example, he suggests, smart thermostats linked to smart meters could monitor variations in the price of electricity—as utilities seek to tailor demand and supply. "There are greater opportunities as more of these things integrated together." He added "maybe their value isn't initially that high but it gets higher over time."

The long-run prospects of the IoT look solid, according to Lineback of IC Insights. According to him, some of the major elements required to make the IoT a success include the technical standards. We have failed so far in this area to develop a common standard. Data analysis can mature quickly, and according to the most recent IoT equipment market forecast projection, the IoT will grow from US$75 billion in 2016 to US$128 billion in 2020. These numbers may not seem so big, but it is thought that this will be an important industry game changer. The IoT is beginning to feel more like reality as we move forward.

References

[1] P. Suresh et al. A state of the art review on the Internet of Things (IoT) history, technology and fields of deployment, in: 2014 International Conference on Science Engineering and Management Research, ICSEMR 2014, 2014. doi:10.1109/ICSEMR.2014.7043637.

[2] J. Romkey, Toast of the IoT: the 1990 interop internet toaster, IEEE Consum. Electron. Mag. (2017). Available from: https://doi.org/10.1109/MCE.2016.2614740.

[3] L. Yehia, A. Khedr, A. Darwish, Hybrid security techniques for Internet of Things healthcare applications, Adv. Internet Things (2015). Available from: https://doi.org/10.4236/ait.2015.53004.

[4] E. Montón, et al., Body area network for wireless patient monitoring, IET Commun. (2008). Available from: https://doi.org/10.1049/iet-com:20070046.

[5] L. Catarinucci, et al., An IoT-aware architecture for smart healthcare systems, IEEE Internet Things J. (2015). Available from: https://doi.org/10.1109/JIOT.2015.2417684.

[6] M. Stočes, et al., Internet of things (IoT) in agriculture – selected aspects, Agris Online Pap. Econ. Inform. (2016). Available from: https://doi.org/10.7160/aol.2016.080108.

[7] FarmBot, FarmBot genesis, FarmBot [Online], 2016. Available: <https://farmbot.io/> (accessed 05.02.16).

Chapter 9

Internet of Things: the new Rx for pharmaceutical manufacturing and supply chains

Deepak Kumar Sharma, Prachi Gupta and Priety
Department of Information Technology, Netaji Subhas University of Technology (formerly known as Netaji Subhas Institute of Technology), New Delhi, India

9.1 Internet of Things architecture

The Internet of Things (IoT) is not only consumer devices connected to the Internet, it is much more than that. It is, in fact, a technology that helps in the building of systems, that have the capability of autonomously sensing and responding to stimulations from the real world, without any human interference. There are four fundamental components to the IoT architecture as described next [1,2].

9.1.1 Stage 1: sensors/actuators

Sensors or actuators are the key components, equipped in devices, which help in the collection of live data from the surrounding environment. Devices thus are capable of accepting, emitting, and processing signals. Sensors collect data from an object or the environment under consideration and transform them into useful information, whereas actuators are equipped to help change the physical conditions that are useful for generating the data, such as adjusting an air flow valve or shutting off the power supply [3].

9.1.2 Stage 2: data acquisition systems

The data received from sensors is in analogue form, and need to be converted into digital data so that they can be processed further, downstream. This data conversion and aggregation is performed by the data acquisition

An Industrial IoT Approach for Pharmaceutical Industry Growth.
DOI: https://doi.org/10.1016/B978-0-12-821326-1.00010-3

systems. This system basically establishes a connection to the sensor network, collects the analogue data from there, and then converts it into digital data. These collected data are then sent to the cloud system, using the Internet, wired LANs, WANs, or the Wi-Fi, for further processing [4].

9.1.3 Stage 3: data processing

The digital data collected from the second stage have to be sent to the data center after further processing. This is where the functioning of edge IT systems comes into play, and helps in this further analysis. Though they may be located in remote locations, the places where sensors are in close proximity are the most suitable places for the installation of these processing systems, such as a wiring closet [4].

After the data are collected, they are processed by the software and sent to the cloud. This processing can include anything, from highly complex jobs such as identifying objects on a video using computer vision, or simply checking the temperature.

9.1.4 Stage 4: cloud analytics

Data that need even further processing are sent to a cloud-based system or a physical data center. Here the data are analyzed, managed, and stored more securely using IT systems which are much more powerful. At this stage, we can combine our sensor data with data from other sources for a deeper understanding, as well as for a more in-depth analysis to be executed; however the trade-off for all this is the longer feedback time, which means that this stage takes a long time to produce the results.

This information now needs to be made available to the end user using various methods, which may be sending text messages, or notifications via emails or even triggering an alarm on their mobile phones; however, this in no way means that this is always a one-way communication. Even the user can perform some actions, dependent on the complexity of the system and the IoT application being used. This could create a cycle, where a user can give commands to the first stage (actuators and sensors), for some actions to be performed. This could be regarded as the fifth stage of the IoT architecture—the user interface (Fig. 9.1) [4,5].

9.2 Internet of Things technologies

9.2.1 Automatic information data collection

Automatic information data collection (AIDC), which stands for automatic identification and data capture, is a technology that can be used to record, verify, identify, communicate, and store information about various items.

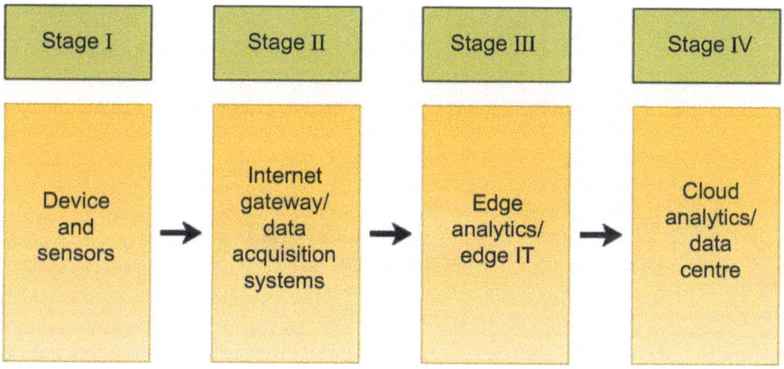

FIGURE 9.1 IoT architecture stages.

One of its main applications is in controlling warehouse and manufacturing operations. This being an automated process, information can be collected accurately and quickly [6].

Throughout computer systems, AIDC identifies and tracks most items. First, with its digital signal, an object is identified by the system. Such data can include machine performance, inventory counts, and supplier data. Then a server gathers the key data. Access to track different stages of the manufacturing process is given to the user [6].

Each development phase in the past included the manual entry of information. This was a system that was very labor intensive for staff. The AIDC innovations of today, by comparison, have simplified the entire operation [7].

This is achievable by software and hardware applications of data collection technology. AIDC tracks dynamic supply chains, quality control, and monitoring of inventories. Thus the work of humans is reduced and the productivity of AIDC provides affordable and high-quality goods for the market. It has made them more productive and meaningful for workers. As such, the digital economy has been shown to be important (Fig. 9.2) [7].

9.2.1.1 Types of data capture technologies

Various techniques are used for capturing and identification of data using AIDC. These include:

- Voice recognition
- RFID tags
- Barcodes
- Optical character recognition
- Magnetic strips [7]

AIDC provides a variety of benefits, including:

- *Visibility*—AIDC aids in monitoring the completion of a processes;

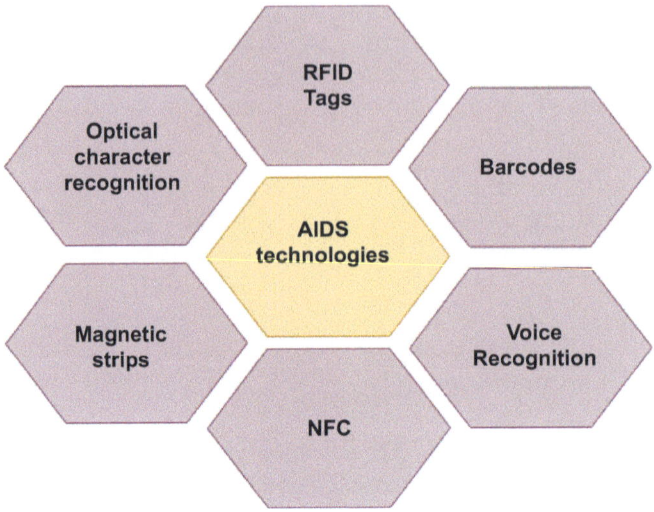

FIGURE 9.2 AIDC technologies.

- *Control*—AIDC systems help maintain inventory control from acquisition to storage and distribution;
- *Picking efficiency*—Workers can fulfill a large number of orders faster, due to the ability of AIDC to eliminate paper-based picking;
- *Picking accuracy*—AIDC eliminates mistakes by checking the reliability of products chosen for orders;
- *Location*—For information regarding location and the open space available for storage of processed items, AIDC could be used to collect data;
- *Inventory accuracy*—As AIDC allows the number of units within the facility to be monitored in real time, an exact inventory count is maintained [8].

9.2.2 Radio-frequency identification

Radio-frequency identification (RFID) is a technology where radio waves are used to capture data which are encoded in smart labels or RFID tags. The three components which form the RFID system are the smart labels, the RFID reader, and the antenna. The tags, composed of an antenna and an integrated circuit, are used for data transmission to the RFID reader, where these radio waves are converted to more usable information. This information is then transferred to a host system via a communication interface. Here the data are stored for further analysis in a database [9].

RFID is used in the pharmaceutical industry for various purposes, such as:

1. Tracking raw materials—Warehouse efficiency can be improved by tracking and identifying those batches of chemicals which are close to

their expiry date and thus help workers to identify which batch to use first, thus helping the company to save money.

2. Controlling temperature—Some medicines require specific temperature conditions for their storage. Therefore, to prevent them from spoiling, temperature sensors are installed and data from them are recorded. If the temperature does not lie within the prescribed range, then this information is sent to the tags and the readers. This makes companies more proactive in resolving these issues.
3. Packaging of pharmaceuticals—Used to prevent tampering and counterfeiting of drugs [10,11].

9.3 Pharmaceutical manufacturing

Although this is a tough and niche industry, pharmaceutical manufacturing has deployed its very own business models, which have been tested over time and helped other businesses to thrive in this field. Many economic decisions, which cannot always be consistent with technical decisions, are made with each model. Moreover, new business models are introduced as times change. To survive in this unpredictable and tough economic environment, each company is searching for methods to reduce costs and as a result a number are disappearing. The pressure to optimize their operations is thus increasing for all disciplines in the pharmaceutical sector. The formulation experiments are not special in this regard, and do not shield a growing organization against significant economic headwinds [12].

Pharmaceutical manufacturing has contributed much to the medical world and has helped in curing a number of catastrophic illnesses and diseases. For the manufacture of important drugs, a series of steps are followed which are equally important, and only when all these production phases are completed successfully, can we say that the entire production process is complete. Production steps include design conception, manufacture, extraction, supply, alteration, packaging, storage, and release. Pharmacists in these manufacturing industries have to undergo all these steps for the development of useful drugs and several other pharmaceuticals. This chapter is about the important steps that are followed in all pharmaceutical plants for the manufacture of essential drugs.

The process of pharmaceutical manufacturing can be broadly divided into two steps, as described next.

9.3.1 Primary processing stage

This stage focuses primarily on the development of drug active ingredients. Research at various centers, conducted by pharmaceutical experts to provide efficient pharmaceutical ingredients, also comes within this step.

9.3.2 Secondary processing stage

The second stage is the unit of final drug processing, where it is crucial to develop all products to be used as medicines in the health organizations for the management of patients suffering from various diseases. This unit mainly includes the conversion of active ingredients in pharmaceuticals into important medicines.

Pharmaceutical companies are constantly under pressure to monitor progress and to speed up the introduction of successful and innovative medicines into the market. Different scientific achievements fuel a fresh wave of innovation in the pharmaceutical industry, and hence it is essential to review the current trends in drug development and their effect on the manufacture of pharmaceuticals [13].

The bulk of the pharmaceutical industry has until recently focused on a batch production process to deliver their drugs to the public. However, in the recent years, continuous pharmaceutical development has been increasingly popular given the proliferation and sustained interest in some drugs. While both continuous and batch processing have proven to be beneficial within the pharmaceutical industry, depending on the needs of consumers and manufactures, both have their advantages and disadvantages. In general, batch processing halts from step to step throughout the production cycle, whereas the continuous method produces a product without stopping until the entire process is finished and the product is completed.

9.3.3 Batch processing

Batch processing takes place when the various components of a drug are brought together through a discrete multistep process, where, before a subsequent batch can be processed, the current batch must be finished as the material goes from step to step. While samples are tested for quality off line, production comes to a halt. During these halt times, to complete the manufacturing process, the material is either stored or shipped to other facilities. Due to the presence of approximately five to six steps in the completion of the whole process, the time taken to manufacture drugs can greatly increase. This then introduces new risks to quality and the continuity of the supply chain [14].

One of the biggest advantages of batch manufacturing is that pharmaceutical companies can produce a limited amount of products and then alter their manufacturing preferences to align with the changing requirements of the market. Also, the economical set-up costs allow one to easily tailor the batches for them to be unique [15].

On the whole, drug manufacturing industries are making huge losses due to the generic inefficiency of the batch manufacturing process—whether this is due to contamination, time constraints, or shipping losses. Most of the

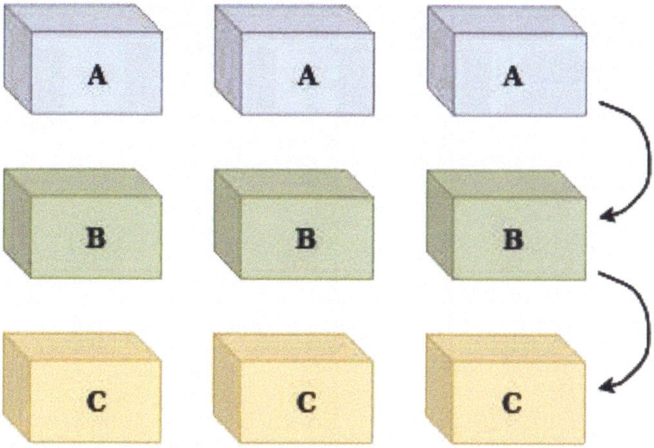

FIGURE 9.3 Batch manufacturing process.

industry's manufacturing giants are exploring technical solutions that can increase process throughput, quality, and yields. One of the best potential solutions to enhance the quality of products and address many of the underlying causes of shortage of drugs is to implement continuous manufacturing (CM) (Fig. 9.3) [14,15].

9.3.4 Continuous manufacturing

CM allows a drug to be taken from its base ingredients to the final product without halting during any of the production steps. Therefore there is no requirement to shutdown equipment and no downtime as the product is manufactured, making it more time efficient, which helps to increase productivity, and also reduces energy needs and the overall amount of waste generated. The risk due to human error is also reduced because here are fewer people involved in the manufacturing process from start to finish [16].

CM eliminates the halt time between various steps because drugs are moved continuously within the same facility and thus this method saves a considerable amount of time invested in manufacturing. Also, as it has fully integrated components and machinery, it reduces the possibility of human error, and thus allows the industry to rapidly respond to market changes. CM plants should be allowed to run for a longer duration of time, so that they can cater to increasing demands, thus reducing the possibility of pharmaceutical shortages (Fig. 9.4) [16,17].

9.3.5 Continuous versus batch processing

CM has proven its worth across the fine chemical industry (among other industries) and is slowly gaining traction as an effective way to produce

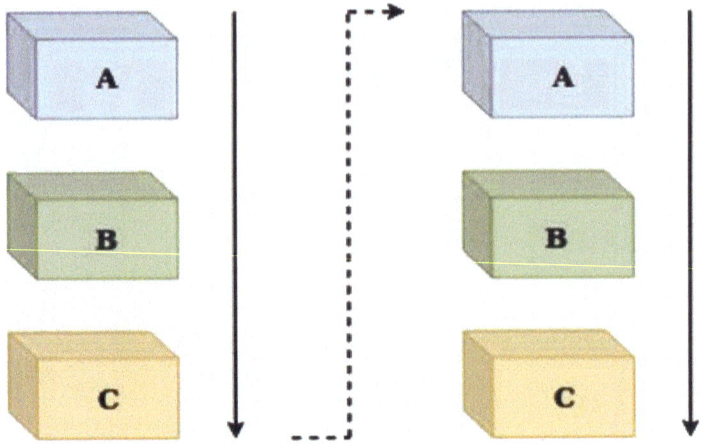

FIGURE 9.4 Continuous manufacturing process.

chemically synthesized APIs. Drugs produced by CM have higher quality and a lower cost of goods sold than drugs made by batch processing.

Four distinct advantages of CM include:

- The same quality control standards apply to both batch production and CM production but monitoring has been automated in CM plants and is usually more common than batch production.
- Like all machines, CM systems are subject to wear and tear, but automatic testing can spot issues well before failure happens.
- The longevity of a piece of equipment can be predicted using monitoring, thus enabling better proactive maintenance.
- CM allows for more flexible tracing and tracking, which is an advantage in the event of a product [14,17,18].

The pharmaceutical industry is vigilant about using new technology and has thus gained a reputation for being slow to adopt changes. This conservatism is at least partly due to regulatory requirements that improvements in the system are shown to not adversely affect product quality. In the last two decades, however, regulatory agencies and businesses have acted in concert to implement strategies and techniques that enhance performance and output efficiency. There is also recognition of the need to better adapt to new industrial strategies like CM, Industry 4.0, personalized pharmacy, and engineering aid, such as product management or automated process control [17].

If an organization aims for development, then their software programs cannot lagged behind. The IoT provides medical companies with a range of new technologies in the same way as artificial intelligence (AI) and machine-learning: better control of drug production, diligent monitoring of infrastructure, and enhanced management of the supply chain [14].

9.4 Pharmaceutical manufacturing challenges and solutions

The pharmaceutical industry is now regarded as one of the most advanced industries due to its fast-paced market growth. Combining both inorganic and organic segments, companies are now able to produce outputs at a better and more improved level. To increase their product advancements, they are also expanding through partnerships and licensing [19].

Pharmaceutical manufacturers, apart from managing and regulating the manufacturing complexities during batch production, also have to ensure complete quality assessment by plugging the gaps in the logistics chain. While control systems and industrial automation have developed in the manufacturing sector of life sciences, management remains unable to make timely decisions, due to the unavailability of crucial data pertaining to the equipment's real-time status. In addition, strict good manufacturing practices (GMP)—which comply to the agencies' guidelines regulating the licensing and authorization of the sale and manufacture of drugs—require strict adherence across all equipment.

The increase in the demand for organic drugs, that is short-term medicines and medicines sensitive to temperature, in the market, requires manufacturers to assure the consistency of temperature and error-free delivery of the drug from the origin to the destination. Because of costly cold-chain storage and failures due to poor handling, the operational costs are high.

In the production and distribution of generic medicines that make up 80% of today's pharmaceutical industry, this problem is accentuated. Very specialized logistical capabilities are also required at the lowest cost with the highest performance to compete with the tough competition on the generics market [20].

Warehousing, an important component in the production of pharmaceutical goods, is costly and its reliability and productivity are essential to the sustainability of a company. Because of the delicate nature of the goods, most companies decide to handle their operations internally. A report revealed that warehousing incurs about 95% of all drug distribution expenses. However, without real-time insight into operations, it is difficult to follow goods in the warehouse and to use workers and distribution equipment efficiently [19].

There is great diversity in the challenges. The continued growth in the pharmaceutical industry as a result of lifestyle changes as well as high demands for good healthcare makes this segment one of the most important contributors to the economies of countries.

Notwithstanding all the challenges, the growth of manufacturing industries is expected to escalate in the coming years. There are also several internal challenges faced by producers, in addition to volatility in national and international economic circumstances.

9.4.1 Forecasting demand for products

Most suppliers still have trouble predicting future demand. The primary issue is that they do not have sophisticated reporting tools to predict the number of products they should produce in the next couple of months or years. As a result, the goods do not fulfill the demands of the consumer and thus decreased sales are observed.

The solution is to foresee consumer demand appropriately, and to do this suppliers must use specific tracking tools that can help them plan their revenues and predict how many and what products they are to generate in the future.

The method can help producers not only evaluate consumer preferences, but also allow them to distinguish quick- and slow-moving goods, what products they would like, when they usually make transactions, what they typically purchase, and how many items they purchase at certain times.

In addition to making reliable predictions using technology, manufacturers should also consider external factors, such as current market trends, inflation rates, and currency fluctuations [21].

9.4.2 Improving efficiency at manufacturing plants

Manufacturers continue to trade off product quality with lower production costs, not realizing that this will also lead to lower profits, since unsatisfied customers will stop purchasing from them.

To date, producers have found efficient ways of reducing costs and improving performance at their factories; but many of them choose to compromise product quality to lower their production costs, which leads to dissatisfied customers avoiding purchasing from them thus overall leading to lower productivity.

The solution to this is the modernization of operations and the systemization of workflows as one of the most efficient methods of maximizing the output of production plants. Manufacturers must reduce time-consuming and laborious operations, maximize the use of machinery by minimizing damage, optimize their supply chains, and reduce the amount of material waste produced. The enterprise resource planning (ERP) system can make this much simpler, so that suppliers can achieve optimal performance [22].

9.4.3 Inventory controlling

The control of inventories remains one of the major problems in manufacturing, but it has become somewhat easier due to the help provided by automated solutions. Nonetheless, several businesses, particularly small manufacturers, still manually manage their inventory.

Tracking the inventory is a laborious operation, which can be simplified with technology. Manual inventory control is highly ineffective and vulnerable to errors that may lead to unidentified damages, overstock, and shortages.

The solution to this is effective inventory management techniques that are essential in order to avoid the unnecessary purchase of equipment or raw materials. In order to identify variations between the figures in the accounts of the organization and the number of items actually on site, regular stock inspections should be carried out. This tracking process can be accelerated using barcode scanners.

An inventory management software should be implemented that allows the managers to assure that, for production, there are ample raw materials and that the finished products are qualified to be delivered to the ultimate consumer. It should also update the stock levels frequently, raising alerts when the stock levels fall below the prescribed danger level and then reordering the correct products at the correct time [22].

In order to accelerate the tracking process and also for other software systems, for example, related to procurement and accounting processes, effective stock management technology can also be combined with barcode scanners, but at the same time it is advisable for small-scale manufacturers to invest in affordable cloud solutions rather than this [23].

9.5 Pharma 4.0 and the industrial Internet of Things in the pharmaceutical industry

Pharmaceutical companies previously reaped high profit margins and had little incentive to improve their industries. Nevertheless, as patents expire and competition grows globally, profits have been eroded.

The boom in personalized medicine—in which medicines are manufactured for specific patient groups in fewer, less costly quantities—also has to pay off. Such aspects have also helped to make producers more open to IoT engineering applications that can increase efficiencies and avoid costly errors [24].

One recent example includes that of a pharmaceutical brand which uses vacuum pumps to freeze-dry drugs—this helps in reducing humidity— however, when a pump breaks, the company could face losses worth millions of dollars. To prevent this, the company has now installed IoT sensors in its pumps and other machines. This collects information, which is then sent to a cloud-based platform.

This system will notice when a failure happens and warn the producers that the equipment should be repaired or replaced before it fails during production. This can be achieved using the historical data in the database.

The transfer of data to a modern platform together with sensor information and then on analysis in the cloud provides visibility for managers throughout the plant. Bottlenecks can easily be detected and recognize over- or underutilized machinery which affects performance [24].

The increasing medical conditions around the globe would certainly make the pharmaceutical industry more important than ever in the future. There has been a change from large-scale production of medications to a larger patient population receiving personalized or specialized products, which is thought to significantly increase the effectiveness of therapies. In addition to solving sustainability problems, it is anticipated that Pharma 4.0 will also provide smarter tools for maintaining drug safety. The emergence of digitization has thus helped in making the operations, people, and even the machineries intelligent.

The digitalization cycle will also play an integral role in providing smooth sharing of information across the drug supply chain. The interconnection of products, people and processes is what defines the *Industrial Internet of Things (IIoT)*. Manufacturing will be the largest benefactor of the IIoT. Thus, pharmaceuticals and healthcare sectors will have a good opportunity to continue to benefit from the IIoT [25].

Investments in IIoT methods that have proven to be effective primarily in other sectors by enhancing the efficacy of their complicated processing activities should be taken into consideration by the pharmaceutical industry. This could help boost organizational and financial performances.

As a consequence, technological developments have become a key part of the manufacturing sector, streamlining procedures, automating activities, and the convergence of different manufacturing departments which are not sufficiently related otherwise.

The IIoT technologies—by making the entire production data available to manufacturers with just the touch of a screen—would provide an overall view of the operations carried out in a plant, along with the advantage of moving down to any level of detail at any stage of production. Thus it gives companies the opportunity to grow from small local centers to global markets [26].

9.6 Pharmaceutical supply chain

Although the pharmaceutical industry is related to healthcare and medical research, the supply chain and logistics play an important role in the industry. Pharmaceutical drugs have some specific requirements with regards to the environment in which they should be handled. Due to this, the pharmaceutical industry's supply chain has some focus points that are different to those of other supply chain networks.

For instance, in the case of healthcare utilities, it is extremely necessary that there should not be any alteration in the products while they are in transit from the manufacturing units to customers. To ensure real-time visibility in transportation and warehouse operations, remote monitoring and network optimization analytics are used by companies. Also, for the distribution, storing, and handling of environmentally sensitive products, certain regulatory requirements should be met. These techniques also ensure that government regulations are complied with and product quality is maintained [27].

Another focus point is to ensure proper cold chain management of drugs by the pharmaceutical supply chain members so as to prevent the quality and effectiveness of the products from being compromised. Therefore cold chains are a necessary component of the pharmaceutical supply chain, and so detection as well as real-time monitoring of cold chains is an essential requirement for improvements to the supply chain network.

Also, biopharmaceuticals form a major component of the pharmaceutical products that are sold worldwide, and the biopharmaceutical market is growing at a faster rate than the traditional pharmaceutical market. This has been one of the reasons for increased management focus on the biopharmaceutical market. Other factors for increasing investments in the area include:

- The ability of a biopharmaceutical drug to target an important unmet need;
- Premium pricing as compared to other pharmaceutical products;
- Limited generic erosion [28].

The fact that biopharmaceuticals tend to be temperature sensitive further contributes to the importance of cold chains in the overall structure of the pharmaceutical supply chain.

Drugs requiring temperature-controlled storage conditions should always be distributed in such a manner as to prevent adverse effects on their quality. Industry regulatory standard practices related to cold chain management are as follows:

- For ensuring that medicines' original quality is maintained, pharmaceutical products' distribution should be handled as per the principles of good manufacturing practice, good storage practice, and good distribution practice.
- Proper evaluation of a drug under the storage conditions should be carried out to test the thermal stability and sensitivity of the product toward moisture or the potential for solvent loss.
- The lengths of studies selected, to establish well-researched storage conditions should cover all the stages in the life cycle of a product—storage, shipment, and the final use.
- Management, control, and oversight of environmental conditions covering the entire supply chain (that is, starting from the manufacturing of the product until it reaches the consumer) for temperature-sensitive drugs should be increased.
- Ensure that protocols and procedures for storage and transport conditions are in place and enforced, such as: light regulation, temperature, turnover of stock, sanitation, humidity, and many other measures necessary to maintain the quality and safe delivery of the pharmaceutical product (Fig. 9.5) [29].

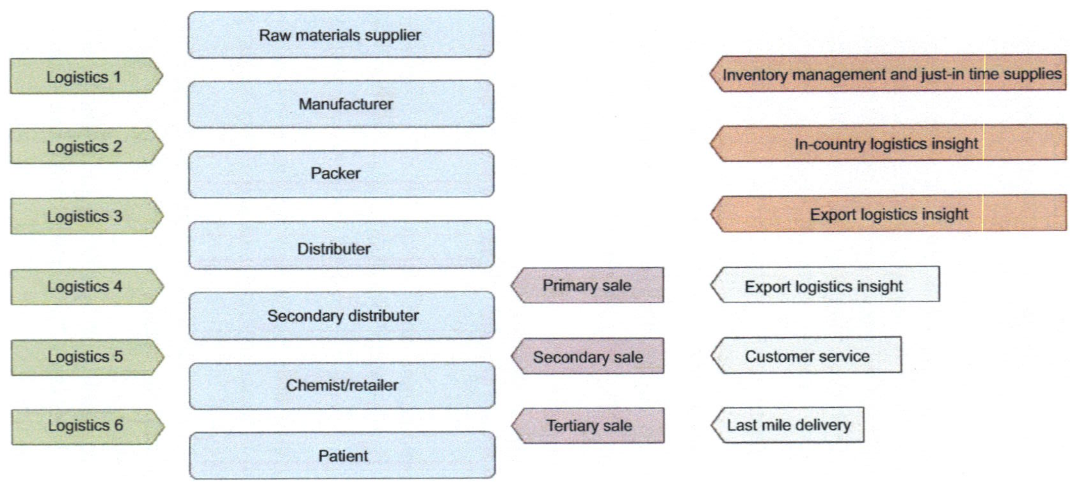

FIGURE 9.5 Pharmaceutical supply chain.

9.7 Supply chain challenges

This section addresses the crucial challenges that the pharmaceutical supply chain and logistics face, so that suitable solutions to resolve these problems can be suggested.

9.7.1 Data handling

Almost every sector is becoming increasingly data intensive and utilizing resources to optimize the use of available information to analyze trends and develop marketing strategies and business models accordingly. This is a tremendous opportunity for the pharmaceutical industry, but at the same time, it is an equally big challenge.

Two major challenges are:

- Development of data centers that have the capability to handle the data;
- Ensure cybersecurity from hackers seeking all sorts of data breach—from highly critical information, intellectual property, to the private health information of the patients from research laboratories.

Data-warehousing integration projects have been initiated by many of the established pharma companies over the past years. The goal of these initiatives is to harmonize data-warehousing resources across international sites to facilitate efficient data storage, processing, and retrieval, as well as to allow the data to be exchanged both internally and externally. Improving information flow both internally within the company and externally to the regulatory agencies and R&D stakeholders would shorten product-to-market cycles significantly [27,30].

9.7.2 Regulatory compliance

Pharmaceutical products worldwide are subject to varying standards, regulations, and quality compliance directives. It is extremely important for pharmaceutical manufacturers and supply chains to conform to and comply with the regulations of the respective regions, which makes supply chain networks more complex. In such a situation, there is added pressure to ensure optimal temperature control, maximum supply chain visibility (SCV), and efficient warehousing [27,31,32].

9.7.3 Warehouse management

The following factors explain the challenges faced during warehouse management in pharmaceutical supply chains:

- Due to the rapid rise in the sales of pharmaceutical products, there is an urgent requirement to increase the warehousing capacities. However, logistics companies are still unprepared to cope with the rise in pharmaceutical demands, as the vast majority of them do not have adequate infrastructure to address the needs of increased visibility and reduced overall cost.

- It is a strategic move by most pharmaceutical companies to keep their warehouse operations in-house, which leads to managing the storage facilities and warehouse processes internally because of the nature of pharmaceutical products. It is often difficult to track goods in the warehouses without real-time visibility into operations and completely utilize transportation equipment and operators.
- Specialized warehouses with temperature-controlled zones to store pharmaceuticals require massive investments.
- In this sector, there are already large companies with a more established set-up, which makes the arrival of new companies very complicated.
- Another major challenge is the reliability of the data shared between the logistic companies and the pharmaceutical products' manufacturers, along with their confidentiality [27,33].

9.7.4 Temperature control

The key factor which sets apart pharmaceutical supply chains from the rest of the supply chain network is temperature regulation. It is highly important to maintain a particular temperature range for most of the pharmaceutical products throughout the storage and transit period.

During the transportation phase, drugs and other pharmaceutical products are exposed to various uncertain temperature conditions, in which case maintaining a particular temperature is a challenge; however, it is required to ensure that the products conform to the quality and regulatory standards.

Special attention is required in the case of products that are especially sensitive to high temperatures and that begin to degenerate if the specific environmental conditions are not met. Drug manufacturers are increasingly dealing with such products like biologics that are particularly sensitive to storage conditions. These are high-end drugs that have a greater proportion of high-value active compounds with shorter shelf-life and are subject to strict temperature requirements. Most need to be maintained at temperatures below 77°F; others demand 35–46°F during cold chain transportation.

Nevertheless, pharmaceutical companies continue losing millions because of spoilage due to temperature fluctuations and are facing an even larger threat in the case that this deterioration goes undetected—leading to severe health risks for patients, as well as further regulatory activities. Especially in emerging markets like South America and Asia, where large temperature fluctuations are highly prevalent, this problem is becoming increasingly critical [19,27,34].

9.7.5 Collaboration issues

The pharmaceutical supply chains are confronted by the challenge of trust deficit, reluctance to share data, and a lack of transparency. The collaboration of this multilayer supply chain involves huge conglomerates. Therefore, the data flow in the supply chain network is huge and complex, and so is prone to delays and is extremely vulnerable. In this situation, the lack of willingness to share data by partner organizations is an obstacle to the successful collaboration and resilience of the pharmaceutical supply chain network [27,31].

9.7.6 Supply chain visibility

Visibility is a complicated issue involving individuals, processes, the flow of information, and technology. SCV is the ability of an enterprise to gather and scrutinize distributed information, produce specific recommendations, and suit tactical insights. Although a rise in supply chain information provides an illusion of visibility, it actually adds to the difficulties of a business.

SCV is essentially concerned with increasing communication between the various nodes in a supply chain. This allows an organization to understand the situation better across the supply chain. From a logistic point of view, SCV is a transparent view of content, time, location, and status, according to the definition put forth by John Fontanella.

Drug thievery and counterfeiting during the transit period are one of the main concerns of every pharmaceutical company. If a pharmaceutical company is unable to avoid these leaks in the supply chain, then it will face difficulties in ensuring compliance with regulations. Therefore the visibility of a supply chain is very important for pharmaceutical industries, but shipment traceability has some challenges that the existing systems cannot overcome [27].

A pharmaceutical supply chain has the following needs with respect to SCV.

9.7.6.1 Inventory visibility

1. Inventory visibility is a key requirement in the pharmaceutical supply chain. Many local hospitals retain a 14-day supply of common drugs, which costs around $4 million. A reduction of the inventory level is a major concern in such cases.
2. The inventory status of hospitals is not sufficiently accessible to distributors and wholesalers in order for them to shorten the replenishment cycles of pharmaceuticals.
3. At present, many hospitals do not have proper facilities to manage drug inventory in wards. A major portion of the working hours of nurses is spent in handling the medication inventory. Another requirement is to

create an environment conducive to the better management of inventories to reduce the nurses' efforts.

4. There is no sample drug management system. Visibility in cases of sample drugs is a necessity.
5. There is a lack of visibility with respect to price changes of drugs. This is a major concern for wholesalers as they are notified about the change in prices with only short notice.
6. Tracking of the expiry date of pharmaceutical products.
7. Real-time inventory tracking is required for controlled drugs.
8. Visibility of the chain of custody can be improved [35,36].

9.7.6.2 *Demand visibility*

1. Wholesalers require information on the customer demand. At present, there is an absence of visibility of demand and they have to estimate the figures on the basis of historical data, such as the level of usage over the past 6 months. The current practice is unable to get rid of this bullwhip effect. They need a mechanism that can provide visibility of demand. After this objective has been achieved, a major challenge would be discovering and monitoring the consumption pattern.
2. Manufacturers require demand forecasts, which should be provided by the wholesalers [35,37,38].

9.8 Supply chain solutions

This section addresses IoT solutions to the above-mentioned problems. The IoT solutions are effective when implemented in the pharma supply chain, but the successful execution of these IoT techniques requires cooperation with various conventional methodologies. The traditional methods include modeling of the transportation network, business models for efficient pharmaceutical supply chain operation, warehouse design concepts, and optimization. The solutions are described here.

9.8.1 Telemedicine

Telemedicine monitoring aims to improve diagnosis and medication, lowering medical bills, and developing patient-based infrastructure by collecting data from past medical programs and lifestyle choices of patients. In this system, the focus of IoT implementation is to facilitate healthcare service providers to attain consistent improvements in the field of medical care.

RFID sensors and tags should be used to gather information that would aid in the development of new pharmaceutical drugs.

Telemedicine has favorable implementation for pharmaceutical companies in research and development which will potentially help them to build new drugs in a shorter timeline and simultaneously comply with regulations, speed up development, and make the pharmaceutical products available to consumers at a lower cost [27,39,40].

9.8.2 Temperature control

The growth of cold chain pharma products has been boosted worldwide. If contaminated medications are overlooked and pass into the supply chain network, the risk of drug expiry and degradation can worsen. It is estimated that 20% of products are lost from the cold chain market during shipping.

- Environmental sensors are used to gather information from storage units to detect pharmaceutical products' conditions in real time to maintain optimal storage conditions of such critical drugs.
- Active or passive log data, when compared with the medicinal thermal stability tables, can be used to produce warnings in the case of temperature excursions. This may help avoid premature degradation and expiry of drugs.
- In addition to regulating the temperature spikes across shipments, it also is extremely important to preserve the product quality during transit. The temperature of the shipped products can also be monitored to make sure that they stay within the defined stability range.
- Environmental sensors are incorporated and program sensors are used to trigger an alert if the storage conditions of the pharma products vary from the specified temperature range.

  Such strategies also provide various other advantages in addition to the above:
- They help in ensuring conformity to the local and international target market standards and regulations;
- Wastage of medicines is minimized;
- It helps in ensuring the effectiveness of pharmaceutical products.

9.8.3 Shipment traceability

Due to shipment theft, pharmaceutical companies suffer significant losses in inventory. However, on a positive note, the average loss rate of the drug industry has dropped by more than 50% in recent years due to the adoption of smart methods for shipment monitoring and information gathering. This clearly indicates the significance of monitoring shipments in the pharmaceutical industry, largely because of the sensitive nature of health concerns. This becomes even more important owing to the involvement of medicine.

In order to make sure that the quality and integrity of drugs remains unaffected, product verification and bidirectional communication are required at each transit point. The following IoT techniques are employed to help with shipment traceability:

- Condition monitoring and the global positioning system (GPS) also provide electronic real-time alternatives for tracking the safety of products during transportation. One such example of real-time tracking of shipments includes a model which would also provide updates on the environmental and temperature conditions.
- Furthermore, AIDC integrated with digital chips in packaging for smart serialization helps track the incorporation of drugs into the healthcare system, and is offering another level of security to ensure safe product delivery [41].
- IoT-enabled packaging monitors goods constantly to prevent counterfeiting and to stop fraud as the products pass through the transit network. Therefore, RFID tags with smart labels, 2-D barcodes, and near-field communication devices can be used for labeling [27].

9.8.4 Real-time visibility

As discussed in Section 9.9.3, for improving such operations, real-time visibility is invaluable. Through the incorporation of IoT systems, real-time visibility, and 3-D visualization of current warehouse operations, operators can monitor all the relevant information, which then allows them to regulate the sensitive areas of drugs.

- The key idea behind the connection of equipment for real-time updates for improved process yield is connecting the equipment sensors to a back-end database or cloud network.
- Shop floor visibility, that is, visibility across equipment and goods helps in dynamic scheduling of the shop floor activities.
- Tracking of overall equipment effectiveness can be achieved by the automated triggering of incidents using a ticket management system.
- Sensor packs can be added to collect the device information, and to build a back-end platform using IIoT platforms like Amazon web services, which could be implemented to reduce equipment downtime and back-up equipment.
- Access to the human and machine interaction on the shop floor can be utilized to determine the causes of variability in the process by conducting statistical analysis on the collected data, which in turn would result in increased production yield.
- Furthermore, utilizing the obtained information enables technicians to develop machine learning models to perform diagnostic services for a firm's machine portfolio. Which could be used to increase availability and minimize interruptions in the supply chain.

The degree of automation already existing in the company determines the value addition from the connected production process and accordingly a set of solutions should be recommended for the specific company [27,42].

9.8.5 Smart warehousing

Warehouse operations are a vital part of the pharmaceutical operations, and cost about 90% of the logistics costs.

With real-time updates on inventory, floor space can be utilized to its maximum potential.

- Sensors can be used to track and analyze the inventory flow, in which case information will be captured by the wireless readers and used for aggregate planning.
- Embedded sensors on the forklifts are used to track the asset movement.
- RFID tags are used to track location and consumption rate. Inconsistencies in the processes are captured by these sensors and transmitted for analysis.
- Data from arriving pallets are also captured with the help of wireless readers.

These solutions altogether contribute toward higher productivity and maximum floor space utilization. The data of the inventory flow from warehouses can be fed to production planning to optimize the inventory levels, reduce stock outs, and improve demand fulfillment [43,44].

9.8.6 Network of information transfer in the pharma supply chain

Due to the problems of mismanagement and counterfeiting of drugs, the government also faces difficulties in complying with the regulatory norms and to guaranteeing the safety of customers. The IoT model in the network of drug manufacturers and healthcare providers is explained and illustrated here.

- Outside of the line, RFID tags with EPCs, that is, electronic product codes, are attached to medicines. All the necessary information can be associated to those RFID tags or EPC-IS (the electronic product code information service) database, such as manufacturing date, chemical composition, SKUs (stock keeping units), expiry date, and safety instructions.
- As medicine batches are associated with EPC codes, the company can use EPC codes as keywords to update and query medicinal information. By means of this updated network of IoT devices, real-time updates of medicines can be transmitted to concerned parties [45].
- On the other hand, RFID readers and analytics and location engine servers are employed at drug-querying counters in hospitals and local pharmacies [46].
- As medicines with RFID tags enter into the magnetic field of these counters, connected data will be read and processed. The information will then be queried at the local and root object naming services, so that

consumers can identify whether the medicines they are buying are counterfeit or not.

- The link between ERP and WMS represents the internal information system integration within the ERP and warehouse management system (WMS). In this way customers can protect themselves even if counterfeit drugs enter the supply chain network (Fig. 9.6) [27,44,46–48].

9.9 Revolutionizing the industry using the Internet of Things

The IoT has the ability to completely reform the pharmaceutical manufacturing processes, from discovery of the drug to remote patient access and monitoring. In their manufacturing plants, many major pharmaceutical companies worldwide are rapidly leveraging IoT devices to simplify and boost system performance. Previously data access and understanding presented a significant barrier to effective communication because of various data formats for different processes in a drug manufacturing plant.

In a pharmaceutical fabrication plant, IoT technologies enable the standardization of a network, equipment, and systems effectively throughout the plant. In addition, pharmaceutical companies can access real-time data and operations visibility throughout the production process using IoT.

When a company is pursuing innovation, its software cannot lag behind. The IoT system offers the pharmaceutical industry a wide variety of new opportunities, which includes improved control of drug production and supply chain management, preventive maintenance of equipment, etc. IoT sensors and trackers in particular are able to offer optimum conditions for the handling of biomaterials and chemicals, ensuring a flawless system.

A large majority of manufacturing companies have been urged to use this technology, given the unprecedented strength of the IoT, to unlock unlimited potential. Along with advanced analysis and big data, the IoT can be used to improve manufacturing efficiency by scrutinizing huge amounts of information.

In production facilities, pharmaceutical companies use IoT technology to help with the CM process. The IoT contributes to data integrity along with the standardization of the production process. The IoT provides visibility in a manufacturing plant from production to distribution, leading to more feasible and better analysis, thus enhancing and improving operational efficiency.

IoT technologies are mainly adopted by pharmaceutical companies to increase the efficiency of machines and optimize their processes. Older methods of data synchronization were highly prone to human error as well as being time consuming. The presence of different data formats, for different processes in a manufacturing plant, made data interpretation and its access a barrier to effective communication. By connecting systems, networks, and equipments, to share data, IoT technology facilitates standardization.

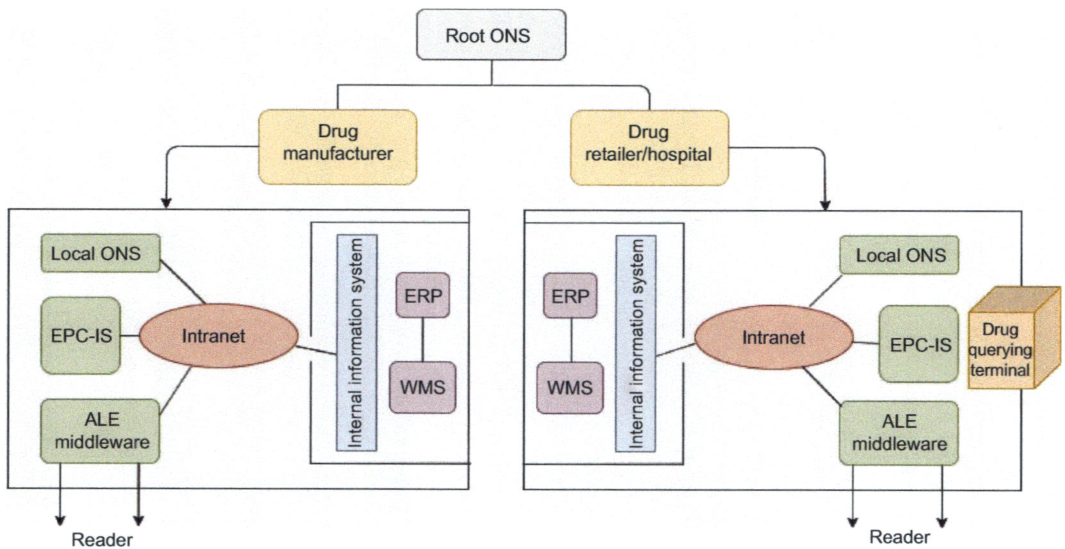

FIGURE 9.6 Network model of the IoT in the supply chain.

There are a large and diverse number of IoT implementations. At present, IoT solutions can be used for the following purposes:

- *Smart metering*: Smart meters could be used to monitor resource consumption such as fuel, water, and electricity. By using IoT sensors, manufacturers will be able to keep track of consumption. Operational expenditure can be substantially reduced through effective management.
- *Enhanced industrial safety*: To ensure better workplace conditions, focal performance indicators of safety and health could be continuously monitored by effectively combining the IoT with big data analysis, and thus issues like an increasing number of accidents could be immediately addressed.
- *Facility management*: Maintenance based on condition is an effortless task. Equipment manufacturing is highly susceptible to wear and tear and therefore sensors are installed which help to monitor vibration, temperature, and other factors leading to nonoptimal operational conditions, and thus help improve facility management efficiency.
- *Inventory management*: In order to reduce the risk of errors in inventory management, IoT solutions can be very useful. Events across the supply chain are much easier to monitor, giving companies a broader view of the inventory. Estimates of supplies and materials available can be made more accurate, preventing slowdowns.
- Real-time logistics visibility uses RFID and sensors to capture and report parameters.
- Smart serialization through AIDC, which involves automatically identifying objects, collecting information, and feeding it into the computer system is all done without human involvement.
- IoT sensors help in providing insights on power use and balancing energy utilization during peak times, which reduces costs for the industry [49].

Most companies are currently adopting an extremely simple mode of operation—if something breaks they fix it. However, the use of the correct software and intelligent sensors will enable the proper prediction of future failures. Thus it helps improve the ways in which equipment is maintained and replaced, therefore reducing breakdown risks which may cause the industrial processes to halt. The IoT has greatly helped them in the following spheres of manufacturing processes [50].

9.9.1 Real-time visibility in warehouse operations

For all pharmaceutical manufacturing companies, one of the critical components is warehousing. Shortages of some drugs can have severe effects, while overstocking can contribute to wastage of goods and higher retention costs of stocks. As pharmaceutical companies maintain many warehouses spread over vast areas, the majority of pharmaceutical logistics costs are due to warehousing operations. The monitoring of warehouse goods helps smart warehouses to improve real-time awareness of activities and optimally utilize employees and distribution machinery. In the pharmaceutical supply chain, IoT technologies are vital because they can report inconsistencies, optimize the floor space of the warehouse, and monitor the storage of sensitive medicinal products [51].

9.9.1.1 Industrial mechanics and maintenance

While there is still significant use in the pharmaceutical industry for industrial monitoring devices, data on the real-time state remain widely unavailable. IoT monitoring sensors can warn supervisors of any abnormal conditions or pressing maintaining needs. The IoT also allows sensitive situations to be treated remotely in drug manufacturing.

The IoT expands visibility across all business areas from development via production, consumption, dispensing, transportation, and distribution. On the floor of a factory, real-time sensor data would allow transparency in every area of work, leading to better performance and quality, and lower cycle time and cost of output.

Overall, IoT-enabled smart warehouse management systems will give greater transparency, include real-time data for monitoring and documenting anomalies (e.g., storage temperature), and ensuring that the correct information is made available at the correct time. In real time, tracking the movements of drug inventory can help the company to save huge amounts in logistics. Intelligent drug labeling will ensure reliable monitoring of imports and drugs to make the supply chain open, efficient, and cost-effective. Packaging pharmaceuticals smartly can help ensure that medications and shipments are tracked accurately, and hence the supply chain remains fluid, cost-effective, and efficient [19,52].

9.9.2 Remote fleet management

With digital technologies, such as GPS units, the shipping cycle can be automatically tracked and the supply chain transparency assured. There are a range of details on the position of the truck, operator experience, and vehicle performance, that can be analyzed and decisions can be taken on the remote side to improve fleet management. This ensures that the time to market can be drastically reduced, spoilage of the product is avoided, and thus prices are minimized.

Thus, IIoT technologies may contribute to the development of computer-enhanced vehicles which can help in the transmission of relevant information and geo-intelligence, thus improving fleet management [24,52].

9.9.3 Connected equipment

One potential pitfall in drug manufacturing is rooted in equipment. Lack of maintenance, unstable environment, chemical deterioration, excessive voltage, and mechanical damage are some of the many reasons for an asset's failure. Pharmaceutical producers are largely self-contained and rely heavily on batch processing. This therefore demands materials and men move continuously on the shop floor. While controlling technology and industrial automation are used extensively, equipment status is not readily available in real time for informed decisions which can help enhance efficiency of equipment.

Pharmaceutical companies require real-time monitoring status of the equipment 24/7, because they simply cannot afford to discard batches due to accidental shutdown of equipment.

Through increased insight into shop floor operations, IoT technology can help ensure the efficacy and safety of pharmaceutical items.

This is achieved by connecting the machine parts across the company's facilities and monitoring the data on the state of these components constantly, such as:

- Vacuum pumps
- Pressure gauges
- Air compressors
- Multimedia filters
- pH probes
- Heat exchangers
- Sterilizers

In order to plan repair and maintenance, maintain workplace safety, minimize downtime, and prevent critical problems, sensors can be used to collect the information, which provides a complete picture of the status of the equipment. This overview can be used to reduce waste and optimize performance. For even better maintenance of equipment, this technology can further be combined with AI.

In continuous control of connected factory equipment and personnel, the IIoT plays an important role. This real-time equipment monitoring may contribute to the efficacy of production processes by requiring informed decision-making through data as a result of connectivity. In pharmaceutical production, the IIoT has another important advantage, namely the modular automation of pharmaceuticals that bring them to market quicker [53].

9.9.4 Controlling the drug manufacturing environment

During the manufacture of biological goods, producers are required to be extremely cautious, given the fragile nature of pharmaceuticals. Being generally of high value and having short shelf-lives, they therefore require highly sensitive storage conditions, with strict requirements for temperature. Suboptimal environmental conditions can prove to be catastrophic to the drugs.

IoT technologies help in bringing transparency to the storage environment and drug production using connectivity and sensors—which helps in real-time monitoring of the drugs' condition, by tracking of environmental data such as:

- CO_2 level
- Light
- Temperature
- Radiation
- Humidity

All this collected information is updated continuously on the systems, allowing operators to quickly obtain an analysis of the environment and take any required measures. This information can also be used to activate automated changes, when combined with climate control systems. For example, in the event of a disaster, such as leakage of toxic material, the system would immediately alert staff to ensure prompt evacuation [54].

9.10 Challenges in Internet of Things implementation

In order to maximize the profit promised by the IoT and reap the benefits across their supply chain, pharmaceutical organizations should first invest in a robust IoT network with the requisite ability to meet heavy-duty specifications. Because security is of utmost importance, businesses should also be willing to invest in security solutions focused on the IoT (Fig. 9.7).

We have to take care of a few challenges faced while implementing IoT solutions, including the following:

- The accessibility and appropriateness of production equipment to customize and reinforce IoT solutions is a major obstacle.
- Finding workarounds can be expensive for IoT applications.
- For proper evaluation of operational efficiency, the project demands need to be predominantly aligned with the company's supply requirements and key performance indicators. This can be an overwhelming task. If the goal of the IoT solution is to support the GMP best practices, the solution may require appropriate validation. This could lead to the extension of project goals and timelines.
- Another challenge is to deal with different suppliers and maintain common vocabulary and practices, in cases which involve integration with existing systems.
- In addition, the production floor usually contains a wide range of machines and related software systems, many of which are no longer under manufacturer warranty, although they are perfect for manufacturing purposes.
- IoT solutions like visibility on shop-floors enhance and improve organizational efficiency. Simultaneously, they question change

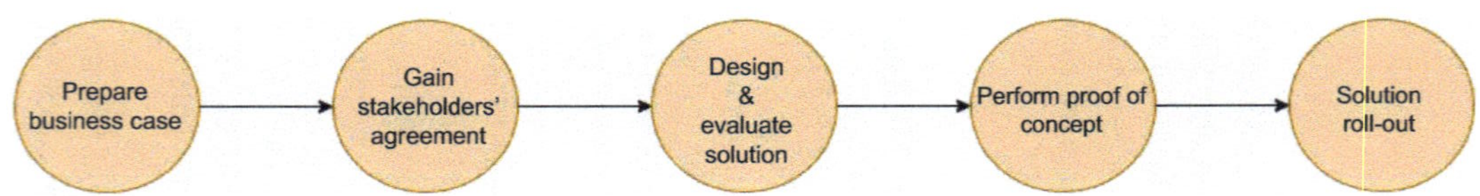

FIGURE 9.7 An IoT roadmap: from planning to execution.

management—a complex issue that should take procedures, people, and obligations into consideration.

- In order to achieve a seamless transition, organizations must establish and explain market drivers and develop communication techniques in order to address the concerns of staff and stakeholders.

Pharmaceutical companies need to look at their IoT initiatives from the point of view of facilitating innovation, increasing efficiencies, optimizing processes and operational efficiency, and bringing more value to the company, its customers, and its suppliers. While conforming to IoT industry standards and learning from other organizations that have successfully deployed IoT systems and strategies, pharma companies would be well prepared to face today's ultracompetitive, superconnected global economy's rising demands [19,55,56].

References

[1] El Hakim, A. (2018). Internet of Things (IoT) system architecture and technologies, White Paper. 10.13140/RG.2.2.17046.19521.

[2] B.N. Silva, M. Khan, K. Han, Internet of Things: a comprehensive review of enabling technologies, architecture, and challenges, IETE Tech. Rev. (2017) 1–16.

[3] D. Bandyopadhyay, J. Sen, Internet of Things – applications and challenges in technology and standardization, Wireless Pers. Commun., April 9, 2011.

[4] J. Gubbi, R. Buyya, S. Marusic, M. Palaniswami, Internet of Things (IoT): a vision, architectural elements, and future directions, Fut. Gener. Comp. Syst. 29 (2013) 1645–1660.

[5] D. Miorandi, S. Sicari, F. De Pellegrini, I. Chlamtac, Internet of things: vision, applications and research challenges, Ad Hoc Netw. 10 (2012) 1497–1516.

[6] S. Hodgson, F. Nabhani, S. Zarei, AIDC feasibility within a manufacturing SME, Assembly Autom. 30 (2) (2010) 109–116.

[7] A.K.N. Parlikad, D.C. McFarlane, M.G. Harrison, A.J. Thorne, The role of AIDC technologies in product recovery – and information quality perspective, Int. J. Prod. Lifecycle Manag. 4 (2009) 129–145. ISSN 1743-5110.

[8] V. Agarwal, Assessing the benefits of auto-ID technology in the consumer goods industry, published September 1, 2001. Cambridge, MA: White paper, Auto-ID Center, MIT.

[9] M. Bhuptani, S. Moradpour, RFID Field Guide: Deploying Radio Frequency Identification Systems, Prentice Hall PTR, USA, 2005.

[10] P.J. Sweeney, RFID for Dummies, Wiley Publishing Inc, Hoboken, NJ, 2005.

[11] S.A. Weis, S.E. Sarma, R.L. Rivest, D.W. Engels, RFID systems and security and privacy aspects of low-cost radio frequency identification systems, in: D. Hunter, G. Müller, W. Stephan, M. Ullmann (Eds.), Security in Pervasive Computing. LNCS, vol. 2802, Springer, Heidelberg, 2004. Available from: https://doi.org/10.1007/978-3-540-39881-3_18.

[12] B. Mittal, Business Acuity, in: B. Mittal (Ed.), How to Develop Robust Solid Oral Dosage Forms from Conception to Post-Approval, Academic Press, 2017, pp. 155–163. Available from: https://doi.org/10.1016/B978-0-12-804731-6.00008-X.

[13] T. Sandle, Data handling and trend analysis, in: T. Sandle (Ed.), Biocontamination Control for Pharmaceuticals and Healthcare, Academic Press, 2019, pp. 225–247. Available from: https://doi.org/10.1016/B978-0-12-814911-9.00014-6.

[14] A. Garg, P. Mhaskar, Utilizing big data for batch process modeling and control, Comput. Chem. Eng. 119 (2018) 228–236. Available from: https://doi.org/10.1016/j.compchemeng.2018.09.013.

[15] Z. Xiong, J. Zhang, X. Wang, Y. Xu, Tracking control for batch processes through integrating batch-to-batch iterative learning control and within-batch on-line control, Ind. Eng. Chem. Res. 44 (11) (2005) 3983–3992. Available from: https://doi.org/10.1021/ie049000o.

[16] K. Plumb, Continuous processing in the pharmaceutical industry: changing the mind set, Chem. Eng. Res. Des. 83 (6) (2005) 730–738.

[17] P. Poechlauer, J. Manley, R. Broxterman, B. Gregertsen, M. Ridemark, Continuous processing in the manufacture of active pharmaceutical ingredients and finished dosage forms: an industry perspective, Org. Proc. Res. Dev. 16 (2012) 1586–1590.

[18] S.L. Lee, T.F. O'Connor, X. Yang, C.N. Cruz, S. Chatterjee, L.X. Yu, et al., Modernizing pharmaceutical manufacturing : from Batch to Continuous Production, J. Pharm. Innov. 10 (2015) 191–199. Available from: http://dx.doi.org/10.1007/s12247-015-9215-8.

[19] Cognizant, Retrieved from the Internet of Things: the new Rx for pharmaceuticals manufacturing & supply chains [white paper]. <https://www.cognizant.com/whitepapers/the-internet-of-things-the-new-rx-for-pharmaceuticals-manufacturing-and-supply-chains-codex2437.pdf>, 2017.

[20] N. Yadav, A. Mazumder, R. Mazumder, N. Rana, Pharma-IoT: its opportunities and challenges in digital India, 2018.

[21] S. Yerpude, T.K. Singhal, Impact of Internet of Things (IoT) data on demand forecasting, Indian. J. Sci. Technol. 10 (15) (2017). Available from: https://doi.org/10.17485/ijst/2017/v10i15/111794. ISSN (Print): 0974-6846; ISSN (Online): 0974-5645.

[22] W. Ding, Study of smart warehouse management system based on the IOT, in: Z. Du (Ed.), Intelligence Computation and Evolutionary Computation. Advances in Intelligent Systems and Computing, vol. 180, Springer, Berlin, Heidelberg, 2013.

[23] Guidelines, Guidance for Industry Q10 Pharmaceutical Quality System, FDA, ICH, 2009, p. 1. <http://www.fda.gov/cber/guidelines.htm>.

[24] R. Geoff, Cold chain shipping loss in pharmaceuticals, Retrieved from <http://www.cargosense.com/cold-chain-shipping-loss-in-pharmaceuticals.html>, 2015.

[25] G. Dutton, Genetic engineering & biotechnology news, August 2019. Ahead of print <https://doi.org/10.1089/gen.39.08.04>.

[26] F. Chiacchio, D. D'Urso, L. Compagno, M. Chiarenza, L. Velardita, Towards a blockchain based traceability process: a case study from pharma industry, in: F. Ameri, K. Stecke, G. von Cieminski, D. Kiritsis (Eds.), Advances in Production Management Systems. Production Management for the Factory of the Future. APMS 2019. IFIP Advances in Information and Communication Technology, vol. 566, Springer, Cham, 2019.

[27] Marathe, A., Internet of Things: opportunities and applications in pharmaceutical manufacturing and logistics, 2018.

[28] R. Handfield, Biopharmaceutical supply chains: distribution, regulatory, Systems and Structural Changes Ahead, CRC Press, 2012.

[29] R.H. Bishara, Cold chain management–an essential component of the global pharmaceutical supply chain, Am. Pharm. Rev. 9 (1) (2006) 105–109.

[30] S. Alshawi, I. Saez-Pujol, Z. Irani, Data warehousing in decision support for pharmaceutical R&D supply chain, Int. J. Inf. Manag. 23 (3) (2003) 259–268. Available from: https://doi.org/10.1016/s0268-4012(03)00028-8.

[31] H. Rees, Supply Chain Management in the Drug Industry: Delivering Patient Value for Pharmaceuticals and Biologics, John Wiley & Sons, 2011.

[32] P. Brhlikova, I. Harper, R. Jeffery, et al., Trust and the regulation of pharmaceuticals: South Asia in a globalised world, Glob. Health 7 (2011) 10. Available from: https://doi.org/10.1186/1744-8603-7-10.

[33] D. Jomaa, T. Monteiro, B. Besombes, Design and development of a forecasting module: case of a warehouse management system, IFAC Proc. Vol. 46 (24) (2013) 177–182.

[34] D. Smith, L. Sparks, Temperature controlled supply chains, Food Supply Chain Manag. (2004) 179–198.

[35] N.S. Zhang, W. He, P.S. Tan, Understanding local pharmaceutical supply chain visibility, Oper. Res. 25 (2008) 234–239.

[36] A.N. Zhang, M. Goh, F. Meng, Conceptual modelling for supply chain inventory visibility, Int. J. Prod. Econ. 133 (2) (2011) 578–585.

[37] J. Småros, J.M. Lehtonen, P. Appelqvist, J. Holmström, The impact of increasing demand visibility on production and inventory control efficiency, Int. J. Phys. Distrib. Logist. Manag. 33 (4) (2003) 336–354.

[38] J.M. Lehtonen, J. Småros, J. Holmström, The effect of demand visibility in product introductions, Int. J. Phys. Distrib. Logist. Manag. 35 (2) (2005) 101–115.

[39] C.S. Pattichis, E. Kyriacou, S. Voskarides, M.S. Pattichis, R. Istepanian, C.N. Schizas, Wireless telemedicine systems: an overview, IEEE Antennas Propag. Mag. 44 (2) (2002) 143–153.

[40] M.J. Field, J. Grigsby, Telemedicine and remote patient monitoring, JAMA 288 (4) (2002) 423–425.

[41] M. Papert, P. Rimpler, A. Pflaum, Enhancing supply chain visibility in a pharmaceutical supply chain: solutions based on automatic identification technology, Int. J. Phys. Distrib. Logist. Manage. 46 (9) (2016) 859–884.

[42] S.J. Mason, P.M. Ribera, J.A. Farris, R.G. Kirk, Integrating the warehousing and transportation functions of the supply chain, Transport. Res. E: Logist. Transport. Rev. 39 (2) (2003) 141–159.

[43] A.M. Atieh, H. Kaylani, Y. Al-abdallat, A. Qaderi, L. Ghoul, L. Jaradat, et al., Performance improvement of inventory management system processes by an automated warehouse management system, Procedia CIRP 41 (2016) 568–572.

[44] A. Ramaa, K.N. Subramanya, T.M. Rangaswamy, Impact of warehouse management system in a supply chain, Int. J. Comput. Appl. 54 (1) (2012) 14–20.

[45] U. Barchetti, A. Bucciero, M. De Blasi, L. Mainetti, L. Patrono, RFID, EPC and B2B convergence towards an item-level traceability in the pharmaceutical supply chain, 2010 IEEE International Conference on RFID-Technology and Applications, IEEE, 2010, pp. 194–199.

[46] B. Yan, G. Huang, Supply chain information transmission based on RFID and internet of things, 2009 ISECS International Colloquium on Computing, Communication, Control, and Management, vol. 4, IEEE, 2009, pp. 166–169.

[47] D.C. Wyld, Genuine medicine? Why safeguarding the pharmaceutical supply chain from counterfeit drugs with RFID is vital for protecting public health and the health of the pharmaceutical industry, Compet. Rev. An. Int. Bus. J. 18 (3) (2008) 206–216.

[48] U. Barchetti, A. Bucciero, M. De Blasi, A.L. Guido, L. Mainetti, L. Patrono, Impact of RFID, EPC and B2B on traceability management of the pharmaceutical supply chain, 5th International Conference on Computer Sciences and Convergence Information Technology, IEEE, 2010, pp. 58–63.

[49] S. Jayanth, M.B. Poorvi, M.P. Sunil, Inventory management system using IOT, in: S. Satapathy, V. Prasad, B. Rani, S. Udgata, K. Raju (Eds.), Proceedings of the First International Conference on Computational Intelligence and Informatics. Advances in Intelligent Systems and Computing, vol. 507, Springer, Singapore, 2017, pp. 201–210.

[50] C. Yang, W. Shen, X. Wang, Applications of Internet of Things in manufacturing, in: 2016 IEEE 20th International Conference on Computer Supported Cooperative Work in Design (CSCWD), Nanchang, 2016, pp. 670–5. doi:10.1109/CSCWD.2016.7566069.

[51] D. McFarlane, Y. Sheffi, The impact of automatic identification on supply chain operations, Int. J. Logist. Manage. 14 (1) (2003) 1–17. Available from: https://doi.org/10.1108/09574090310806503.

[52] R. Vargheese, H. Dahir, An IoT/IoE enabled architecture framework for precision on shelf availability: enhancing proactive shopper experience, in: 2014 IEEE International Conference on Big Data (Big Data), Washington, DC, 2014, pp. 21–26. <https://doi.org/10.1109/BigData.2014.7004418>.

[53] M. Vinay, S.K. Shivashankar, Monitoring and controlling of smart equipments in manufacturing industry using IoT applications, Int. J. Tech. Res. Appl. 4 (3) (2016) 96–100. ISSN:2320-8163, www.ijtra.com.

[54] L. Dan, C. Xin, H. Chongwei, J. Liangliang, Intelligent agriculture greenhouse environment monitoring system based on IOT technology, 2015 International Conference on Intelligent Transportation, Big Data and Smart City, Halong Bay, 2015, pp. 487–490. Available from: http://dx.doi.org/10.1109/ICITBS.2015.126.

[55] S.C. Mukhopadhyay, N.K. Suryadevara, Internet of things: challenges and opportunities, Internet of Things, Springer, Cham, 2014, pp. 1–17.

[56] A. Banafa, Three major challenges facing IoT. *IEEE IoT Newsletter*, 2017.

Chapter 10

Smart cervical band: an Internet of Things- and artificial intelligence-based neck pain and cervical spondylosis healing system

Vikram Puri[1], Chung Van Le[1], Sandeep Singh Jagdev[2] and Long Cu Kim[3]

[1]*Duy Tan University, Da Nang, Vietnam,* [2]*Ellen Technology (P) Ltd., Jalandhar, Punjab, India,* [3]*School of Information Communication Technology, Hanoi University of Science and Technology, Hanoi, Vietnam*

10.1 Introduction

Cervical spondylosis (CS) is a general term related to neck spinal disc problems due to improper neck positions and movement. CS is very common among those aged 60 and over, with more than 85% of people affected by CS at some time [1,2]. In the general population, 3.3 per 1000 people are affected by CS [3,4]. Neck pain, also known as cervicalgia, affects 70% of the total population with a physical disability, as well as being the fourth most common of 310 classified physical pain syndromes [5]. Women suffer more through this pain than, men similarly to back pain, it is more common in middle age [6]. The main causes of cervicalgia and spondylosis are injury, compression of the neck, strain, and degenerative disc disease. Spondylosis and back pain are the third most expensive treatment, with an annual cost of \$87.6 billion [7] after diabetes and heart-related diseases. Radiological changes inside the body are explained by structural changes to the vertebrae but they are insufficient to be the cause and severity of the problem [8]. Proper knowledge of the cause and symptoms related to neck pain or CS help to prevent cervical spine problems. Cervical position sensing is essential to maintain stability between the joints under static and dynamic conditions, and spontaneous pain may cause impaired proprioception (IP) [9]. Moreover,

An Industrial IoT Approach for Pharmaceutical Industry Growth.
DOI: https://doi.org/10.1016/B978-0-12-821326-1.00012-7

joint position error (JPE) is used to quantify the degree of IP. Neuronal and muscular control of cervical joints are disturbed due to impaired position sensing that leads to unbalanced muscular force and puts pressure on the joints, hence increasing the risk of trauma [10,11]. CS most often occurs at the C5–C6 and C6–C7 levels, however other levels may be involved (see Fig. 10.1) [12]. From a single area, cervical pain can originate [11] and then rapidly spread to distant areas and also result in chronic fatigue syndrome. The anatomy of the cervical spine [2] is explained next.

The cervical spine consists of seven vertebrae and eight sets of nerve roots, which connect to the upper extremities with the use of nerve plexus. C1 and C2 vertebrae are connected to the alar and cruciate ligaments, which are parts of the ligamentous complex. In addition, a number of facet joints and intervertebral discs are attached from C3 to C7.

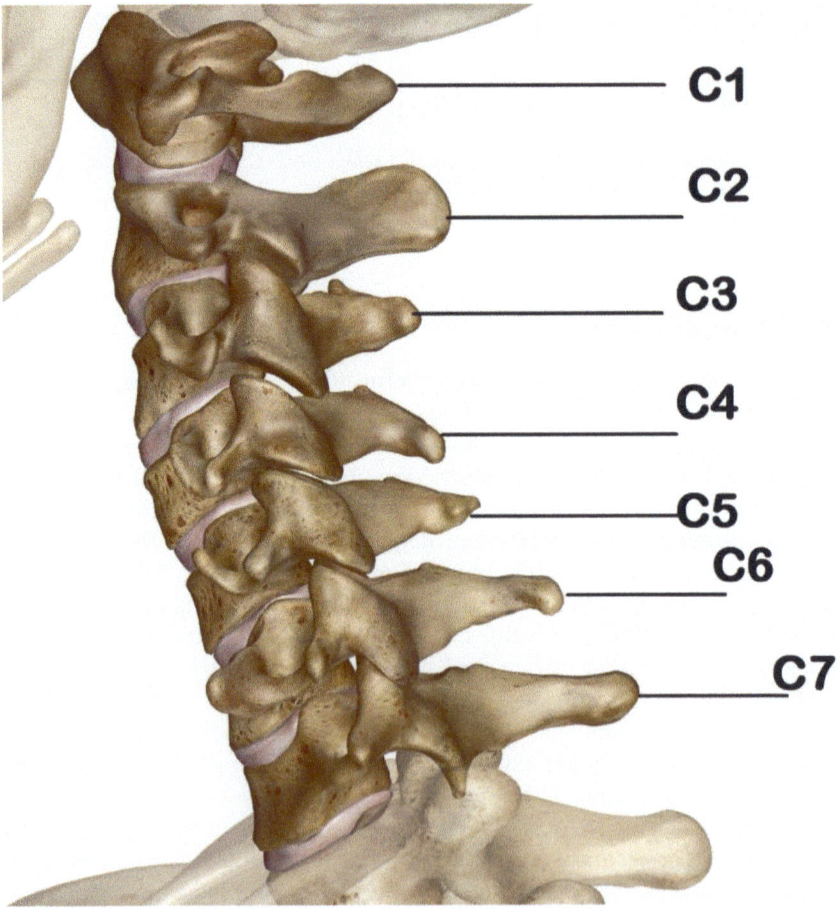

FIGURE 10.1 Cervical spine levels.

The main causes of neck pain are as follows:

- Pain generated due to muscular tension around the neck and shoulder;
- Foraminal stenoses and herniated discs lead to nerve compression.

The Internet of Things (IoT) plays a vital role in the personalized medical system (PMS) and also offer different kinds of efficient solutions such as e-healthcare and Telemedicine [13]. Fruitful integration of the IoT with PMS can make medical devices smarter, reduce the deployment cost, improve medical practices at patient care centers, and also improve sustainability [14]. Integration of the IoT in PMS devices is still challenging due to the shortage of accurate medical sensors and the costs of these sensors, which generate a huge amount of data, also IoT architecture has inefficient standardization. Nowadays, neck pain and CS are very common musculoskeletal disorders [15,16], often caused by working on a computer for a long time, such as IT engineers, researchers, and scientists. Due to this issue [17], many problems arise such as a reduction in work productivity, limited presence at work, and greater reliance on healthcare services. A flexible computer screen, keyboard mouse, and chair can help to improve body posture [18]. In addition, continuous neck monitoring provides the required information that is used to improve the posture position and reduce stress due to continuously work in the same manner [19].

Some previous studies have discussed the various feedback devices that help to improve neck posture. Azrin et al. [20] proposed portable operant treatment instruments to reduce the poor posture problem. Twenty-five adults participated in an experiment wearing the instrument while working and aversive tones generated stimulus warning during the time of slouching. If there is no slouching, tones automatically terminated and postponed. The results of this experiment showed that posture can be improved with the use of behavioral engineering. Yoo et al. [21] designed a chair-based feedback device that continuously monitored body posture including the head, shoulder, and trunk. Twenty subjects participated in this experiment and monitored their posture through the aid of a 3D motion axis sensor. Proximity sensor feedback was shown to help to improve the posture. A paired *t*-test was used to measure the statistical difference between the feedback and nonfeedback devices and the results showed that the feedback system had a greater benefit for body posture. Riandy et al. [22] proposed a wearable device equipped with voice- and vibration-based feedback that helped to monitor poor body posture. Moreover, this device provided a feedback technique to improve bad habits during sitting and was also integrated into a mobile-based graphical user interface via Bluetooth technology. The capability of the proposed wearable device measures between 0 and 90 degrees with an average error of 0.42. Ma et al. [23] designed a system based on a 3D axis that helps to monitor five types of sitting procedure and analysis of the accelerometer via machine learning techniques such as support vector machine and k-means clustering that were used to classify the obtained data. Lou et al. [24] designed a monitoring system to improve body posture which was equipped with an accelerometer, microcontroller, and battery. This system provided feedback to the user on the second day. Thomas et al. [25] designed a user-friendly web application to monitor body posture and provide feedback through an alert. This application showed the results of the current posture and orientation of the body on a 3D graph. Moreover, this system was also able to monitor the heartbeat. Breen et al. [26] proposed a biofeedback system equipped with an accelerometer to monitor the neck position. This system is used to detect the bad posture of the neck via feedback mechanism. Six users participated and either received or did not receive feedback. Alattas [27] designed a system integrated with Arduino and ultrasonic sensors to monitor neck posture. This system provides feedback when bad posture is detected. In Ref. [28], the authors presented an application-based solution with the main aim of monitoring the tilt angle of the mobile phone. An alert is generated when the angle of the phone is greater than 45 degrees. Liu et al. [29] developed a convolutional neural network-based algorithm model that worked on human pressure data. Data are recorded with the aid of pressure data from chair contacts and trained through a body posture algorithm. In this study, the authors declared that results arrived from the system with high accuracy, security, and robustness. In Ref. [30], the author deployed a real-time system for monitoring head and neck conditions as well as to prevent pain in the head and neck. In this proposed system, cloud services were introduced to identify management, data storage, and a setup communication bridge between the cloud and real-time systems. Wang et al. [31] designed a low-cost energy-efficient wearable device to monitor the cervical vertebrae posture with the aid of a three-axis accelerometer. The authors classified seven different types of neck posture and five participants participated in the study. This system is able to monitor the time duration of bending neck and results can be displayed on the mobile application. In the above-mentioned studies, the authors and researchers focused on the offline or online modes of a cervical posture monitoring system. In this chapter, we propose a low-cost, low-power consumption cervical or neck posture monitoring band that works with and without cloud services. The main advantage of the proposed system is that it works with a low-power microcontroller which makes our system energy efficient. Moreover, an artificial intelligence technique is applied to predict the neck posture as well as provide information to reduce the CS or neck pain.

Section 10.2 discusses the applied methodology in the system, Section 10.3 includes the results and a discussion, and Section 10.4 concludes the chapter.

10.2 Methodology

In this section, the architecture, circuit diagram, algorithm workflow, artificial intelligence techniques, and statistical parameters are described (Fig. 10.2).

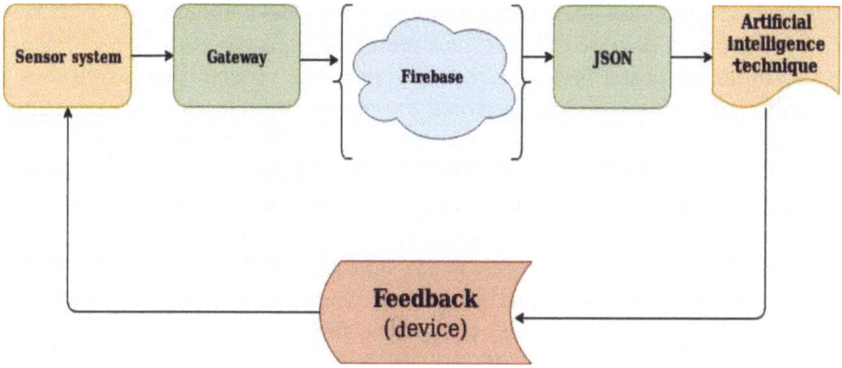

FIGURE 10.2 Architecture of the proposed system.

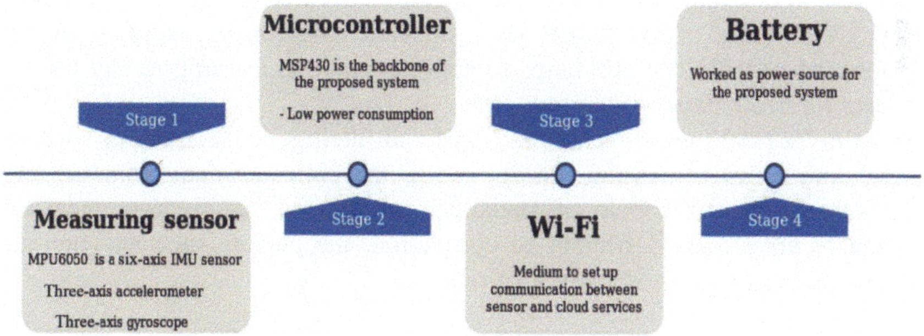

FIGURE 10.3 Sensor system module.

10.2.1 Architecture

To tackle the problem of neck pain or CS, a smart wearable band is proposed. The system workflow is categorized into six module as follows:

1. *Sensor system*: The sensor system module consists of an accelerometer/gyrometer, power-efficient microcontroller, Wi-Fi module, and power source, such as a battery. Selection of the microcontroller is the major part of this module which directly depends on the power consumption of the device (Fig. 10.3).
2. *Gateway*: The gateway is an essential component in any kind of IoT system and provides a solution between the node to node system or node to cloud system. The IoT gateway is also involved in performing some specific tasks, namely buffering, storing of data, data filtering, optimization, and configuration management. In the proposed sensor node, the gateway is used to set up the communication bridge between the sensor device and firebase.

3. *Firebase*: Firebase [32] is a mobile and web development platform run by Google. It provides the services presented in Table 10.1.

   In the proposed system, a real-time database service is deployed to set up a real-time connection between the device and the cloud. A real-time database is based on the cloud-hosted database that is used to store data as Javascript object notation (JSON) and connected to the client on the basis of real-time synchronization. The main feature of the real-time database is sharing the same database with multiplatform devices such as android and iOS. Moreover, it works on online and offline modes. In the offline mode, data are stored on a disk, when connectivity is available it synchronizes with the current state of the firebase server.
4. *Javascript object notation (JSON)*: JSON [33] is an open standard format file consisting of a language-independent data format that uses human-friendly text to share data objects composed of data-type arrays and value pairs. Firebase database stores in the form of JSON objects and creates a JSON tree. Every updating of data in the JSON tree creates a node with a key consisting of the USER ID. Collected data can be exported in the form of a JSON file that helps to apply a machine learning model on the data to create better data prediction.
5. *Artificial intelligence technique*: Artificial intelligence techniques such as random forest regression are applied to the collected data from Google Firebase. With the help of statistical parameters, namely root measure square error and coefficient of correlation, the performance and predictive power of the model are measured.

TABLE 10.1 Services of Firebase cloud.

S. no.	Name of service	Function
1.	Cloud firestore	Data synchronization and data storage for application
2.	Functions for cloud	Without managing server, test and run code at backend
3.	Storage at cloud	Serve and store data at the cloud with the use of Google
4.	Real-time database	Transmission of data between cloud and device in milliseconds
5.	Web hosting and authentication	With the point of data transferring speed and security helps to deliver the web application. Authentication used to maintain the security

6. *Feedback*: The feedback system is one of the most important parts of any system. In this system, the vibration motor is integrated to give an alert warning through vibration. This warning is issued to warn of the wrong neck posture and also helps to improve CS and neck problems.

10.2.2 Circuit diagram

The microcontroller plays a major part in any kind of system. MSP430 [34] offers a wide range of low-power 16-bit microcontrollers, as well as being integrated with the analogue and digital sensing measurement. MSP430 is especially designed for low-power embedded devices. In the idle mode, the current drawn is less than equal 1 μA. MSP430 contains six types of modes for power handling. However, the following three modes are mostly used:

- *Active mode*: MSP CPU and other peripheral devices working at the same time. This mode uses the time of calculation, decision-making, and other related tasks where the CPU is needed.
- *Low-power mode 3 (LM3)*: This mode works as normal mode for the range of applications between the 99% to 99.9%. This mode is also called the done or sleep mode. The LM3 mode is mostly used in the case of battery-powered applications. This mode [35] only disable the CPU usage but still active with the external peripheral devices (see Table 10.2).
- *Low-power mode 4*: This mode is mostly deployed in the case of storage time and is also referred to as the off-mode. The main purpose of this mode is for when there is a requirement for extra low-power supply and it is mostly used in the case of storage or calibration time (Figs. 10.4 and 10.5).

Moreover, the wake up time for the MSP430 is less than 1 μs, and the sleeping time is increased, which help to reduce power consumption. MSP430 directly connects to the MPU6050 to measure the exact position of the neck. MPU6050 is a sensor based on a microelectromechanical system and integrated work with the six-axis accelerometer and gyrometer. Three axes are for the accelerometer and three are for the gyrometer and communication setup with MSP430 via I2C (inter-interconnected protocol). The principle behind the three-axis accelerometer

TABLE 10.2 LM3 active and sleeping modes.

Active mode	Sleeping mode
RAM	Main processor (CPU)
ACLK (crystal clock)	MCLK (CPU clock)
Oscillator	Peripheral devices
Timers	Frequency-locked loop

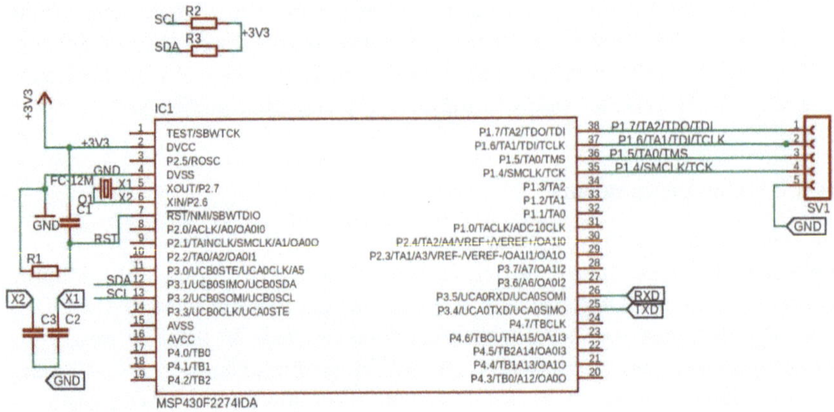

FIGURE 10.4 Circuit diagram of MSP430.

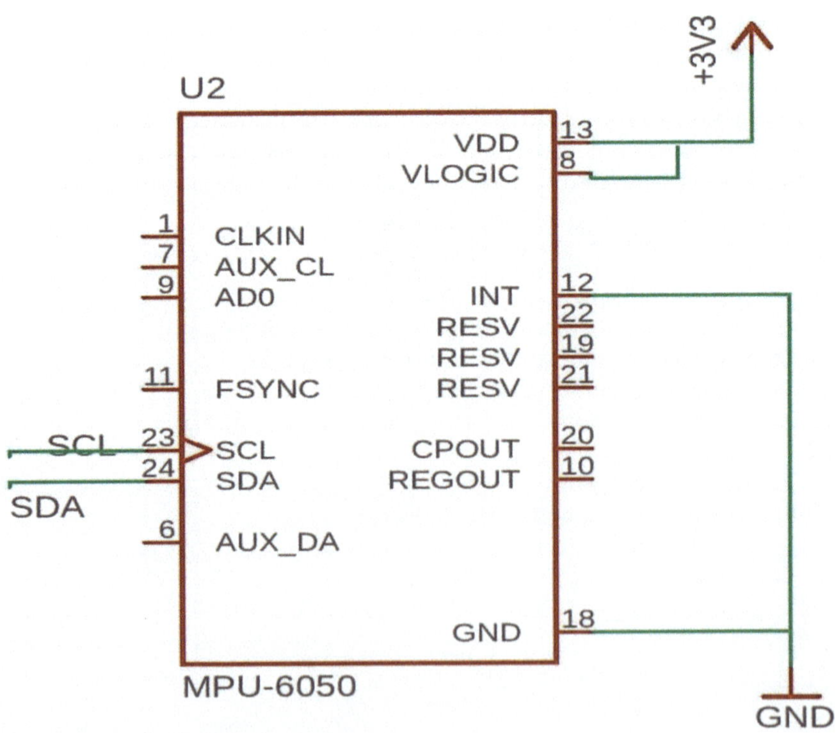

FIGURE 10.5 Circuit diagram of MPU6050.

is the piezoelectric effect. Current from every axis is directly dependent on the determination of the magnitude and direction of inclination. The principal behind the Gyroscope is Coriolis acceleration. In the imaginary way, it is a fork-like structure that constantly experience the motion and placed via Piezoelectric crystal. Piezoelectric crystal experience force in the route of inplimentation that lead to cause inertia of moving fork. Current is generated from solidarity of crystal with the piezoelectric effect. ESP8266 works as a bridge to set up communication between the device and the cloud server. ESP8266 [36] is low-cost low-powered Wi-Fi chip integrated with a TCP/IP stack and onboard antenna. It is able to host an application as well as working with another processor on Wi-Fi functions. ESP works on the Attention (AT) commands due to already uploaded AT firmware. With the aid of general purpose input output pins, it is capable of integrating with the sensors and other related applications. The operating voltage is 3.3 V, and for transferring the data, a logic shifter is required, although the possibility of pin damage is high. Table 10.3 presents the AT commands used for ESP8266.

A vibrator sensor is integrated with the MSP430 for providing feedback on the wrong neck posture. A coin-shaped vibrator motor consists of a shaftless motor packed in a pancake-shaped mounted body. Fig. 10.6 presents the workflow of the proposed system algorithm. In this system, MSP430 is connected to MPU6050 which helps to capture the axis of the neck. Data are sent directly to the Firebase to indicate whether the neck position is correct or not. ESP8266 and the gateway are two media which are helpful for connecting the sensor module to the cloud services. However, if the position is wrong the vibrator is activated for feedback to correct the neck posture. Data are stored at the Firebase on a real-time basis and a JSON file is exported for further processing. After exporting of data, a machine learning model random forest is applied. The performance of the model is measured through a statistical parameter such as root mean square error (RMSE) or R^2. Fig. 10.7 presents the proposed system for neck monitoring.

TABLE 10.3 AT commands for ESP8266.

Command	Function
AT	Test with AT firmware
AT + RST	Reset module
AT + GMR	Print firmware
AT + GSLP	Sleep mode
ATE	Enable/disable ECHO
AT + CWMODE	Modes in the Wi-Fi
AT + CWJAP	Set up connection with gateway
AT + CWDHCP	DHCP: enable/disable

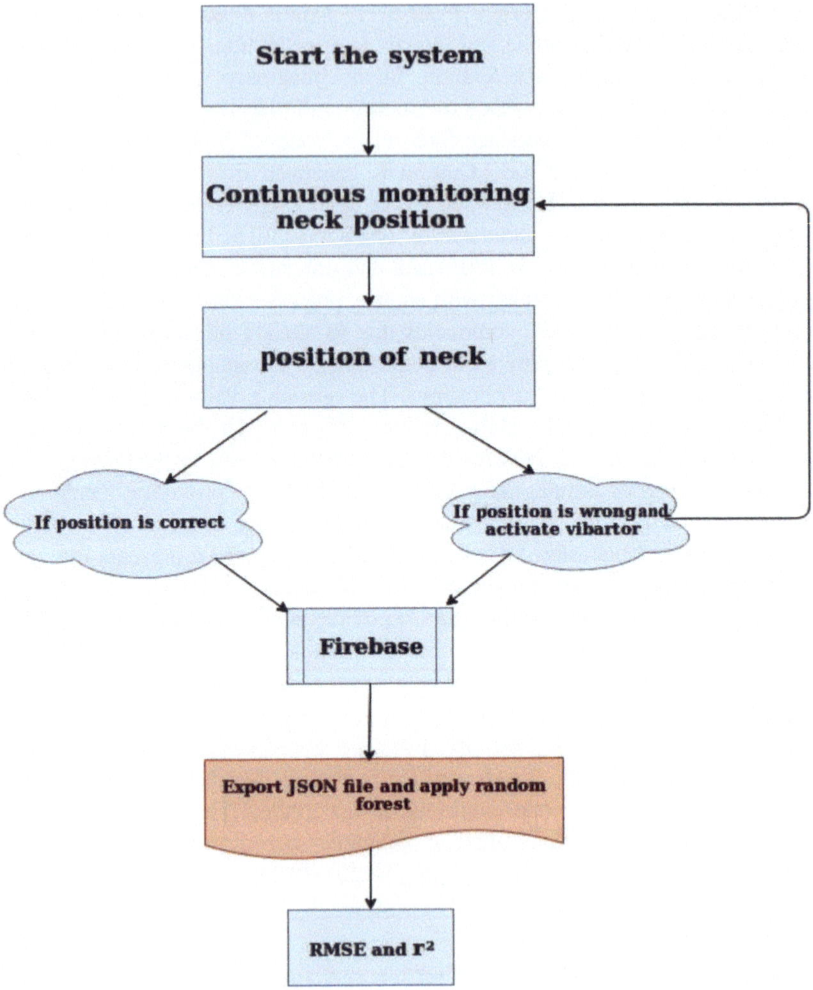

FIGURE 10.6 Proposed system algorithm.

10.2.3 Artificial intelligence technique

10.2.3.1 Random forest regression

Random forest regression is an ensemble technique that is capable of handling the classification and regression task with the aid of a multiple regression tree using a technique described as bagging, commonly called bootstrap aggregation. The main principle behind random forest regression is to determine the multiple tree output and evaluate the final output from this. This approach is much better than relying on a single output (see Fig. 10.8).

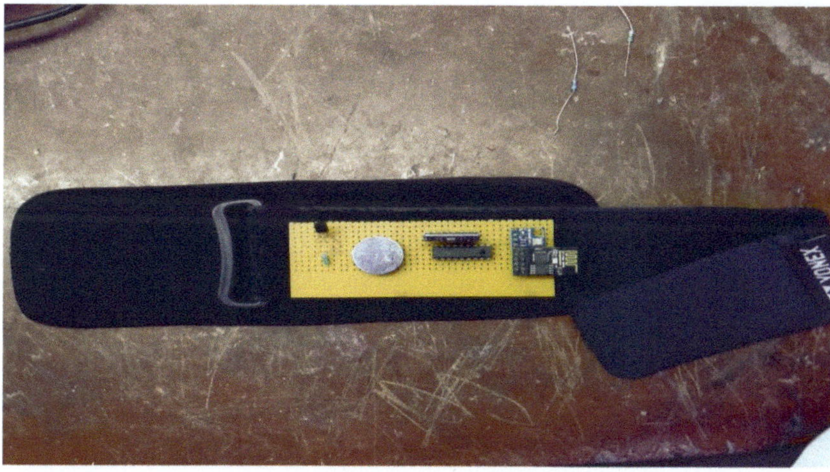

FIGURE 10.7 Proposed system to tackle CS or neck pain.

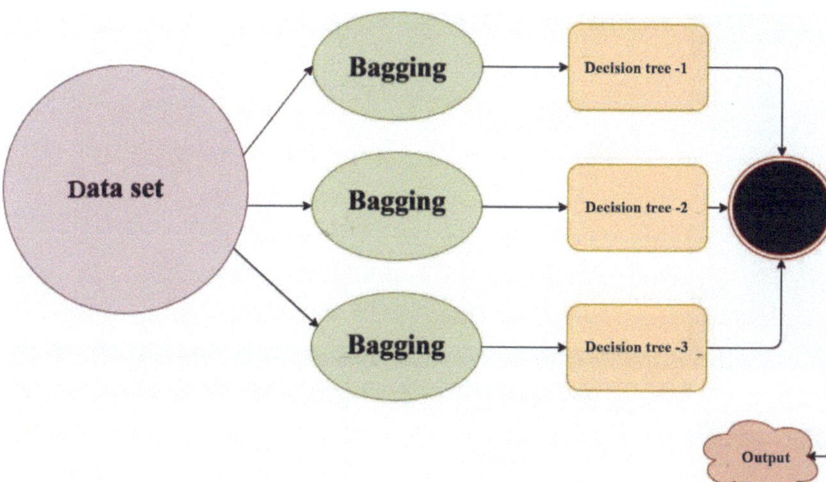

FIGURE 10.8 Random forest regression.

10.2.4 Statistical parameters

In this study, performance evaluation of the models is done with the aid of statistical methods such as RMSE, and coefficient of correlation (R^2). R^2 values vary between 0 and 1 showing the predictive capability of the deployed model. RMSE provides the error of the model used for training. In addition, the statistical value from the evaluation techniques directly depend on the trained model capability.

$$\mathrm{RMSE} = \sqrt{\frac{1}{N}\sum_{K=1}^{N}(X_A - Y_A)^2} \tag{10.1}$$

$$\mathrm{MAE} = \frac{1}{N}\sum_{K=1}^{N}|X_A - Y_A| \tag{10.2}$$

where X and Y are the predicted and actual values.

10.3 Results and discussion

Firebase, a real-time database, helps to set up the communication channel between the cloud and neck monitoring system. Firebase works on the basis

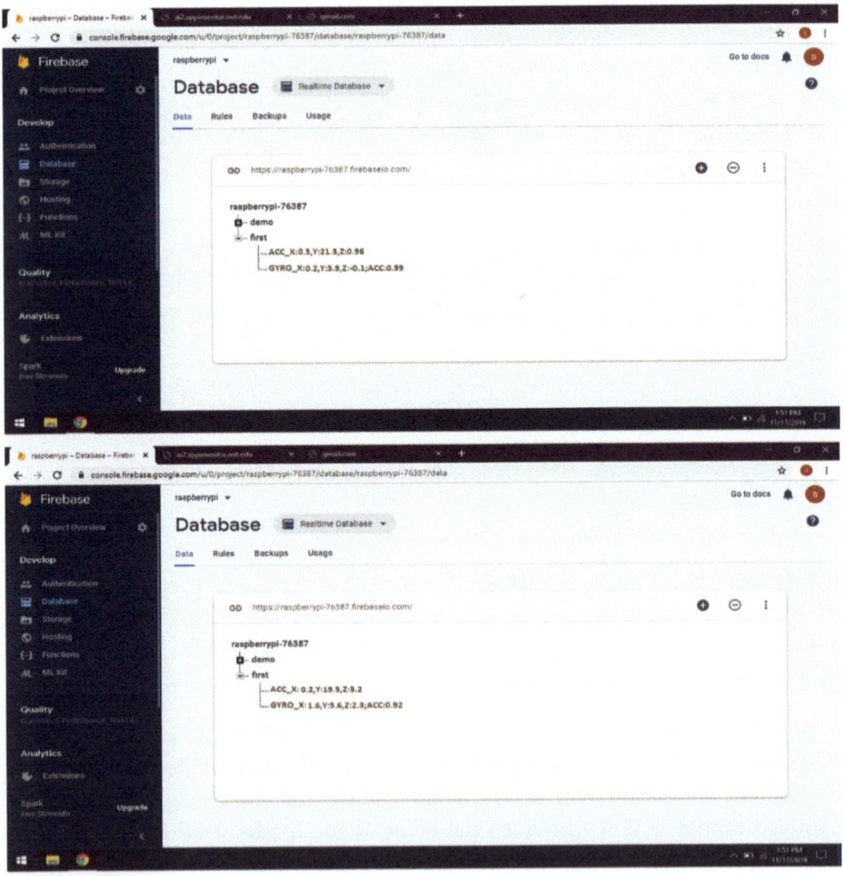

FIGURE 10.9 Data stored at the firebase.

of real-time and stored data on the server. Fig. 10.9 present data stored in the cloud storage.

The proposed system was tested on four neck movement positions (see Fig. 10.10). These movements help to calibrate the sensor for CS or neck pain limits and send data to the cloud server via values processed through MSP430.

To evaluate the performance of the proposed cervical monitoring system, statistical parameters such as RMSE and R^2 are used in the system and validated with the aid of Python. The main motive behind the use of Python is its open source nature. The data set consists of 1875 samples that include posture accelerometer x, y, z axes, gyrometer x, y, z axes, and barometer readings for acceptable and unacceptable postures. Seventy percent of the data set is used for training and 30% for testing. The RMSE and R^2 of the proposed system are 48.646 and 0.87, respectively.

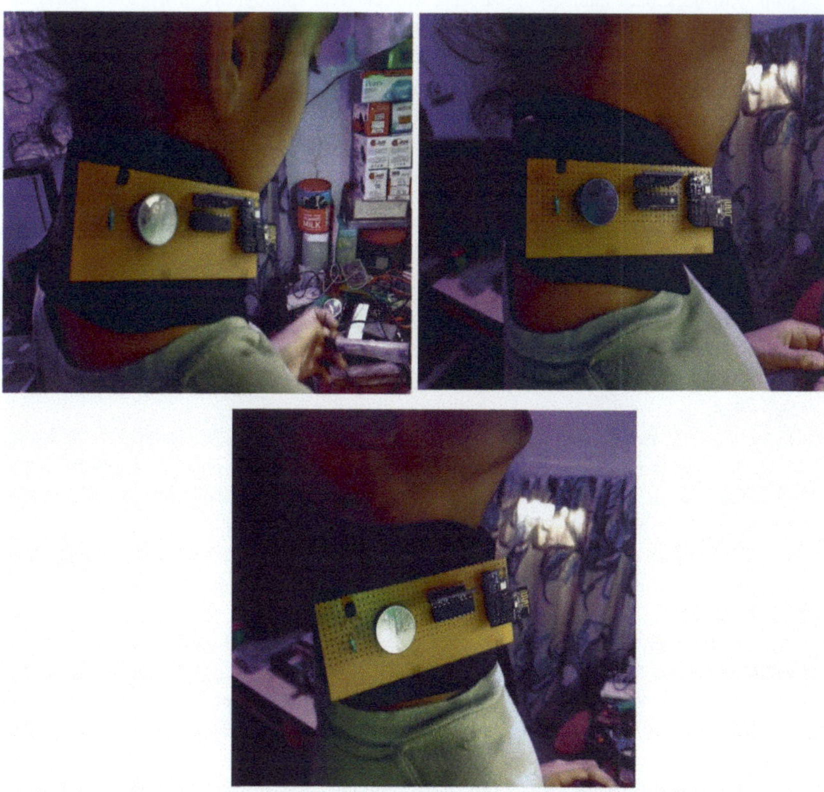

FIGURE 10.10 System calibrated and tested on different neck positions.

10.4 Conclusion

In this chapter, an IoT- and artificial intelligence-based neck pain and CS monitoring system is proposed. Data are collected at a Firebase and exported in the form of a JSON file and validated through the AI model, namely random forest regression. A three-axis accelerometer and three-axis gyrometer help to monitor the posture of the neck and connect to the processor for further processing and send data to the Firebase which acts as a real-time database via the Wi-Fi module and gateway.

For the validation process, the model is developed from real-time monitoring of neck posture and is trained and tested through random forest regression via two statistical parameters, RMSE and R^2. Total input samples are considered to predict the accurate posture of the neck. Result outcomes from the proposed system give RMSE and R^2 values 0.189 and 0.87, respectively.

References

[1] R.S. Reddy, J.S. Tedla, S. Dixit, M. Abohashrh, Cervical proprioception and its relationship with neck pain intensity in subjects with cervical spondylosis, BMC Musculoskelet. Disord. 20 (1) (2019) 1–7.

[2] A. Starkweather, Cervical pain: when it is more than a pain in the neck, Top. Pain. Manag. 34 (8) (2019) 1–8.

[3] S. Hogg-Johnson, G. van der Velde, L.J. Carroll, L.W. Holm, J.D. Cassidy, J. Guzman, et al., The burden and determinants of neck pain in the general population: results of the bone and Joint Decade 2000–2010 Task Force on Neck Pain and Its Associated Disorders, J. Manipulative Physiol. Ther. 32 (2) (2009) S46–S60.

[4] K. Radhakrishnan, W.J. Litchy, W.M. O'Fallon, L.T. Kurland, Epidemiology of cervical radiculopathy: a population-based study from Rochester, Minnesota, 1976 through 1990, Brain 117 (2) (1994) 325–335.

[5] V.L. Feigin, G.A. Roth, M. Naghavi, P. Parmar, R. Krishnamurthi, S. Chugh, et al., Global burden of stroke and risk factors in 188 countries, during 1990–2013: a systematic analysis for the Global Burden of Disease Study 2013, Lancet Neurol. 15 (9) (2016) 913–924.

[6] L.J. Carroll, S. Hogg-Johnson, G. van der Velde, S. Haldeman, L.W. Holm, E.J. Carragee, et al., Course and prognostic factors for neck pain in the general population: results of the Bone and Joint Decade 2000–2010 Task Force on Neck Pain and Its Associated Disorders, J. Manipulative Physiol. Therap. 32 (2) (2009) S87–S96.

[7] J.L. Dieleman, R. Baral, M. Birger, A.L. Bui, A. Bulchis, A. Chapin, et al., US spending on personal health care and public health, 1996-2013, JAMA 316 (24) (2016) 2627–2646.

[8] A.E.W. Brooker, R.W. Barter, Cervical spondylosis: a clinical study with comparative radiology, Brain 88 (5) (1965) 925–936.

[9] H.Y. Lee, J.D. Wang, G. Yao, S.F. Wang, Association between cervicocephalic kinesthetic sensibility and frequency of subclinical neck pain, Man. Ther. 13 (5) (2008) 419–425.

[10] F.E. Valergakis, Cervical spondylosis: most common cause of position and vibratory sense loss, Geriatrics (Basel, Switz.) 31 (7) (1976) 51–56.

[11] M. Revel, C. Andre-Deshays, M. Minguet, Cervicocephalic kinesthetic sensibility in patients with cervical pain, Arch. Phys. Med. Rehabil. 72 (5) (1991) 288–291.
[12] A.I. Binder, Cervical spondylosis and neck pain, BMJ 334 (7592) (2007) 527–531.
[13] J. Qi, P. Yang, G. Min, O. Amft, F. Dong, L. Xu, Advanced internet of things for personalised healthcare systems: a survey, Pervasive Mob. Comput. 41 (2017) 132–149.
[14] J. Qi, L. Chen, W. Leister, S. Yang, Towards knowledge driven decision support for personalized home-based self-management of chronic diseases, 2015 IEEE 12Th Intl Conf on Ubiquitous Intelligence and Computing and 2015 IEEE 12th Intl Conf on Autonomic and Trusted Computing and 2015 IEEE 15th Intl Conf on Scalable Computing and Communications and its Associated Workshops (UIC-ATC-ScalCom), IEEE, 2015, pp. 1724–1729.
[15] S. Bagheri, M. GHaljahi, Ergonomic evaluation of musculoskeletal disorders with rapid office strain assessment and its association with occupational burnout among computer users at zabol university of medical sciences in 2017, Asian J. Water Environ. Pollut. 16 (1) (2019) 91–96.
[16] B.N. Green, A literature review of neck pain associated with computer use: public health implications, J. Can. Chiropr. Assoc. 52 (3) (2008) 161.
[17] D.L. Presti, C. Massaroni, J. Di Tocco, E. Schena, A. Carnevale, U.G. Longo, et al., Single-plane neck movements and respiratory frequency monitoring: a smart system for computer workers, 2019 II Workshop Metrology Ind. 4.0 IoT (MetroInd4.0&IoT), IEEE, 2019, pp. 167–170.
[18] K. Mekhora, C.B. Liston, S. Nanthavanij, J.H. Cole, The effect of ergonomic intervention on discomfort in computer users with tension neck syndrome, Int. J. Ind. Ergonom. 26 (3) (2000) 367–379.
[19] C. Niewiadomski, R.J. Bianco, S. Afquir, M. Evin, P.J. Arnoux, Experimental assessment of cervical ranges of motion and compensatory strategies, Chiropr. Man. Ther. 27 (1) (2019) 9.
[20] N. Azrin, H. Rubin, F. O'brien, T. Ayllon, D. Roll, Behavioral engineering: postural control by a portable operant apparatus 1, J. Appl. Behav. Anal. 1 (2) (1968) 99–108.
[21] W.G. Yoo, C.H. Yi, M.H. Kim, Effects of a proximity-sensing feedback chair on head, shoulder, and trunk postures when working at a visual display terminal, J. Occup. Rehabil. 16 (4) (2006) 631–637.
[22] R. Riandy, M.N. Suya, A.N. Ratna, The development of a wearable biofeedback trainer to increase quality of sitting posture, In: J. Phys. Conf. Ser. 1192 (1) (2019) 012067. IOP Publishing.
[23] S. Ma, W.H. Cho, C.H. Quan, S. Lee, A sitting posture recognition system based on 3 axis accelerometer, 2016 IEEE Conference on Computational Intelligence in Bioinformatics and Computational Biology (CIBCB), IEEE, 2016, pp. 1–3.
[24] E. Lou, M. Bazzarelli, D. Hill, N. Durdle, A low power accelerometer used to improve posture, Canadian Conference on Electrical and Computer Engineering 2001. Conference Proceedings (Cat. No. 01TH8555), vol. 2, IEEE, 2001, pp. 1385–1389.
[25] P. Thomas, P. Manikonda, S. Josyula, S.V. Boppana, J. McManus, E. Chawla, Sensor Technology to monitor body postures, 2016.
[26] P.P. Breen, A. Nisar, G. ÓLaighin, Evaluation of a single accelerometer based biofeedback system for real-time correction of neck posture in computer users, 2009 Annual International Conference of the IEEE Engineering in Medicine and Biology Society, IEEE, 2009, pp. 7269–7272.

[27] R. Alattas, Postuino: bad posture detector using arduino, Int. J. Innov. Sci. Res. 3 (2) (2014) 208–212.

[28] S. Elnaffar, A. El Allam, An app approach to correcting the posture of smartphone users, 2018 *Adv. Sci. Eng. Technol. Int. Conf. (ASET)*, IEEE, 2018, pp. 1–4.

[29] W. Liu, Y. Guo, J. Yang, Y. Hu, D. Wei, Sitting posture recognition based on human body pressure and CNN, in: AIP Conference Proceedings, 2019, vol. 2073, no. 1, p. 020093, AIP Publishing.

[30] D.Y. Liao, Design of a secure, biofeedback, head-and-neck posture correction system, IEEE First International Conference on Connected Health: Applications, Systems and Engineering Technologies (CHASE), IEEE, 2016, pp. 119–124.

[31] Y. Wang, H. Zhou, Z. Yang, O.W. Samuel, W. Liu, Y. Cao, et al., An intelligent wearable device for human's cervical vertebra posture monitoring, 40th Annual International Conference of the IEEE Engineering in Medicine and Biology Society (EMBC), IEEE, 2018, pp. 3280–3283.

[32] Google Firebase Services. <https://firebase.google.com/> (accessed 21.10.19).

[33] D. Crockford, The application/json media type for javascript object notation (json), 2006.

[34] MSP430 offer 16 bit low power consumption microcontroller. <http://www.ti.com/microcontrollers/msp430-ultra-low-power-mcus/overview.html#%20> (accessed 25.10.19).

[35] MSP 430 features. <http://www.ti.com/sc/data/msp/databook/chp1.pdf> (accessed 25.10.19).

[36] ESP8266 overview. <https://www.espressif.com/en/products/hardware/esp8266ex/overview> (accessed 26.10.19).

Chapter 11

Medical data analysis in eHealth care for industry perspectives: applications

Chandrakanta Mahanty and Brojo Kishore Mishra
Department of CSE, GIET University, Gunupur, India

11.1 Introduction

Over the last decade, unusual developments have been observed in the variety and volume of digital data related to research and development, patient self-tracking, and the collectively identified big data. In many nations, health care is a booming economic sector [1]. With its growth, its cost growth, inefficiency, poor quality, and increasing complexity have created challenges [2]. US spending on health has grown from $26 trillion to $32 trillion between 2010 to 2015. Researchers have found that about 251,454 US patients die annually from clinical errors [3]. Better decision-making based on the available information could alleviate these challenges and promote the transition to a value-based healthcare sector [4].

11.1.1 Data mining

Data mining is defined as a nontrivial extraction process from the data stored in a database of implicit, previously unexplored, and potentially useful information. Health databases have a large amount of data, but there are insufficient tools available for analyzing this knowledge in depth. Adequate machine-based data and decision-making systems can help scientists make less costly clinically similar decisions in their practice. Efficient and reliable computerized system implementation requires comparison of many available techniques. Prediction of disease plays a crucial role in data mining. In the field of medicine, data mining is efficiently used to predict diseases such as heart disease, skin cancer, breast cancer, and lung cancer.

The eHealth record is quickly becoming the clinical toolkit's most convincing tool, with data stored in the cloud because it is estimated that

An Industrial IoT Approach for Pharmaceutical Industry Growth.
DOI: https://doi.org/10.1016/B978-0-12-821326-1.00013-9

the extent of the e-documents carrying entire patient records may be as high as 6 TB.

To allow the practice of precision medicine, a large data file is required. This is a new health revolution that is capable of specifically focusing on a person's health care.

The following data forms are derived from the human anatomy:

Social information: personal activities, for example, living and employment.
Genome: full gene array "printed" in the patient's DNA.
Computer data: knowledge obtained from exercise and sleep measurement applications, digital inhalers, etc.
Metabolome: bodily chemical products, along with the enzyme response, that are created, modified, and broken.
Transcriptome: DNA-generated messages that produce protein stencil (mRNA).
Clinical data: patient's medical records.
Microbiome: 100 trillion microscopic files living within us, combined with their names.
Exposures: outdoor results, such as tobacco smoke and pollution, etc.
Proteome: a structure of proteins that are the building blocks of your body, including chemicals.
Images: medical images, X-rays, MRI scans, etc.
Epigenetic: modifications in the human genome to the structure of nucleic and methylation.

Patient information from IoT technologies, such as health trackers, blood pressure monitors, and metric scales, is also routinely provided with basic data on a person's daily lifestyle readings.

Healthcare industries now operate from a volume-based business to a value-based business that requires greater performance from doctors and nurses to be extra productive and efficient contributing to increased longevity and preventing illnesses, infections, and diseases. A data-mining system is presented in Fig. 11.1.

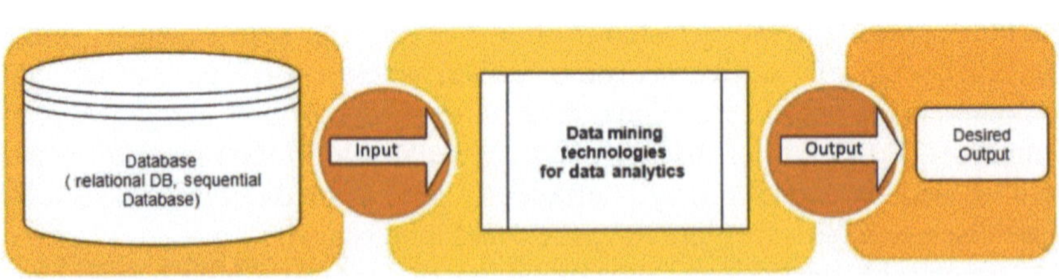

FIGURE 11.1 Data-mining system.

11.1.2 Data-mining applications in the pharmaceutical industry

Due to its excellent efficiency, a number of medical companies, hospitals, and pharmaceutical production units use data-mining tools. The recognition and quantification of pharmaceutical data can be extremely helpful for clinicians, medical doctors, pharmacists, health departments, assurance firms, lawyers, regulatory agencies, investors, drug-testing companies, and the pharmaceutical industry.

11.1.2.1 Data requirements of the pharmaceutical industry

See Ref. [5].

1. Data description and summaries: For example, historical hospital data analysis over a period of changes in medicine indicates an increasing demand from the hospital for medicinal products, etc.
2. Data characterization summary: This is the target class summary. For example, more than 40% of sales, or sales of more than 20% of consumers as the target class for recognition of desires of key customers and production and customer service are improved for the development of the most lucrative customer markets.
3. The cluster analysis means that we may collect a great deal of information in order to properly interpret the sales data of main customers without predetermined interpretation of the premise, such as, which customers or which regions favor which drugs. The isolated point analysis should be taken particular care of; these customers are to be cautiously dealt with as their loyalty cannot be guaranteed.

   Simultaneously with a view to maintaining current consumers and developing new clients, customers should be measured on the basis of historical data, including the analysis of data linked to time.
4. Prediction: To predict future drug sales based on the previous drug sales period in order to decide the future medicines to be purchased.
5. Data relevance discovery: To assess which drug a person has obtained and which medications he/she can also buy.

Depending on the current situation, the pharmaceutical company is considering the forms of drug sales over the last century in order to decide on the next purchase scenario. The problem that needs to be resolved in this example is the relationship between drug sales and pharmaceuticals and the development of a relationship model. In this case, data mining is used to identify the relation between sales volume and the role of the drug using a selection tree classification.

11.1.2.2 Data sorting

Here basic information about drugs and drug sales are used. The drug's basic information includes its name, description, region of origin, measuring unit, expiration date, dose, approval number, type of dosage, and other characteristics. Drug advertising includes sales information, sales unit, sales volume, wholesale price, retail price, date of sale, etc. In addition, related fields and data-mining information from the original repository should be removed. In order to make it easier to run more effective data-mining applications, they are linked up with ID on two columns and recombined on a data-mining database.

11.1.2.2.1 Data analysis

Among several attributes, Song [5] is concerned more about the label, the manufacturer's usage intent, the purchasing system, and the validity, dose, dosage type, batch number, and other data. The author does not pay attention to the connections between drug sales. Therefore during data mine work, to boost data-mining performance and reliability, these features can be disabled. There are several fundamental variables to be determined in data mining: name, manufacturer, purpose of use, and buying unit.

11.1.2.3 Selection tree model

The selection tree model is a subset of the decision tree (DT) system. Its core principle and evaluation approach is the same as the DT, except that the selection tree gives more branch choices.

A set of data with class labels is the input for DT building and results in a binary tree or multitree. The inner nodes of a binary tree are usually represented as a logical choice, such as a logical decision of the ai = vi type where ai is the attribute and vi is the attribute's property value. The inner node in a multitree is the property, the edge of the property is the value, which has various features and there are several sides. Tree Leaf tags select the classifier tree places each record in a class. Tree leaf nodes are category tags.

This example uses the choice tread model instead of the DT model, but certain tree branch data attributes, such as product name, vendor sale volume, task purpose, purchasing level, and other knowledge attributes, are net specific to their current situation. In the same scenario, the nodes have net sales value. Other parameters such as different variables can be chosen as the major revenue component to simplify the calculation. Each record is categorized according to its sales volume into the following categories.

The DT remains the basic structure of the grouping. Although the value of producing a choice tree is higher than that of a DT, there are two significant advantages for the selection tree: branches that are more reliable on a level of background knowledge are able to be identified; there is more precise than the DT, and, in many situations, a combination of several options leads to more consistent and less unpredictable predictions.

In this case, the selection tree model is adopted instead of the DT model. In addition to the above advantages, some of the tree branch information attributes are net unique in real situations, for example, the sales volume of the nodes includes the name of the product, the manufacturer, the purpose of use, purchase units, etc. To facilitate the analysis, choose sales as the target attribute, and other attributes as independent variables. The sales volumes of all records are divided into categories such as 0, 1−50, 50−100, and more than 100.

Based on these categories, a data set is used to generate the complete selection tree. A rule base is created in the resulting selection tree. A rule base contains a set of rules, each rule corresponding to a different path to the selection tree, and this path represents a connection that passes under the conditions represented by the node. Finally, the author a selection tree classifier is designed to demonstrate the importance of data mining in customer relationship management (CRM) of pharmaceutical industry.

11.1.3 Applications of data mining in the pharmaceutical industry

A review of medical data integrates and streamlines a clinical data review from vast amounts of information. Data mining helps in the development and launching of a product by illustrating trends, irregularities, and risk. Data mining helps to examine consumer behavior in terms of renewal of prescriptions and purchases of products [6].

Consumer analysis uses customer profiles that rely not only on goods but also on the ability to pay for them by analyzing recent health patterns associated with demographics that can be built utilizing data mining.

Sales data are combined with client groups and information on consumer networks to determine what leads patients to complete orders more rapidly or what leads doctors to prescribe other medications more rapidly.

Identify and approach people and groups that may be deemed to be "undiagnosed" through outreach programs aimed at encouraging such persons to be screened and checked for future issues.

Marketing and sales analysis provides a mobile analytics to the regularly disconnected sales force, so that they can not only respond to detailed questions about drug information but also to questions about history and trends.

In order to make market size changes, financial analysis analyzes prescription behavior in a particular area.

Product analysis is the analysis of purchasing patterns and treatment outcomes to build more combinations of drugs and goods targeted specifically to different age and risk factors.

Analysis of the supply chain improves production schedules by analyzing which pharmaceutical products remain on the shelves the longest and how well each pharmaceutical product sells.

11.2 Big data

Big data mining is now a most compelling and fast growing field, enabling the collection, discovery, and modeling of the overwhelming amount of medical data to assist clinical decision-making, avoid medication errors, and improve the outcomes of patients. Big data relate to data collection which cannot be collected, managed, and processed within a certain time by conventional software tools. The modern processing model requires massive, high-growth, and diversified knowledge to provide better decision-making, perspective, and system optimization skills [7].

In Cooke's Big Data Era, big data means that all data are used for analysis rather than random analysis. The 5V big data (IBM) characteristics [8] include volume, speed, variety, value, and accuracy. Gartner, a big data research organization, defined "big data" as a massive, high-growth, and diverse information asset that requires a new processing model to have greater decision-making, insight discovery, and process optimization capabilities [9].

11.2.1 Big data in the pharmaceutical industry

The pharmaceutical sector is a data-intensive enterprise that produces or uses different types of data. With additional sources, the volume of data increases exponentially daily. Raw data are produced from different internal and external sources in the healthcare and pharmaceutical industry, including research, R&D, clinical trial processes, physicians, academics, medical equipment, and patients. Effective use and analysis of these data will enhance the identification and rapid development of active ingredients that have the desired therapeutic effect. The production of drugs is an expensive and time-consuming process. Big data can be used at different stages of the process of drug development by offering more information, new techniques, accurate decisions, and minimizing time and cost.

It also offers the potential for increasing the participation of all players in the fields of drug research and development, marketing, and supply. Through time, real-world findings can be interpreted by both insurers and insurance payers. As beneficiaries, hospitals and even customers tend to anticipate the need which could become a prerequisite for real-world data in the future, the profit pricing paradigm for pharmaceutical firms is becoming more relevant. Pharmaceutical companies may respond to this cost–benefit stress by pursuing big data, to mine text literature or monitor social media, etc., showing definitive real-world results and separating themselves from the rest of the competition.

Regulators are reviewed in order to identify any signs of adverse effects or health negligence in this broad set of information. These data insights can improve patient safety and make drug manufacturers or distributors more accountable, among other things.

Focus groups produce active listening results from the collection, processing, and correlation of millions of online customer interactions using algorithms that feed computer visualization so that what consumers really want is clarified.

This offers companies an opportunity to show their concern for the general public and big data can also contribute to better monitoring. Enterprises can react quickly to the situation. It may also help companies store or displace medicines in unfortunate events such as a natural disaster, or predict a pandemic that could save many lives in a short time.

As there are already thousands of used or prescribed products and therapies on the market, it is becoming all the more important to validate their efficacy, drug safety, to compare product differentiators, to draw up marketing strategies, or to protect patents, etc. The use of big data also helps businesses to develop goods or processes and creates new ideas.

This presents a challenge, however in the drug sector, big data potential is huge and the rewards can be immense for successful companies. Big data analysis, which includes data integration, data extraction, and analytics, will help the pharmaceutical industry to make smarter decisions to help companies fulfill their business goals and overcome challenges.

11.2.2 Big data in pharmaceutical R&D

Due to constantly increasing costs while overall output stagnates, over the past 20 years, productivity in the pharmaceutical industry has decreased. Productivity remains a problem in the market, despite many attempts at its improvement. At the same time, providers of health care understandably expect better value for money and clear proof that new medications are better than the current standard of care, making a difficult problem even more challenging. With its introduction in life science, companies have established a new frontier with big data efforts that incorporate data from different data sources and backgrounds, which can provide an insight into turning this around and enabling the industry to return to sustainable growth.

Pharmaceutical companies must provide real evidence that new medicines that are on the market are better placed to obtain profits than current products or those of their rivals. Productivity is no longer only a feature of R&D efficiency; drug companies have an enormous range of internal data, ranging from basic scientific testing to sophisticated clinical trial programs, that are not completely analyzed and that lie idle in corporate data silos. Many companies are starting to use outside investigators to further study some of their clinical data.

Biomarkers will help verify risk targets and identify appropriate patient populations for development programs in the drug R&D community and provide early signs of safety issues and effectiveness to promote "go/no-go" decisions. The use of biomarkers can also provide early signs of real-world efficacy, which can help in assessing the commercial viability of medications at an early stage.

The pharmaceutical sector has only begun to utilize big data, and has a long way to go to make full use of it. The industry has nevertheless recognized that it must concentrate on its main assets: its own data and other data. This will help to better understand the mechanisms of diseases, define true medical needs, and provide an increasingly extended healthcare system with more affordable products for patients.

11.2.3 Use of big data in drug development for precision medicine

In pharmaceutical companies large data-driven drug discoveries have became an increasingly popular strategy. Currently, many therapeutic companies have incorporated an analysis of genetic expression, mobile screening systems, and computing software to identify the chemical structures of oncology drug discovery attributes of interest [10].

In contrast, a higher quality cell-based phenotypic screening method was used by pharmaceutical companies to integrate existing components of various illnesses, for example, inflammation, digestion, metabolic, and infectious diseases, in order to treat multiple dysfunction pathways.

The eTOX project involved the development of a comprehensive toxicity database and reliable modeling of drug safety endpoints between 13 pharmaceutical companies, 11 academic institutions, and 6 companies [11]. Quantitative structure activity relationship (QSAR) is a commonly used method of forecasting planned and nontarget events in the pharmaceutical industry. QSAR data sets often involve a large number of compounds ($>$ 100,000) and descriptors ($>$ 1000), which is why it is often computationally intensive to give priority to QSAR drug compounds, and many of its sensitive parameters need to be adapted to achieve good prediction [12].

11.2.4 Big data in the healthcare sector

Big data in health care refers to large and complex electronic health data sets that are difficult (or impossible) to deal with using conventional software and/or hardware; nor can it easily be handled using traditional or common data management tools and techniques. Big data is characterized as very large data sets that are computer based for searching for patterns, trends and associations, visualization, queries, privacy, and predictive analysis for large-scale data collection. The expanded use of telehealth will further test patient data storage capacity, and doctors' innovative use of Google Glass will also contribute to big data's social and behavioral dimensions. Big data is daunting in the healthcare sector not only because of its size, but also due to the diversity of data types and the pace at which it has to be delivered. There is a wide variety of data types available from electronic medical records such as patient demographic, biological, and clinical data that are renowned for assisting healthcare streams. Data analytics can be used for monitoring of these data in order to make them usable. This allows the concept of "big data" to be applied. They are currently viewed as a method for analyzing and making decisions on a large-scale and in enabling cheap health care. However, one of the most important questions is the growing divide between costs and results, and efforts are under way in many advanced countries to bridge that divide. The gap between the cost and outcome of medical services has been analyzed as the result of poor research understanding, poor use of available evidence, and poor acquisition of healthcare experience, leading to missed opportunities, a waste of resources, and potential harm to patients [13].

11.2.5 Features of medical big data

See Ref. [14].

11.2.5.1 Wide variety

Diversification characterizes the types of medical big data, which can be defined as follows.

1. Data collected or stored in hospitals by internal basic fund processes and information systems, such as auxiliary decision statistics, financial data, administrative information, market data, material data, supply chain data, patient data, call records, etc.
2. Data gathered by different facilities and equipment such as sensor readings, administrative data, physical data, hospital statistics, meter readings, video data tracking, etc.
3. Information on human behavior, information on behaviors, preferred details, data on communications, etc.
4. Basic geographic location information, radio frequency identification (RFID) read information, global positioning system (GPS) mapping data, image files, hospital information, time and location data, currency data, high-resolution images, vectors, remote sensing and dynamic data monitoring, etc.
5. Data on Internet messages, data on movement analysis, data on e-commerce purchases, data on digital activity consumers, etc.
6. Machine usage data and mobile device usage data, etc.
7. Documentation, science data, survey data, public data, and public information.
8. CRM, flow control, search software, analyzers, and other system data.

11.2.5.2 Complex sources

Health big data comes mainly from the following eight data sources.

1. Hospital-assisted information technology, such as the Internet of things (IoT), currency networking, RFID, GPS, etc.

2. Sensor networks, interconnecting devices, smart terminals, real-time monitoring equipment, etc.
3. Social media, Internet and mobile Internet, e-commerce, etc.
4. Professional research reports, industry information, business records, etc.
5. Business intelligence (BI), hospital-assisted decision-making system, etc.
6. Computers, tablets, mobile phones, other mobile devices, mobile storage, etc.
7. Cloud computing, third-party data processing technological development, third-party platform applications, etc.
8. Other large-scale interactions, large-scale data transaction sources.

11.2.5.3 Multidimensional structure

It is possible to deconstruct health big data from multiple dimensions.

1. Internal and external data. Health big data covers hospital services internal transaction data, CRM information, and other outside outlets' data. Health big data has dual characteristics, both public and private. The public nature of external data is more apparent than private internal data because they are closely linked to industry standards and company secrets.
2. Data structure: semistructured and unstructured. Health big data contains all structured data stored in the cloud as well as semistructured data such as log files, XML documents, JavaScript Object Notation (JSON) documents, and emails. Other types of data are unstructured files, including office documents, text, photos, XML, HTML, various reports, photographs, and audio/video information. Semistructured and unstructured data constitute 75%–85% of the big data total.
3. The format of a big data hospital is diverse, in accordance with the multidimensional features of the data structure. It includes digital text, video, SMS, audio, video, fax, storage data, configuration of the file, photographs, files, etc. Diversity and incompatibility of data formats and random access make it difficult to obtain, store, analyze, and use data.

11.2.6 Big data-mining applications in the medical industry

The application of data mining has been implemented and could lead medical care in the business and marketing field. This is not the case currently, however. In the medical field, effective mining applications have been updated, and some of these are described below [15].

11.2.6.1 Patient monitoring

The use of big data reduces the work of doctor and staff and allows them to work more effectively. Prolonged monitoring of the pulse, blood pressure, and respiratory rate uses sensors placed near the beds of patients.

Any changes are immediately detected and staff notified. One way to help monitor the lives of patients with big data is through the use of electronic health records (EHRs). These electronic data enable specialists to accurately trace the history of patients and to foster a improved communication of human services. Another important benefit is that patients can easily monitor their medications and laboratory tests. Thus, patients can significantly reduce the frequency of hospital visits by coordinating EHRs across restorative offices [16].

11.2.6.2 Health risk identification

We can identify patients with high-risk health conditions by providing medical data and robust mining methods and model construction solutions. Doctors and medical professionals should help patients to identify any health issues so that they can take steps to improve their healthcare quality and prevent future health problems [17]. Cancer, for example, is a severe disease that can be prevented and cured using big data analytics. Cancer is a devastating worldwide disease. Big data are extremely effective in the war against disease. In their early periods of its use, health providers enhanced the capacity to identify and analyze infections, assign more appropriate treatments to the genetic make-up of individual patients, and direct medication measures to limit the symptoms and improve the viability of medications. It is also aimed to chart the 3 billion database sets of DNA in a unique way [18].

11.2.6.3 Early epidemic breakdowns and recognition require constant monitoring

They can use the assistance of computer-assisted monitoring analysis to classify high-risk patients and potential incidents, and to track anomalies when predefined events occur. Efficient mining tools can be used to extract trends and associations with the clinical big data already in our possession to explain the health behavior in a particular area. In order to identify new and interesting trends in infection control data, the surveillance system utilizes data-mining techniques. The program also includes rules theory and patient care data obtained from the knowledge from the laboratory management systems in accordance with monthly averages, which are reviewed by an infection control specialist. The system developers conclude that enhancing the data-mining system's infection control is more sensitive and significantly more specific than traditional infection control monitoring. This was built at the University of Alabama [17].

11.2.6.4 Strategic planning and predictive analytics

In the treatment of patients with various complex conditions, quick and intelligent decisions should be taken. In this particular circumstance, the role played here by big data is essential. Big data provides an insight to make intelligent decisions and to improve the general system of diagnosis. These individual pieces of knowledge help in advanced testing, because it becomes clearer in predicting which patients are at risk of diabetes or cardiovascular diseases. Health fraud is a domestic problem, predominantly in government and public–private protection programs. Over the last decade, the misrepresentation of human services has increased and billions of dollars have been paid to fraudulent claims. The National Association for Anti Fraud Healthcare estimates that the medical services misrepresentation is costs 3% of all human services expenditure or 60 billion dollars, annually.

11.2.6.5 Telemedicine

Digitalization, cell phones, remote gadgets, and online video gatherings have set the stage for remote clinical administration transmission. Today, thanks to these mechanical developments, telemedicine is a very important and useful tool. The data collected from these gadgets can be easily shared, which simplifies diagnosis. Big data also helps in predicting severe medical conditions and to diagnose patient conditions, as well as enabling remote patient checks. Another important way for big data to modify telecommunications is through the provision of continuous information from remote regions using robots [19].

11.3 Data analytics

Data analytics is the application of mathematical methods to gain information from data, which can then be used to optimize processes or support decisions [20]. Here we subdivide data analytics into data analytics methods and techniques. Thus, this method describes which general type of data analytics has been used (descriptive, for example). The term "technique" represents the specific approach used for the data analytics method. Different methods can therefore be defined as follows.

Descriptive analytics implies data analytics applications in the explanation and comprehension of circumstances based on past and present data [21].

Diagnostic analytics is a data analytics program used to analyze the causes and effects of situations.

Predictive analytics is an application of data analytics using information and computational methods to view data relationships to predict future outcomes based on data set changes [21].

Prescriptive analysis defines the use of mathematical models to create a set of complicated options from existing data to prescribe the optimal solution. This analytics describes the data analysis.

11.3.1 Machine learning in health care

Machine learning is a multidisciplinary field that can be used extensively for medical purposes. In general it exists to find deep pathways among symptoms and diseases in four types of generic machine learning algorithms [22].

Healthcare supervised learning: This is carried out through a data set of symptoms and diseases, where we already know the disease in each case and generate a model [23]. Random forest, support vector machine (SVM), K nearest neighbor (KNN), Naive Bayes (NB), etc. are supervised learning algorithms.

Unsupervised learning in health care: For each instance of different symptoms, no class label (disease) is available in this learning [24]. A community or cluster is established based on symptom similarities. Each cluster is a single disease. The clustering algorithm consists of fuzzy clustering, hierarchical clustering, k-means clustering, density-based spatial clustering of applications with noise (DBSCAN), etc.

Semisupervised healthcare learning: This is a method to find the best identification of illness from marked and unlabeled information [25]. It uses unlabeled data to carry out high-performance classification. The algorithms for semisupervised learning are self-training, generative model, graph-based methods, etc.

Healthcare reinforcement learning: This form of learning offers a complex environment for learning from the symptoms data set and provides feedback in terms of punishments [26]. When the model confirms the diagnosis, the punishment will be the solution, and the models will try to classify other types of symptoms to forecast the disease. Thus, prediction accuracy will be increased after much iteration. The algorithms for improving learning are Q-learning, rule gradients, etc.

Some common techniques such as SVM, naive Bayes classification, DT, KNN, fuzzy logic, classification and regression tree, and also deep learning (such as Bayesian network, contrastive divergence) have been applied in health care to uncover possible clues between symptoms and diseases, such as heart disease. Such algorithms have already been used to assess diseases from clinical results in practical health systems.

11.4 Various platforms for clinical decision-making

11.4.1 Smart healthcare portal for clinical decision-making

Nalluri et al. [27] developed the health information gateway and exchange application (HINGE) software research tool to collect data from radiotherapy delivery systems, such as linear accelerators, treatment management systems (TMSs), patient outcomes reports, and digital medicinal reporting systems in order to enhance cancer care by collecting reliable information.

1. Research/clinical practice tier: Comprises all IT and medical equipment systems related to various patient treatment phases. Treatment scheduling systems provide information from doctors on the prescribed radiation therapy treatment plan. These plans are used as an input by the TMSs

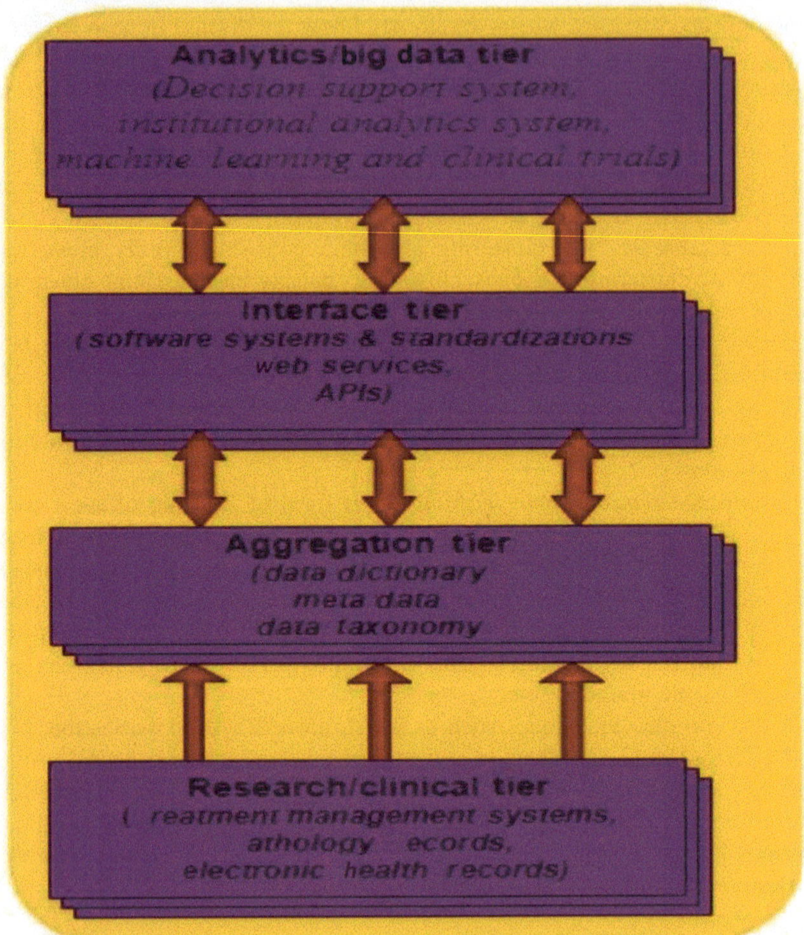

FIGURE 11.2 Smart analysis tools for medical big data analysis.

and to provide the patient with the required radiation dose. These systems are often proprietary software that document and record this information.

2. Aggregation tier: This gathers disparate data from a clinical practice level and organizes it into a standard template in accordance with the established data taxonomy and data dictionary produced by clinical domain experts. The information is organized coherently (Fig. 11.2).
3. Interface tier: This provides end users with resources to interact with patient information. The doctors and clinical team have a place in which all the patient information containing all the documents can be viewed, edited, verified, and shared.

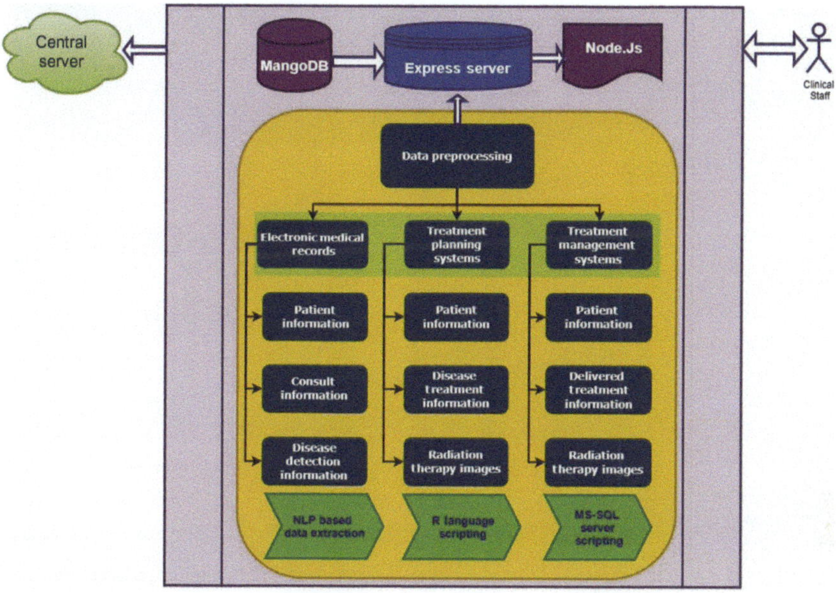

FIGURE 11.3 HINGE system design.

4. Analytics/big data tier: This contains several analytical tools for model-based predictions, including a decision support system, institutional analytics system, machine learning, and clinical trials. These tools can conduct intelligent and predictive analysis of patient information, as well as of several patients or illnesses in a cohort. They can also extract clinical parameters, define data elements, and create statistical models to respond to various outcomes and hypotheses for treatment and research. The HINGE system design is illustrated in Fig. 11.3.

A platform such as HINGE can address some unanswered questions in forecasting the results of radiation therapy of cancer patients in an unpredictable way. The HINGE data can be utilized for the creation of decision support models in clinical systems that enable clinicians to improve patient quality of life and safety. The HINGE platform strives to integrate clinical practice, data collection, automation, and clinical workflow streamlining with the aim of deriving clinically meaningful measures in patient care.

11.4.2 Big data for medical service platforms

The MedicalDataMart (MDataMart) data processing, conversion, aggregation, and extraction, and the isolation of business data from analytical data are used by BDMSP in the construction of data from hospitals, healthcare centers, private clinics, and medical exam centers for the purposes of

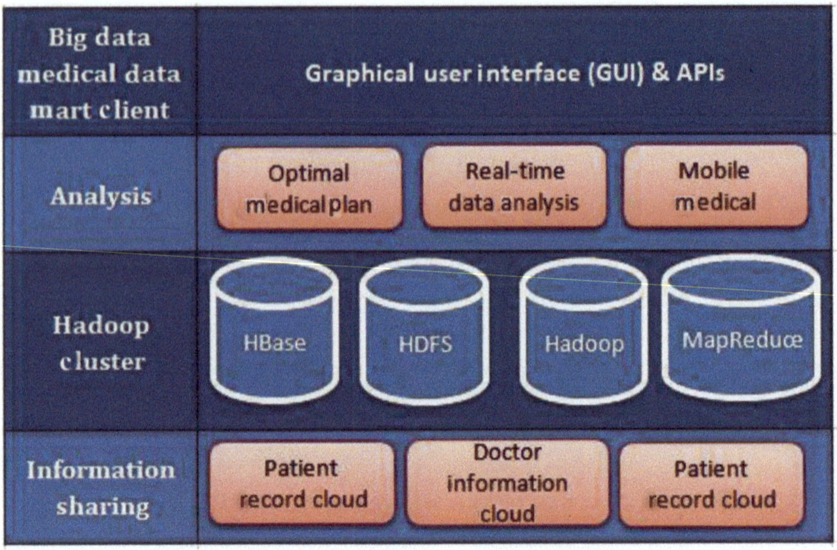

FIGURE 11.4 Key technologies for system implementation BDMSP solution architecture.

medical data accumulation. The medical description model and predictive model are constructed on the basis of the data mart MDataMart using data-mining technology. As shown in Fig. 11.4, BDMSP [28] architectures are divided into three parts.

The first part consists of the medical information that gives raw data for BDMSP. It handles first-level information and loads the data into the MDataMart server based on medical data. Data-mining software is central to and evaluated based on information mart MDataMart. The results are stored in the business model library, and a secondary development interface for the model algorithm is provided to clients on the application server. The third component consists of customer software called BDMSP, which provides a graphical user interface.

This chapter presents a framework on the clinical data mining and knowledge management of big data based on MapReduce's parallel storage and processing architecture. MapReduce is the Hadoop architecture model for data processing programming. This is the basic methodology of big data analysis and the foundation of data extraction and knowledge management system. The code can be executed in the framework of the Hadoop-cluster in different languages, for example, Java, Python, C, C++, Ruby, etc. Google proposed MapReduce for the first time. MapReduce is a summary of Google's major data-processing problems and needs, which decouples computing complexity into two parts: (1) MapReduce abstract machine tasks design computation; (2) distributed MapReduce assisted computational architecture. MapReduce implies that the data to be interpreted by the user are a key value column pair. Two functions are defined, Map and Reduce. The Business Logic Developers with different requirements can define and implement these two tasks. Hadoop MapReduce is an open source Google MapReduce computing model.

11.4.3 Big data predictive analytics and radio frequency identification in the pharmaceutical industry

Recent advances in technology have been able to make significant use of supply chain performance (including RFID) in combination with big data predictive analytics (BDPA). In order to better understand the performance of the supply chain, there is an urgent need to accept RFID technology and BDPA to investigate them empirically. Through this analysis the relationship between the acceptance of BDPA, RFID acceptance, and supply chain quality (SCP) is developed and empirically studied. This study includes the Chinese medical logistics industry. The test was collected using a standard, random sampling process from employees. A personalized questionnaire was used to collect data. For structure equation simulation analysis, the program SmartPLS 3.0 has been used. For the partially smallest square, the path coefficient value was 0.700 in structural equation modeling (partial least squares−SEM). This reveals a direct positive impact on the performance of the supply chain of big data predictive analysis. In addition, the "variance accounted for" mediation method also has tested RFID acceptance. Partial arbitration between BDPA and SCP was shown at 57%. This research, in theory, fills the knowledge gap by defining connections between BDPA and SCP through an RFID mediation function. It would be feasible to use BDPA and RFID technologies to improve the performance of supply chains for supply chain executives in the pharmaceutical industry [29].

Throughout China, barcode scanners are used in many medical distribution organizations for the production and maintenance of inventories. These are old technologies with limited features that prevent them from tracking products in real time. Moreover, the processing speeds are slow. Moreover, it took more than 4 days to buy online from e-commerce platforms to deliver an item to the customer. In fact, consumers do not know where their goods are precisely housed, they receive information about a few central locations that are updated by barcode scanners. This creates the need for information about products in real time, on the one hand, in order to determine the product status, while pharmaceutical companies must track their inventories in real time, on the other hand. In this context, the use of advanced technologies is crucial for pharmaceutical companies to improve performance in their supply chains [30].

In order to improve pharmaceutical logistics performance, BDPA together with RFID technology are necessary. Therefore, the main aim of this research is to use state-of-the-art technology such as BDPA and the detection of radio frequencies and their functions to conduct the supply chain in pharmaceutical companies. This demonstrates that the time is desperately needed to fill technical and organizational gaps in the logistics sector by analyzing the role of BDPA and RFID technology in improving supply chain efficiency.

The RBT's focus is on organizational resources and skills, as well as concepts of competitive advantage.

It has been found that resource-based theory (RBT) is appropriate for examining BDPA and RFID roles in pharmaceutical organizations.

In order to enhance pharmaceutical corporate efficiency, the theory of RBT is based on the four characteristics of a VRIO model: quality (V), rare (R), imitables (I), and organized (O). Therefore it would be possible to improve their supply chain performance if pharmaceutical logistics organizations use BDPA and RFID technology.

The big data analytical process includes five phases, including data generation, data acquisition, storage, analytics, and visualization of data. Pharmaceutical organizations can improve their supplies through the visualization of big data, known as big data analytics.

11.4.3.1 Radio frequency identification

From electronic product codes, radio frequency identification technology was developed. This software is used to improve the efficiency and performance of the network channel for drug companies in supply chain management.

RFID uses radio frequencies to transmit data using a wireless technology. RFID technology consists of three main components: (1) tags; (2) readers; and (3) reader-interacting software applications. RFID technology is thus a combination of software and hardware that aims at identifying an object, collecting data from tags, and tracking them quickly and accurately [31].

Without question, RFID software helps to increase and enhance an organization's efficiency in the supply chain. The benefits to pharmaceutical logistics companies are that the RFID is used to monitor medical products that are forged, medical transcriptions are made, patients are registered and discharged, and products are tracked. Acceptance of RFID technology is thus critical if medical companies are to study their systems empirically.

Application software, including a server, antenna, transponders, and readers, are the essential components of radio frequency identification technology. The technology for identification of radio frequency signals is a small antenna-based microchip. This is used for sending signals to RFID readers, who receive these signals on their servers, through wireless communication [32]. The software code in this application makes it possible to link distributors, vendors, producers, and suppliers of pharmaceutical logistics to their server in order to obtain real-time goods data on the supply chain.

Radio frequency recognition technology has made it possible for pharmaceutical companies to assess the flow of their products. It also measures the overall time needed from pharmaceutical organizations to the manufacturer to the dealer and the seller to the consumer in "all the way" time. It helps producers, distributors, retailers, and consumers to have greater trust with one another.

11.4.4 Big data technologies

New technologies that are typically used in the processing of big data are described next [33].

11.4.4.1 Data collection

Sqoop is an open source platform that runs on the edge of the shared file system Hadoop (HDFS), together with SQL and Hadoop. It is a command line interface that ensures that data between HDFS and relation databases such as Enterprise Data Warehouses, Oracle, Postgres, and Teradaten are imported and exported.

Flume is the decentralized and secure open source platform for the assembly, compilation, and transition from an external computer to HDFS for storing huge volumes of batch files, log files, and data sharing. It provides a simple and scalable architecture based on constant data sharing. Flume has many benefits including fault resistance, endurance, and horizontal scalability.

Kafka is an open source Apache Software Fund's framework. With its distributed system and high throughput, it can continuously collect data from many outlets, including data centers and social media networks. Kafka's main users are LinkedIn and Wikipedia. It has been written in Scala and its scalability and tolerance to faults are characteristic [34].

11.4.4.2 Data processing

MapReduce is an efficient and cost-effective programming method that simplifies and speeds up data processing significantly.

This system has three main functions.

Function map to ensure that data are separated into separate key/value pairs;

- Function shuffles: collect and store key pairs that are then grouped.
- Reduce function: parallel multiplication of pairs by predefined programs. Input is a series of keys with corresponding value.

Yet another resource negotiator is more generic than MapReduce. It is a professional HDFS resource manager that allows the operation of several applications concurrently. Therefore it processes batch and flow. The scalability and security of this structure are also known.

Storm is a platform for open source computations that are distributed in real time. This tool is capable of handling stream data, unlike HDFS which is designed for batch processing. Its scalability, defect sensitivity, and performance are characteristic.

Flink is an open source platform which can use its distributed architecture to process stream data.

Spark is an in-memory cluster computing technology developed by Zookeeper, and is a distributed service designed to synchronize the configurations across a cluster, for rapidly calculating data via its sophisticated libraries; it ensures high data access [34].

Oozie is a task manager in an open source system that runs and handles the Hadoop system workflows. Oozie is an extensible device that is modular, stable, and extensible.

11.4.4.3 Big data storage

Hadoop uses the distributed HDFS and a nonrelational database called HBase to ensure storage of large volumes of data: HDFS is a primary component of a Hadoop cluster, that is, a set of connected computers, which can support up to hundreds of cluster nodes. It is inexpensive and has robust processing, good scalability, and fault sensitivity. HDFS can also manage organized and unstructured data. HDFS is intended for high-latency batch processing.

HBase is an open source project designed for low-latency operations, built on top of HDFS. This nonrelational database has the capacity to hold massive table numbers with trillions of rows and millions of columns, and has been established since the Google Big Table. HBase is a modular, distributed, and customizable database that can be used to process and analyze data in real-time queries.

11.4.4.4 Big data analysis

Pig is a Yahoo forum for open source data analysis that is uses data flows. It uses a high programming language called "Pig Latin" to develop data analysis systems. Developers then have to write Pig Latin scripts in order to analyze data, and then convert them to MapReduce tasks with the Pig Engine module. There are many benefits to Apache Pig. The number of codes is shortened first, because of its multiquest approach. Second, when coding MapReduce jobs, Pig Latin substitutes for the use of Java, traditionally considered to be more complicated. Pig Latin is SQL-like and easy to learn.

Hive is a Facebook-based data warehouse system that enables Hadoop to be used. The collected data are saved to all users in a structured database. Apache Hive server is managed using the same HQL syntax as SQL.

Mahout is an open-source library for the learning of machines and data. It works on the HDFS to perform algorithms through MapReduce.

11.4.5 Big data applications in the healthcare sector

11.4.5.1 Healthcare prediction

Physicians use predictive analytics from large-scale information to forecast and provide targeted medication to high-risk patients, in order to lower medical costs due to high numbers of patients with chronic diseases, such as diabetes [35]. A US company called Asthma polis was developed in order to provide therapeutic follow-up for asthmatic patients to track and follow the inhalation use of sensors placed in the tops of inhalers. The place and time of use of the inhaler are obtained by means of the GPS in real time.

For an asthma attack, the data recorded are transmitted via a smartphone or laptop to an app that can be downloaded by the patient and doctor. On the one hand, the main objective of the sensors is to help patients decide whether or not to visit other places as well as to forecast an asthma attack. The data aggregation, on the other hand, leads practitioners to develop a customized patient care plan, identify times when an asthma crisis is highly likely, and treat them by increasing or decreasing the doses of medication.

One hundred and two of 1000 patients suffering from metabolic syndrome were selected by Li [36]. Data from 600 laboratory tests and 180 reports have been compiled by researchers. A customized treatment schedule was drawn up from patient medical records. This application produced positive results: over the next 10 years the rate of morbidity was reduced by 50%. Different alternative solutions have been discovered: statin treatment, weight loss, and increasing the average blood triglyceride levels if the blood sugar level exceeds 20%.

Google has developed two real-time monitoring technologies in order to track the progress of epidemics around the globe: Google flu trends and Google dengue trends. Data were collected from Internet searches in Google to find information about symptoms, medicines, side effects, etc. Reports also reported that Google's influenza rate was 140% higher than predicted [37].

11.4.5.2 Performance enhancement

King Faisal Specialist Hospital & Research Center successfully managed a large clinical data analysis project with a view to maximizing the performance of emergency rooms (ERs) and reducing crowding in ERs. Medical details were compiled from the hospital's data warehouse. In order to gather positive results, the data analysis made improvements to the workflow of the ERs. The ER waiting time decreased from 140 to 62 min in 2014. Treatment time fell from 17.5 h in 2014 to 10.8 h in 2016. The ER stay length ranged from 20 h in 2014 to 12 h in 2016. This demonstrated the effectiveness of large data analysis in identifying deficient areas and recommending valuable solutions to improve performance [38].

11.4.5.3 Recommendation systems

In general, software tools and technology are recommendation systems designed to offer a series of suggestions to help users make better decisions on a product or service. Recommendation systems are currently increasingly used in the healthcare sector to offer medical recommendations on drugs, diagnoses, and treatment plans.

In accordance with their health status, Katzenbeisser and Milan [39] developed a clinical recommendation system that is useful to obtain reliable recommendations from caregivers for patients. In order to improve the system's performance, patients can also contribute by adding their private notes and assessments of doctors on the basis of health information for different conditions. It must, however, be protected from dishonest and malicious users due to the sensitivity of such data. This program also guarantees that their integrity is safeguarded by care providers. The system provides a list of the most suitable physicians for each patient.

In comparison, an advisory program that forecasts users' points of view about an article that corresponds to the preference of a broad range of users is a form of collective filtration [40]. The Collaborative Assessment and Recommendation Engine was then used in the health sector when it was developed to forecast disease risk.

A patient's initial appointment with a doctor provides basic data on the patient's medical background. Therefore researchers will classify patients suffering from a disease on the basis of health records for that disease (j). The use of collaborative filters produces p(a, j), which indicates that the patient is likely to develop a disease j in the future. Finally, the system gives a sorted list of possible diseases ranging from the highest to the lowest risk for each patient.

This system is useful for the elimination of readmission rates, reliable predictions, and continuous enhancement in treatment ratings. The system was validated on a database of 13 million patients who made 32 million visits over 4 years to ensure its feasibility and effectiveness. In addition, the system has been validated.

11.4.5.4 Healthcare knowledge system

Boldominos et al. [41] proposed a smart healthcare management system called DataCare for the purpose of monitoring clinical key performance indicators and identifying situations in hospitals and health centers. Finally, all information on dashboards was displayed in the visualization module to make interpretations more precise and efficient. In a medical center in Spain, DataCare was tested, and interesting conclusions have been obtained with the results.

The ability for big data analysis to automate hospital services, enhance efficiency, reduce cost and levels of readmission, save lives, and increase quality of care can be extracted from these various technologies in the healthcare sector [34].

The system's architecture consists of three main components: data recovery and storage, data processing and interpretation, and visualization of data. The first module uses advantage software to collect and aggregate data to monitor patient–physician communication. The researchers used Apache Spark for the storage of a huge amount of streaming data, because of its high scalability and fault sensitivity. The data were then put on the MongoDB database for further analysis. The stored data were analyzed to provide valuable insights into security forecasts, medical advice, and alerts.

In reality, on the basis of actual information in the future, DataCare will predict KPIs. In addition, there are thresholds to generate early and real-time warnings when an indicator exceeds its permitted value. DataCare recommended that the quality of care be improved. Depending on the experience, DataCare has 52 guidelines established by physicians.

11.4.6 Heart disease prediction system using the K nearest neighbor classification technique

The major cause of death in developing countries is heart disease and it is one of the major contributors to the burden of disease in developing countries such as India. Several Andhra Pradesh (AP) studies have shown cardiovascular diseases as high as those in Canada and the United States were the leading cause of death, accounting for 33% of deaths. A clinical decision support for the classification of cardiovascular disease must be designed and developed.

Deekshatulu and Chandra [42] proposed a classification algorithm which combines KNN and a genetic algorithm, to predict heart disease in patients in the AP population. KNN is a simple algorithm, which stores all cases and classifies new cases based on similarity measures.

KNN algorithm is also called (1) case-based reasoning, (2) KNN, (3) example-based reasoning, (4) instance-based learning, (5) memory-based reasoning, and (6) lazy learning [3]. KNN algorithms have been used since 1970 in many applications such as statistical estimation and pattern recognition. KNN is a nonparametric classification method which is broadly classified into two types:

1. Structureless NN techniques;
2. Structure-based NN techniques.

In structureless NN techniques, all data are classified into training and test sample data.The distance from the training point to the sample point is measured and the point with the lowest distance is considered the nearest neighbor.

Structure-based NN techniques are based on data structures like orthogonal structure tree, ball tree, k-d tree, axis tree, nearest future line, and central line [43]. The nearest neighbor classification is used mainly when all attributes are continued. A KNN algorithm is presented in Fig. 11.5.

Deekshatulu and Chandra [42] combined KNN and genetic algorithms to increase the accuracy of classification of cardiovascular data (Fig. 11.6). As a measure of goodness they used genetic search to prune redundant or irrelevant characteristics and to place classifications in the attributes. The least classified attributes are deleted and classifying algorithms are based on assessed attributes. This classifier is trained to classify a heart disease data set as either healthy or sick. The authors proposed an algorithm consisting of two parts:

1. The first part deals with evaluating attributes using genetic search;
2. The second part deals with building a classifier and measuring the accuracy of the classifier.

Accuracy is computed as the number of samples correctly classified in test data divided by the total number of samples in the test data.

Six medical data sets and one nonmedical data set have been used for the analysis. Six of the seven data sets were selected from UCI repository, and some cardiovascular diseases data were obtained from various company hospitals in AP. The accuracy of AP heart disease from the identification algorithm without genetic algorithm (GA) is increased fivefold. Integrating GA into KNN increases the reliability of identification for various data sets and particularly for AP cardiac disease.

This model helps doctors to diagnose efficiently cardiac diseases with less attributes. In India and AP, cardiac disease is the most common contributor to death. The recognition of major risk factors and design of the decision support system and effective control mechanisms and services for health education will decrease deaths from heart disease.

11.5 Efficient large-scale medical data (eHealth big data) analytics in the Internet of things

The aim of the eHealth big data platform is to collect large-scale (big data) medical (eHealth) data in real time from sensors and actuators to patients

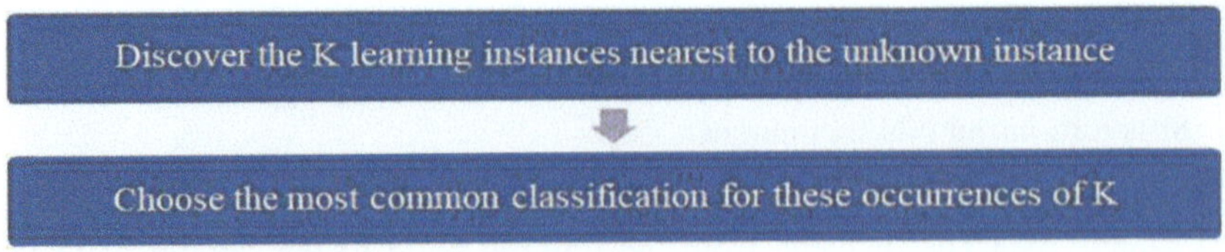

FIGURE 11.5 KNN algorithm.

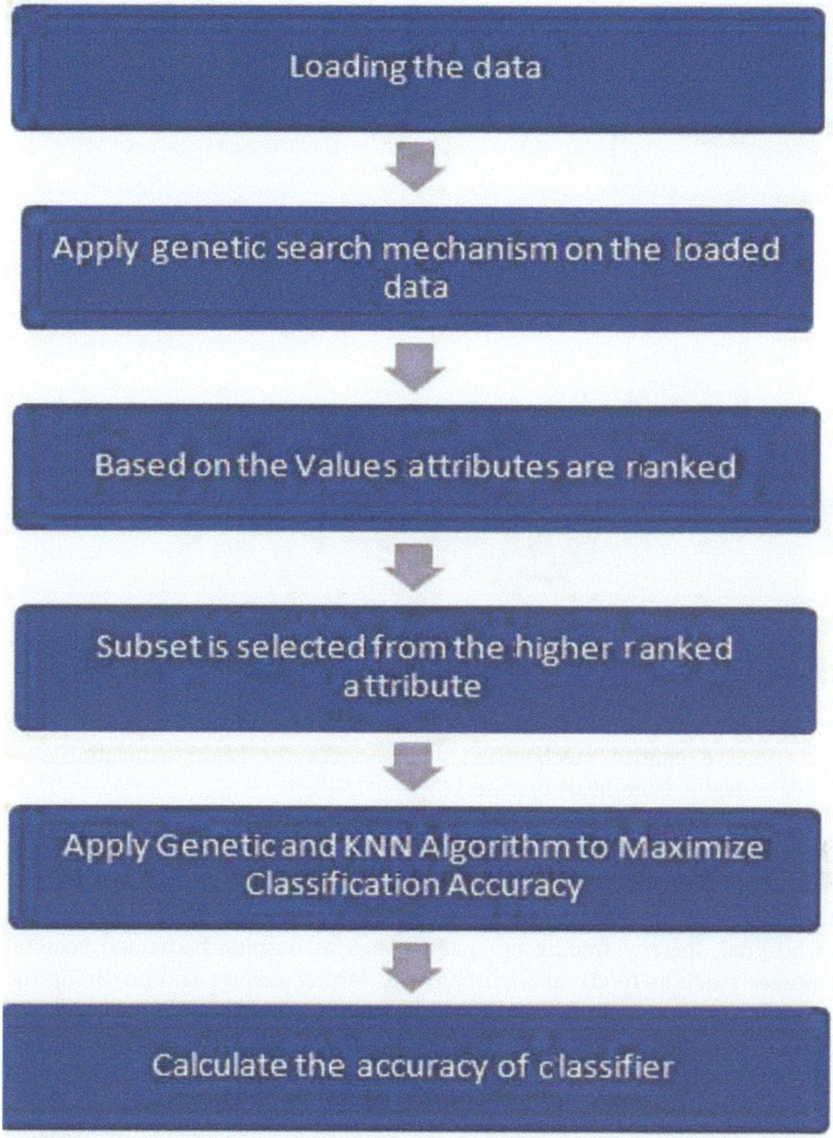

FIGURE 11.6 Combination of genetic and KNN algorithms.

with a chronic, rare, or hereditary disease. The data collected are then (in real time) transported to a "cloud server" over a network and then processed in the cloud computing, which then analyzes these IoT data to obtain knowledge which has no meaning unless analysis has taken place. This model helps doctors to diagnose cardiac diseases with less symptoms efficiently.

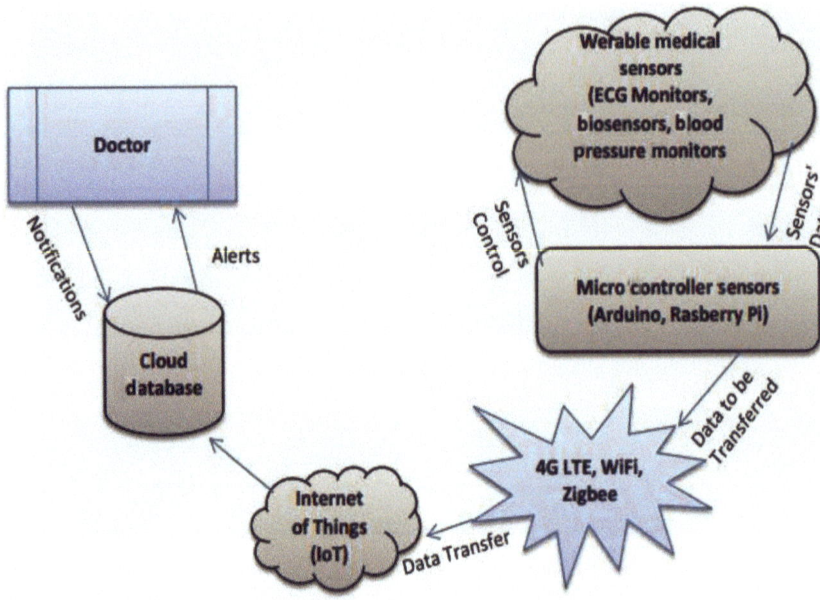

FIGURE 11.7 eHealth Big Data Architecture.

In India, including AP, cardiac disease is the most common cause of death. The recognition of major risk factors and design of a decision support system and effective control mechanisms and services for health education will decrease deaths from heart disease [44].

In this study, given the growing problems in the healthcare sector, the authors proposed an intelligent system, as shown in Fig. 11.7. This provides medical information related to the patient's health to the relevant people in real time, via sensors and other devices, in order to monitor health outwith the hospital, thereby freeing up spaces (such as hospital beds) and hospital resources (such as food), and helping with further savings and providing the patient with a more comfortable environment.

At the same time, in this way, the patient's direct health information, timely information on the patient's health, remote patient monitoring, and timely information from the doctor-in-charge of emergencies are provided.

Specific conversations between the patient and the doctor in order to avoid the manipulation of evidence by playing a "broken telephone" and strengthening the organizational history of the doctors, so that the details can not be changed over time [45] to update and immediately plan for an emergency.

This research proposal is designed to meet the challenges of these technologies and to create a system that has advantages over the systems proposed by other researchers [46], in addition to the improvement of patient health care. The researchers presented a built-in system in which they used the IPv6 via the low-powered, scalable, fault-tolerant wireless personal area network technology (6LoWPAN) for sensor nodes to maintain their connection. Moreover, a "pervasive health system" was used which provides patients with the ability to lead a normal life without qualified medical personnel, accurate medical data, and an emergency alarm system [47].

In this study, two systems used to monitor patients remotely included a wireless sensor network based on ZigBee technology, and a network of wireless sensors that uses Wi-Fi (IP). These systems were complemented by internet protocol (IP).

The system author proposed this as a priority over both systems due to the combination of wireless technology, such as 6LoWPAN, ZigBee, Wi-Fi, LoRaWAN, or Bluetooth low energy technologies.

The focus of this system is on bioinformatics and IoT. The innovation of this system was noted as the two regions being combined. A hybrid wireless technology for transferring data, taking into consideration all modern wireless data transmission methods, will be investigated in the future to transmit sensitive medical data. They also proposed new protocols for data transfer, taking into account the special features of each wireless network protocol and combining the most up-to-date data transmission technology. Methods for data sent for this research include ZigBee, Bluetooth low energy, LoRawan, Wi-Fi 802.11ac, and 4G/LTE. Data transmission is also included.

With the aid of a revised data transfer protocol, which relies on the method of the constrained application protocol sensor data transmitting protocol, the Bluetooth low energy and ZigBee protocols are used in limited energy sensors.

If the distance between the client and the Internet access point is greater than 50 m and less than 2 km, the information is sent to determine whether the LoRaWAN protocol is being used. We use the 4G/LTE mobile network if the patient is in an environment not protected by these protocols.

The safety of medical data, which constitutes personal data by applying innovative methods of coding in the future, is emphasized. For patients and doctors it is important to be supervised, so that the communication and exchange of data are secure. Security relies on the development and implementation of a communication protocol focusing on the secure transfer and encryption of medical data using encrypted algorithms [45].

Therefore the peak upload speed in each protocol is explored by a range of future researches. We also examine which medical information can be sent in real time and which can be sent later. One wireless technology provides a smooth transition to another, reducing delays and errors in packets. In order to reduce the transmission speed, data compression algorithms are suggested, particularly if the network is congested. They also discuss ways of ensuring the reliability of the data transmission system. In the hybrid wireless data network, maximum data security is ensured.

Finally, on the subject of security, this system is aimed at creating a new communication protocol to ensure the secure transfer of medical data, which uses encryption algorithms to provide the best possible protection of personally identifiable patient information to encrypt these data.

The main objective of this system are to examine the IoT, cloud computing, and big data technologies with the aim of solving the complex health-related issues faced. The primary objective of this system is to collect real-time clinical (eHealth) information.

A collection of sensors and actuators for those who have suffered from a variety of conditions (e.g., chronic, inherited, and rare diseases) have been used by patients. These data are then transferred to a cloud server via a network. In turn, these data are stored and interpreted in the cloud to obtain value for the user.

The data-mining method is carried out through the review of these results. In order to tackle the different health problems, the transfer of medical data already analyzed is done with devices that are being worn by those people.

As a future work, the authors developed and implemented a communication protocol, providing secure health data transmission and encryption through the cryptographic algorithms. The above eHealth Big Data Architecture can be used in telecommunications, finance, health care, etc. for the betterment of society.

11.6 Conclusions

As the data sources used in healthcare organizations are well known for their size, heterogeneous nature, and high dynamism, the healthcare sector has gained dominance through the effects of big data. Big data analytics include the application and evaluation of large volumes of complex heterogeneous data in medicine and health. The healthcare domain has been updated with the impact of big data and offers advanced remedies to handle diverse large data sources ranging from patient health data to medical images. In this chapter we have described: (1) a data-mining system, data requirements of the pharmaceutical industry and applications of data mining in the pharmaceutical industry; (2) big data in the pharmaceutical industry, R&D, the healthcare sector, and big data-mining applications in the medical industry; (3) machine learning in health care; (4) various platforms for clinical decision-making (HINGE system, BDPA and RFID in the pharmaceutical industry, and a heart disease prediction system); (5) efficient large-scale medical data (eHealth big data) analytics architecture. This chapter presents a comprehensive system design for smart healthcare systems supported by medical big data analytics.

References

[1] E. Simoes, Health information technology advances health care delivery and enhances research, Mol. Med. 112 (1) (2015) 37.

[2] J.W. Cortada, D. Gordon, B. Lenihan, The value of analytics in healthcare, Report No. GBE03476-USEN-00, IBM Institute for Business Value, Armonk, NY, USA, 2012.

[3] M. Durairaj, V. Ranjani, Data mining applications in healthcare sector: a study, Int. J. Sci. Technol. Res. 2 (10) (2013) 29–35.

[4] K. Chalkidou, J. Appleby, Eliminating waste Healthc spending (2017) j570.

[5] G. Song, Application of data mining technology in the CRM of pharmaceutical industry, in: 2018 International Conference on Intelligent Transportation, Big Data & Smart City (ICITBS), Xiamen, 2018, pp. 61–64.

[6] S. Joshi, H. Joshi, Applications of data mining in health and pharmaceutical industry, Int. J. Sci. Eng. Res. 4 (4) (2013).

[7] X.-J. Tao, X.-F. Hu, Y. Liu, Overview of big data research, J. Syst. Simul. 25 (260S1) (2013) 142–146.

[8] X. Zhou, S. Chen, B. Liu, R. Zhang, Y. Wang, P. Li, et al., Development of traditional Chinese medicine clinical data warehouse for medical knowledge discovery and decision support, Artif. Intell. Med. 48 (2–3) (2010) 139–152.

[9] Y. Wang, L.A. Kung, T.A. Byrd, Big data analytics: understanding its capabilities and potential benefits for healthcare organizations, Technol. Forecast. Soc. Change 126 (2018) 3–13.

[10] Y. Cha, T. Erez, I.J. Reynolds, D. Kumar, J. Ross, G. Koytiger, et al., Drug repurposing from the perspective of pharmaceutical companies, Br. J. Pharmacol. 175 (2) (2018) 168–180.

[11] F. Sanz, F. Pognan, T. Steger-Hartmann, C. Díaz, M. Cases, M. Pastor, et al., Legacy data sharing to improve drug safety assessment: the eTOX project, Nat. Rev. Drug Discov. 16 (12) (2017) 811.

[12] J. Ma, P.S. Robert, A. Liaw, G.E. Dahl, V. Svetnik, Deep neural nets as a method for quantitative structure–activity relationships, J. Chem. Inf. Model. 55 (2) (2015) 263–274.

[13] S.L. Garapati, S. Garapati, Application of big data analytics: an innovation in health care, Comput. Intell. 14 (1) (2018) 15–27.

[14] Li, Y.L. Guijie, H. Cai, Research on application of big data in medical industry, 3rd International Conference on Smart City and Systems Engineering (ICSCSE), IEEE, 2018, pp. 763–765.

[15] L. Elezabeth, P.M. Ved, J. Dsouza, The role of big data mining in healthcare applications, 7th Int Conf Reliability, Infocom Technol Optim (Trends Future Directions) (ICRITO), IEEE, 2018, pp. 256–260.

[16] J. Luo, M. Wu, D. Gopukumar, Y. Zhao, Big data application in biomedical research and health care: a literature review, Biomed. Inform. Insights 8 (2016). BII-S31559.

[17] M.K. Ross, Wei Wei, L. Ohno-Machado, Big data" and the electronic health record, Yearb. Med. Inform. 23 (01) (2014) 97–104.

[18] H. Kaur, S.K. Wasan, Empirical study on applications of data mining techniques in healthcare, J. Comput. Sci. 2 (2) (2006) 194–200.

[19] J. Roski, W.B.-L. George, A.A. Timothy, Creating value in health care through big data: opportunities and policy implications, Health Aff. 33 (7) (2014) 1115–1122.

[20] W. Raghupathi, V. Raghupathi, Big data analytics in healthcare: promise and potential, Health Inf. Sci. Syst. 2 (1) (2014) 3.

[21] R.H. Goldstein, Analytics and data warehousing infrastructure and services, U.S. Patent 7,191,183, issued March 13, 2007.

[22] M.D. Islam, L. Daizong, K. Wang, P. Zhou, L. Yu, D. Wu, A case study of HealthCare platform using big data analytics and machine learning, Proceedings of the 2019 3rd High Performance Computing and Cluster Technologies Conference, ACM, 2019, pp. 139–146.

[23] M. Mozaffari-Kermani, S. Sur-Kolay, A. Raghunathan, N.K. Jha, Systematic poisoning attacks on and defenses for machine learning in healthcare, IEEE J. Biomed. Health Inform. 19 (6) (2014) 1893–1905.

[24] J.G. Dy, E.B. Carla, Feature selection for unsupervised learning, J. Mach. Learn. Res. 5 (2004) 845–889.

[25] S. Zhou, Q. Chen, X. Wang, Active deep learning method for semi-supervised sentiment classification, Neurocomputing 120 (2013) 536–546.

[26] A. Esteva, A. Robicquet, B. Ramsundar, V. Kuleshov, M. DePristo, K. Chou, et al., A guide to deep learning in healthcare, Nat. Med. 25 (1) (2019) 24–29.

[27] J.J. Nalluri, K. Syed, P. Rana, P. Hudgins, I. Ramadan, W. Nieporte, et al., A smart healthcare portal for clinical decision making and precision medicine, Proceedings of the Workshop Program of the 19th International Conference on Distributed Computing and Networking, ACM, 2018, p. 9.

[28] S. Guijuan, Y. Wen, Y. Jia, H. Liu, Research on medical service system based on big data technology, International Conference on Intelligent Transportation, Big Data & Smart City (ICITBS), IEEE, 2019, pp. 302–304.

[29] M.N. Shafique, M.M. Khurshid, H. Rahman, A. Khanna, D. Gupta, The role of big data predictive analytics and radio frequency identification in the pharmaceutical industry, IEEE Access 7 (2019) 9013–9021.

[30] J.K. Visich, S. Li, B.M. Khumawala, P.M. Reyes, Empirical evidence of RFID impacts on supply chain performance, Int. J. Oper. Prod. Manag. 29 (12) (2009) 1290–1315.

[31] P.G. Ranky, An introduction to radio frequency identification (RFID) methods and solutions, Assembly Autom. 26 (1) (2006) 28–33.

[32] D. McFarlane, S. Sarma, J.L. Chirn, C.Y. Wong, K. Ashton, Auto ID systems and intelligent manufacturing control, Eng. Appl. Artif. Intell. 16 (4) (2003) 365–376.

[33] S. Bahri, N. Zoghlami, M. Abed, J.M.R.S. Tavares, Big data for healthcare: a survey, IEEE Access 7 (2019) 7397–7408.

[34] A. Oussous, F.-Z. Benjelloun, A.A. Lahcen, S. Belfkih, Big data technologies: a survey, J. King Saud Univ.-Comput. Inf. Sci. 30 (4) (2018) 431–448.

[35] K. Sharmila, S.A. Vetha Manickam, Efficient prediction and classification of diabetic patients from big data using R, Int. J. Adv. Eng. Res. Sci. 2 (9) (2015).

[36] G. Li, Big Data Relat Technologies, Chall Future Prospect (2015) 283–285.

[37] V. Rajaraman, Big data analytics, Resonance 21 (8) (2016) 695–716.

[38] M. Khalifa, I. Zabani, Utilizing health analytics in improving the performance of healthcare services: a case study on a tertiary care hospital, J. Infect. Public Health 9 (6) (2016) 757–765.

[39] S. Katzenbeisser, M. Petkovic, Privacy-preserving recommendation systems for consumer healthcare services, Third International Conference on Availability, Reliability and Security, IEEE, 2008, pp. 889–895.

[40] W.S. Lee, E.J. Han, S.Y. Sohn, Predicting the pattern of technology convergence using big-data technology on large-scale triadic patents, Technol. Forecast. Soc. Change 100 (2015) 317–329.

[41] A. Baldominos, F. De Rada, Y. Saez, DataCare: big data analytics solution for intelligent healthcare management, Int. J. Interact. Multimed. Artif. Intell. 4 (7) (2018).

[42] B.L. Deekshatulu, P. Chandra, Classification of heart disease using k-nearest neighbor and genetic algorithm, Procedia Technol. 10 (2013) 85–94.

[43] M.-L. Zhang, A k-nearest neighbor based multi-instance multi-label learning algorithm, 22nd IEEE International Conference on Tools with Artificial Intelligence, 2, IEEE, 2010, pp. 207–212.

[44] A.P. Plageras, C. Stergiou, G. Kokkonis, K.E. Psannis, Y. Ishibashi, B.-G. Kim, et al., Efficient large-scale medical data (ehealth big data) analytics in Internet of things, IEEE 19th Conference on Business informatics (CBI), 2, IEEE, 2017, pp. 21–27.

[45] T.N. Gia, N.K. Thanigaivelan, A.-M. Rahmani, T. Westerlund, P. Liljeberg, H. Tenhunen, Customizing 6LoWPAN networks towards Internet-of-Things based ubiquitous healthcare systems, NORCHIP, IEEE, 2014, pp. 1–6.

[46] M. Hassanalieragh, A. Page, T. Soyata, G. Sharma, M. Aktas, G. Mateos, et al., Health monitoring and management using Internet-of-Things (IoT) sensing with cloud-based processing: opportunities and challenges, IEEE International Conference on Services Computing, IEEE, 2015, pp. 285–292.

[47] C.A. Tokognon, B. Gao, G.Y. Tian, Y. Yan, Structural health monitoring framework based on Internet of Things: a survey, IEEE Internet Things J. 4 (3) (2017) 619–635.

Index

Note: Page numbers followed by "*f*" and "*t*" refer to figures and tables respectively.

N

O

S

FSC
www.fsc.org
MIX
Papier | Fördert gute Waldnutzung
FSC® C083411

Kolibri 360 GmbH
Ferdinand-Jühlke-Straße 7
99095 Erfurt, Deutschland
produktsicherheit@kolibri360.de